Lecture Notes in Applied and Computational Mechanics

Volume 21

Series Editors

Prof. Dr.-Ing. Friedrich Pfeiffer
Prof. Dr.-Ing. Peter Wriggers

Efficient Numerical Methods and Information- Processing Techniques for Modeling Hydro- and Environmental Systems

Reinhard Hinkelmann

Springer

Professor Dr, REINHARD HINKELMANN
Technische Universitat Berlin
Institut fiir Bauingenieurwesen
Fachgebiet Wasserwirtschaft und Hydroinformatik
StraBedes 17.Juni 144
10623 Berlin, Germany

With 179 Figures

Library of Congress Control Number: 2004116864

ISSN 1613-7736

ISBN 3-540-24146-9 Springer Berlin Heidelberg New York

Springer is a part of Springer Science+Business Media

springeronline.com

Printed in Germany

Cover design: design & production GmbH, Heidelberg
Typesetting: Digital data supplied by author

Printed on acid-free paper 62/3141/Rw-5432 10

Preface

In recent years, *numerical simulation models* have become indispensable in *hydro-* and *environmental sciences and engineering*, mainly for making predictions and improving process understanding. For many problems, a physically correct and mathematically accurate simulation of the coupled complex processes requires powerful numerical methods as well as high resolutions in space and time. In information processing, many techniques have been developed for setting up systems with complex geometries and parameter distributions, and high-performance computer systems are available for fast computations. Overall, there is an urgent need for the further development and application of *efficient numerical simulation models* in environment water which consist of *efficient numerical methods* associated with *efficient information-processing techniques.*

After a general introduction to numerical simulation in environment water, the basic equations for groundwater flow and transport processes, for multiphase / multicomponent flow and transport processes in the subsurface as well as for flow and transport processes in surface waters are briefly deduced from the general form of the balance equation. The state of the art of *discretization and stabilization methods* (e.g. Finite-Difference, Finite-Element and Finite-Volume Methods), *parallel methods* and *adaptive methods* as well as *fast solvers* (e.g. Conjugate Gradient, Multigrid Methods) is presented with particular focus on explaining the interactions of the different methods. Parallel, adaptive and Conjugate Gradient or Multigrid Methods can collaborate well and should be chosen to solve large-scale problems. An overview of various *information-processing techniques*, which must work efficiently together is provided, briefly dealing with *preprocessing* (e.g. Computer Aided Design, databases, mesh generation, geostatistical methods), *processing* (e.g. High-Performance Computing, modeling system MUFTE-UG) and *postprocessing tools* (e.g. advanced visualization) as well as *Internet-based Collaborative Engineering.* Eventually, the interactions of the numerical methods with the information-processing techniques in order to achieve efficient numerical simulations are demonstrated for a wide range of applications in environment water. The work includes contributions to an improvement of process understanding, to further developments of numerical methods and information-processing, as well as to new application fields for existing simulation models. Finally, future perspectives are pointed out.

This book summarizes about a decade of my work in the field of numerical simulation of flow and transport processes in hydro- and environmental sciences and engineering carried out at three different institutes in Germany, the *Institute for Fluid Mechanics and Computer Applications in Civil Engineering*, University of Hannover, the *Institute for Computer Applications in Civil Engineering*, Technical University of Braunschweig, and the *Department of Hydromechanics and Modeling of Hydrosystems*, Institute of Hydraulic Engineering, University of Stuttgart. This book is based on the similarly entitled habilitation which I made at the Faculty of Civil and Environmental Engineering at the Universtiy of Stuttgart. As things change with time, I switched to the *Department of Water Resources Management and Hydroinformatics*, Institute of Civil Engineering, Technical University of Berlin.

I would like to thank most specially my former 'boss' and friend, Prof. Rainer Helmig, for numerous fruitful discussions, permanent support and great freedom in carrying out my work.
The discussions with the colleagues of the three aforementioned institutes contributed a great deal to the progress of this work. Their helpfulness, qualifications and collaboration as well as the longstanding good working atmosphere will always remain in my mind. I specially want to thank several colleagues from Braunschweig and Stuttgart for their excellent help and support: Andreas Bielinski, Thomas Breiting, Dr. Holger Class, Dr. Ken Kobayashi, Dr. Lina Neunhäuserer, Steffen Ochs, Ulrich Ölmann, Dr. Maren Paul, Dr. Hussam Sheta and Björn Witte. Many, many thanks to Prudence Lawday for correcting my English drafts and to Brigitte von der Ohe for designing many figures, improving the layout etc.
Finally, I very specially thank my partner Christine Barlag and my children Timon and Schirin Barlag for their understanding and support, and I apologize my frequent absence from home in the last years.

Berlin, Hannover
September 2004

Reinhard 'Phillip' Hinkelmann

Contents

Nomenclature

terms with Latin letters

symbol	dimension	definition
b	$[m]$	constant aperture
c	$[m]$	correlation length
c	$[mg/l]$	tracer concentration
c	$[m/s]$	wave velocity
c		unknown variable
$\hat{c}$		nodal value
$\tilde{c}$	$[mg/l]$	approximated concentration
c_D	$[-]$	wind-stress coefficient
d	$[m]$	characteristic length on the microscale
e		error
e	$[-]$	scalar entity
e_p	$[\%]$	parallel efficiency
f	$[-]$	damping function
f		quantities water level h
g	$[-]$	damping function
g	$[m/s^2]$	gravity
g		exact solution of the partial differential equation
$\overline{g}$		numerical solution of the discretized equation
$\underline{g}$	$[m/s^2]$	gravity vector $(0, 0, -g)^T$
gx		exact solution of the discretized equation
h	$[m]$	piezometric head
h	$[m]$	water level
$\tilde{h}$	$[m]$	approximated piezometric head
$k_{r\alpha}$	$[-]$	relative permeability
l_i	$[m]$	discretization length of element i
l_m	$[m]$	mixing length

m	$[Mbit]$	amount of data transmission
n	$[mol]$	number of molecules
n, m	$[-]$	*Van Genuchten* (VG) parameter
$\underline{n}$	$[-]$	normal vector
ne		nugget effect
p	$[Pa]$	pressure
p		number of processors
p_c	$[Pa]$	capillary pressure
p_d	$[Pa]$	*Brooks-Corey* (BC) parameter, entry pressure
p_g	$[Pa]$	pressure of gas phase
p_g^w	$[Pa]$	partial pressure of water vapor in the gas phase
q_c	$[kg/(m^3 s)]$	sink or source of concentration
q_w	$[kg/(m^3 s)]$	sink or source of water
q_n	$[kg/(m^3 s)]$	flux of non-wetting phase
q_w	$[kg/(m^3 s)]$	flux of wetting phase
r		sink or source
r		round-off error
r, s	$[m]$	natural coordinates
r_{tr}	$[s/Mbit]$	rate of data transmission
ser	$[-]$	serial (non-vectorizable) part
s_p	$[-]$	parallel speedup
s_v	$[-]$	vector speedup
t	$[s]$	time
t_{com}	$[s]$	communication time
t_{st}	$[s]$	start-up time
v	$[m/s]$	flow velocity
v	$[-]$	variance
var		variogram function
vr	$[-]$	vectorization rate
vv		variance
$v_{z,ave}$	$[m/s]$	average gas velocity
$v_{z,i}$	$[m/s]$	velocity of the gas phase at node i
$\underline{v}^*$	$[m/s]$	smoothed velocity
$\tilde{\underline{v}}_f$	$[m/s]$	approximated velocity
$\underline{v}_a$	$[m/s]$	pore velocity
$\underline{v}_f$	$[m/s]$	filter or *Darcy* velocity
$\underline{v}_{f\alpha}$	$[m/s]$	*Darcy* velocity of phase α
$\underline{w}$		diffusive flux
$\underline{x}$	$[m]$	local vector
x_g^w	$[mol/mol]$	mole fraction of water vapor in the gas phase
x_α^κ	$[mol/mol]$	mole fraction of component κ in phase α
x, y, z	$[m]$	coordinates in the global coordinate system
z	$[m]$	geodetic head

A		system matrix
B_i	$[m^3]$	control volume
B		matrix of the right-hand side
C, D		matrices
C_1, C_2		error indicators
Cr	$[-]$	*Courant* number
D		diagonal matrix
D_{ij}		dispersion matrix
D_L	$[m^2/s]$	longitudinal mechanical dispersion coefficient
D_T	$[m^2/s]$	transversal mechanical dispersion coefficient
$\underline{\underline{D}}_{hyd}$	$[m^2/s]$	hydrodynamic dispersion tensor
$\underline{\underline{D}}_{mol}$	$[m^2/s]$	molecular diffusion tensor
$\underline{\underline{D}}_{mech}$	$[m^2/s]$	mechanical dispersion tensor
E	$[J]$	energy
E		total error
F	$[N]$	force
F		functional
$\underline{F}$		flux term
H_w^m	$[Pa^{-1}]$	*Henry* coefficient
$\underline{\underline{I}}$		unit tensor
J_{ij}, J		*Jacobian* matrix
$K1, K2, K3$		variables
$\underline{\underline{K}}$	$[m^2]$	permeability tensor
$\underline{\underline{K}}_\alpha$	$[m^2]$	effective permeability tensor
K_f	$[m/s]$	hydraulic conductivity
K_m		arithmetic average
$\underline{\underline{K}}_f$	$[m/s]$	hydraulic conductivity
L	$[Mflops]$	processor performance
L		strictly lower triangular matrix
M_{ij}		mass matrix
N	$[-]$	interpolation function
N		number of nodes
Ne	$[-]$	*Neumann* number
P	$[Pa]$	air pressure
$P0, ..., P7$		processor
Pe	$[-]$	*Peclet* number
R	$[J/(mol\,K)]$	universal gas constant, $(R = 8.31451)$
Re	$[-]$	*Reynolds* number
Ri	$[-]$	*Richardson* number
S	$[^o/_{oo}]$	salinity
S	$[-]$	saturation
S		Schur Complement
S_0	$[1/m]$	specific storage coefficient
S_α	$[-]$	saturation of phase α

S_e	$[-]$	effective saturation
S_n	$[-]$	non-wetting phase saturation
S_{gr}	$[-]$	residual gas saturation
S_w	$[-]$	wetting phase saturation
S_{wr}	$[-]$	residual water saturation
T	$[\,^\circ C, K]$	temperature
T		square matrix
T_J		polynomial
U		strictly upper triangular matrix
V	$[m^3]$	volume
W	$[-]$	weighting function
W		partial differential equation
$\overline{W}$		discretized equation
X		vector of unknowns
$\overline{X}$		exact solution
$X^{(0)}$		initial solution
Z	$[Nm/(kg\ K)]$	real gas coefficient

terms with Greek letters

symbol	dimension	definition
α	$[-]$	upwind parameter
α	$[1/Pa]$	*Van Genuchten* (VG) parameter
α_i, α_j	$[-]$	weighting factors
α_L	$[m]$	longitudinal dispersion length
α_T	$[m]$	transversal dispersion length
δ		discretization error
ε		residuum
η		error estimator
θ	$[-]$	*Crank-Nicholson* factor
κ	$[-]$	*Karman* constant
κ	$[-]$	condition number
κ		component
λ	$[-]$	*Taylor*-friction coefficient
λ	$[-]$	*Brooks-Corey* (BC) parameter
λ	$[-]$	mobility
λ_α	$[-]$	mobility of phase α
λ_m		eigenvalue
μ	$[kg/(ms)]$	dynamic viscosity
ν	$[m^2/s]$	kinematic viscosity, $\nu = \mu/\varrho$

$\underline{\underline{\nu}}_t$	$[m^2/s]$	molecular diffusivity tensor
ν_{th}	$[m^2/s]$	horizontal turbulent diffusivity
ν_{tv}	$[m^2/s]$	vertical turbulent diffusivity
ν_{wh}	$[m^2/s]$	horizontal turbulent viscosity
ν_{wv}	$[m^2/s]$	vertical turbulent viscosity
ρ	$[kg/m^3]$	density
ρ_a	$[kg/m^3]$	density of air
ρ_w	$[kg/m^3]$	density of water
ϕ	$[-]$	porosity
ϕ	$[°]$	geographical latitude
ω	$[1/s]$	rotation of the earth
Γ		boundary of domain Ω
Γ_C		*Cauchy*-boundary condition
Γ_D		*Dirichlet*-boundary condition
Γ_N	$[-]$	*Neumann*-boundary condition
Δt	$[s]$	time step
$\Delta x, \Delta y$	$[m]$	element length
Ω		solution domain

indices

symbol	definition
a	air
b	brine
c	capillary
com	communication
f	filter
f	freshwater
g	gas phase
hyd	hydrodynamic
i, j, k, l	node
m	mixing
$mech$	mechanical
mol	molecular
n	NAPL
n	non-wetting phase
n	number of time step
sat	saturation
st	start-up time
tr	transmission
w	water
w	wetting phase

x, y, z	coordinate directions
α	phase

exponents

symbol	definition
adv	specifies the result from the advection step
dif	specifies the result from the diffusion step
k	iteration index
m	methane component
n	current time level
$n+1$	new time level
w	water vapor
w	wetting phase
κ	component

abbreviations

symbol	definition
adv	advection step
dif	diffusion step
div	divergence
fcp	free surface-continuity-pressure step
$grad$	gradient
mab	meter above bottom
mvp	matrix-vector product
$preco$	preconditioner matrix
sp	scalar product
AMG	Algebraic Multigrid Method
ART	Almost Regular Triangulation
BC	*Brooks-Corey*
BD	backward differencing
$BiCGSTAB$	Biconjugate Gradient Stabilized
BJ	*Block-Jacobi*
$BSCW$	Basic Support for Cooperative Work
CAD	Computer Aided Design
$CASE$	Computer Aided Software Engineering

CAVE	Cave Automatic Virtual Environment
CD	central differencing
CG	Conjugate Gradient
CPU	Central Processing Unit
CSG	Constructive Soild Geometry
CVFEM	Control-Volume Finite-Element Method
CVS	Concurrent Version System
DBMS	data base management system
DM	distributed-memory
DNS	Direct Numerical Simulation
DTM	digital terrain model
EBE	Element-By-Element
ELLAM	Eulerian-Lagrangian-Localized-Adjoint Method
ENO	Essentially Non-Oscillatory
ER	Entity-Relationship
FCT	Flux-Corrected Transport
FD	forward differencing
FDM	Finite-Different Method
FEM	Finite-Element Method
FMG	Full Multigrid Method
FVM	Finite-Volume Method
GIS	Geographical Information System
GMRES	Generalized Minimal Residual
HPC	High-Performance Computing
HPF	High-Performance Fortran
IFDM	Integral-Finite-Different Method
IMPES	IMplicit Pressure Explicit Saturation
KL	Kernighan-Lin heuristic
LES	Large Eddy Simulation
MG	Multigrid Method
MIMD	Multiple Instruction Multiple Data
MINRES	Minimal Residual
MPI	Message Passing Interface
MPP	Massively Parallel-Processing
NAPL	Non-Aqueous Phase Liquid
OODBMS	object-orientated DBMS
PCG	Preconditioned Conjugate Gradient
PG	Petrov-Galerkin
PVM	Parallel Virtual Machine
RCB	Recursive-Coordinate Bisection
RDBMS	relational DBMS
REV	representative elementary volume
RGB	Recursive-Graph Bisection
RIB	Recursive-Inertial Bisection
RSB	Recursive-Spectral Bisection

SIMD	Single Instruction Multiple Data
SISD	Single Instruction Single Data
SM	shared-memory
SMP	Symmetric Multi-Processing
SPMD	Single Program Multiple Data
SQL	Structured Query Language
SUPG	Streamline-Upwind-Petrov-Galerkin
TIN	triangular irregular network
VG	*Van Genuchten*
VR	Virtual Reality

1 Introduction

Water is essential to life on earth. It is of paramount social and economical value and its availability and use will considerably influence the development of our societies. A *sustainable management* and *protection* of water in the *environment* is one of the key problems of the 21st century, and numerical simulation models will contribute considerably to its solution. Numerical models bridge the gap between systems or domains and processes. They can simulate the processes concerning the flow and transport of water as well as other fluids and substances in different hydro- and environmental systems, and make predictions. To face these pressing problems, there is an urgent need for the development and application of efficient simulation models which consist of *efficient numerical methods* associated with *efficient information-processing techniques.* This chapter introduces systems and scales, numerical process simulation as well as model concepts and modeling systems in hydro- and environmental sciences and engineering, and it describes the objectives of this work.

1.1 Systems and scales

Hydro- and *environmental systems* cover the major part of the whole *hydrological cycle.* Different kinds of systems occur, and a large bandwidth of space and time scales must be considered. The systems can be divided into *subsurface* and *surface-water systems* and the *atmosphere* (see fig. 1.1). Depending on the geological structures, *pore aquifers, fractured aquifers* and *karstic aquifers* are distinguished in the subsurface; this is discussed further in section 2.1.2. Moreover, the subsurface is subdivided into the *saturated* and *unsaturated zones.* In the saturated zone, the pore space of the aquifers is mainly filled with water, while it is mainly filled with soil gas in the unsaturated zone. Depending on the dominant processes, surface-water systems can be subdivided into the groups *rivers / estuaries etc., coastal waters / seas etc.* and *lakes / reservoirs etc.*; this is explained further in section 2.1.3. One can also distinguish between *standing waters*, such as lakes / reservoirs, and *flowing waters*, such as the others mentioned. Alternatively, surface-water systems can be subdivided into *marine and coastal waters* on the one hand and *inland waters*, which include the rivers / estuaries and lakes / reservoirs, on the other hand. Finally, hydro- and environmental systems are in more or less strong *interaction* with one another. Subsurface and surface waters are coupled by *surface-water infiltrations*, also called *leakage*, or *groundwater springs* (see sec. 5.3). Subsurface and surface waters interact with the atmosphere via *rainfall, runoff* and *evaporation*. The interaction processes are very important for integrated modeling.

The systems surface water and subsurface are *multiscale* systems. In surface-water systems, the space scales range from one hundred kilometers and more via meters to micrometers, while the time scales range from more than days via minutes to milliseconds, depending on whether *currents, waves* or *turbulences* are of major interest (see fig. 1.2, sec. 2.1.3). If, for example, the *morphodynamics* of a coastal zone are being investigated, all space scales and even larger time scales must be taken into account.

In principle, the situation is similar for subsurface systems. Space scales range from more than kilometers to micrometers, i.e. from the catchment area of a groundwater reservoir via the hydrogeological aquifer structure and local heterogeneities to the pore scale and the molecular scale (see fig. 1.3, sec. 2.1.2). If, for example, a contamination infiltrates into the ground, its spreading may be relevant on a regional scale, while the biodegradation processes of the contamination take place on the microscale.

Finally, different time scales in the area of interaction between surface water and groundwater are pointed out in figure 1.4 (see sec. 5.3). A typical flow velocity is in the range of $1m/s$ in a river, $1m/d$ in groundwater and $1m/h$ in a surface-water infiltration zone. As already mentioned before, interaction processes must be taken into account for integrated investigations. First, numerical process simulation is considered.

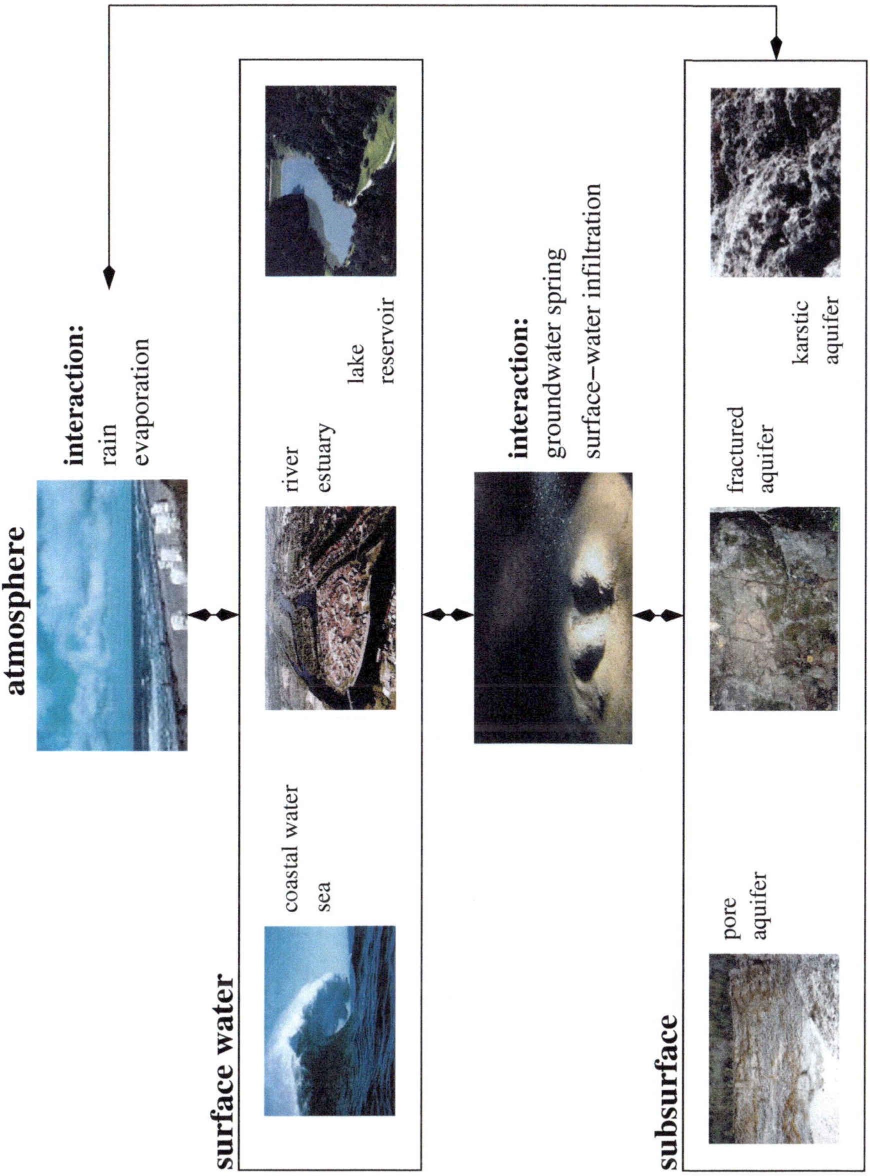

Fig. 1.1. Classification of systems

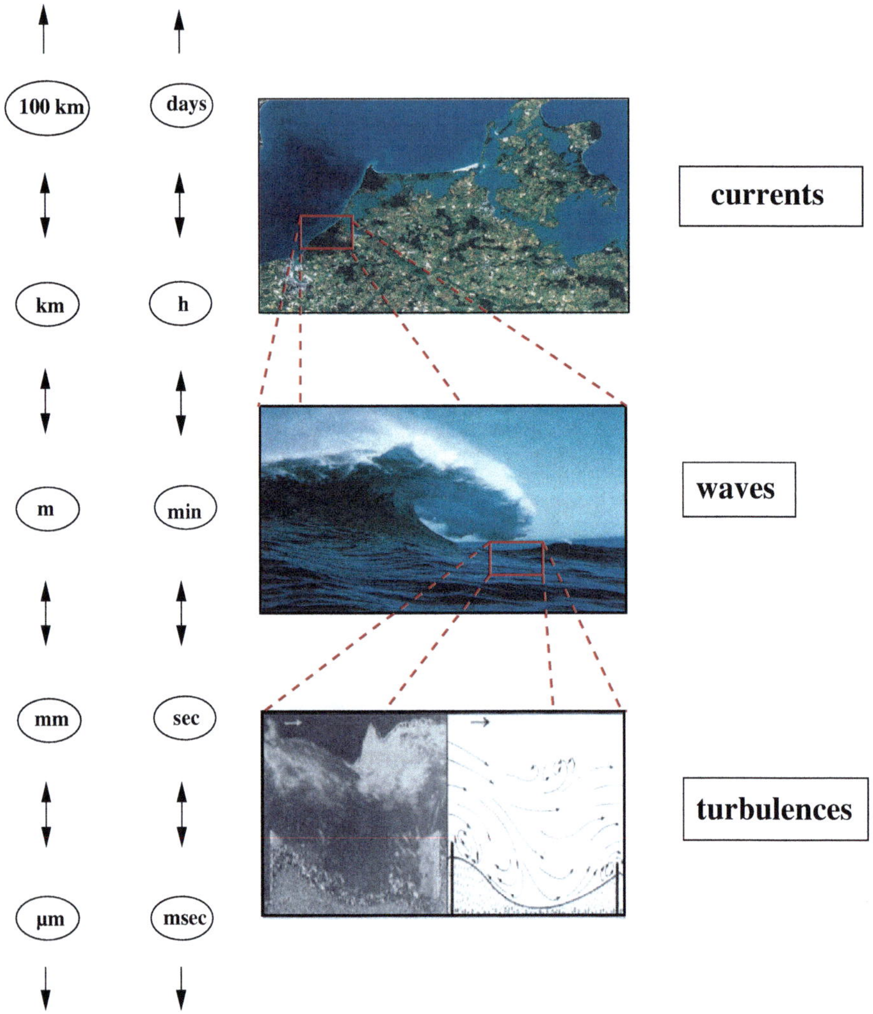

Fig. 1.2. Scales in surface-water systems, bottom after JIRKA (1999 [134])

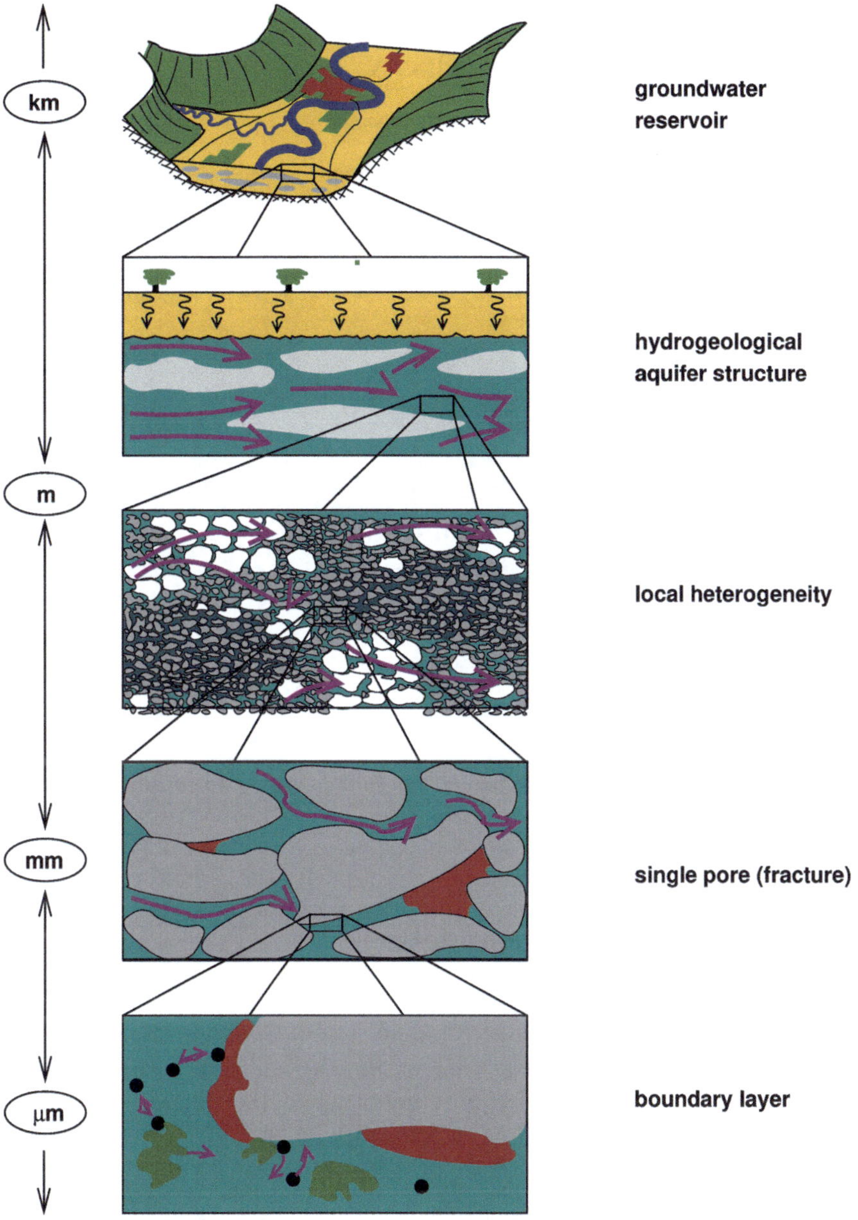

Fig. 1.3. Space scales in subsurface systems, after KOBUS, DE HAAR (1995 [149])

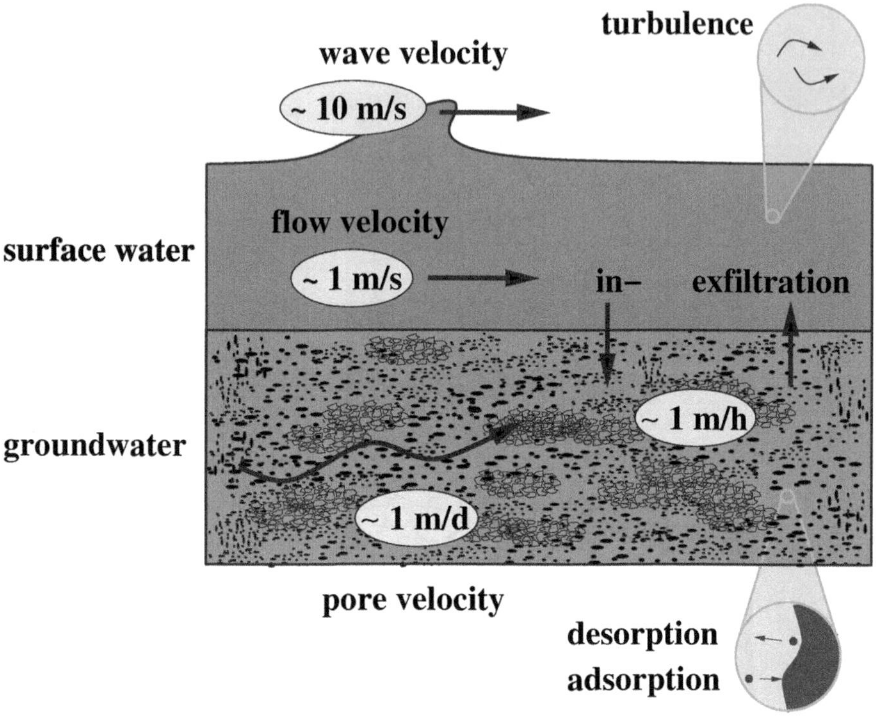

Fig. 1.4. Time scales in the interaction area of surface-water and subsurface systems

1.2 Numerical process simulation

The processes which are considered in *hydro-* and *environmental systems* (see figs. 1.1 - 1.4) deal with the flow of liquids and gases, taking the transport of further substances and heat as well as the reaction of substances into account. Of all these, *water* is the most important. Other liquids treated are *hydrophobic* contaminants such as mineral fuels or solvents. They are called *Non-Aqueous Phase Liquids (NAPLs)* because they are not or only slightly soluble in water. Of the gases, *air* plays an important role. It is the dominant fluid in the soil or the unsaturated zone as well as in the atmosphere. Further gases dealt with are, for example, the greenhouse gases *methane* and *carbon dioxide*. Gases occur as a separated phase or dissolved in the liquids, especially in water. Among the substances transported are, for example, *sediments* suspended in water or as bed load, nutrients such as *nitrogen* or *phosphorus* as well as radioactive substances such as *radon*. Moreover, (micro)biological, chemical and decay reaction processes, for example biodegradation or adsorption, are treated as well as mass-transfer processes between phases, for example evaporation or dissolution.

Nowadays, numerical simulation methods or models are one very important tool in a spectrum of approaches for predicting the consequences of changing situations and conditions, including parameter studies and sensitivity analyses, and for understanding new processes; other tools are, for example, laboratory and field experiments. Models describe flow processes alone or together with transport and reaction processes. They must always be considered critically in the range of their limitations which are discussed in the context of the model concepts (see sec. 1.3.1). A prerequisite for a prediction is the previous calibration and validation of the model (see sec. 1.3). In principle, it is possible to predict the consequences of an interference in a system, for example the changes of the flow and water-level distributions caused by the deepening of a waterway or the reduction of sediment erosion if groynes are constructed along river beds. The effects of expected climate changes, such as the rise of the mean sea level, storm-surge intensity and its frequency, on flood-protection measures can be estimated. The spreading of a real or possible contamination in the groundwater can be predicted as well as their effects, for example, for waterworks. Moreover, numerical simulations enable or simplify the understanding of processes. For example, it is possible to consider single processes in a strongly coupled multiphase subsurface system and to investigate their effects on remediation measures. Furthermore, numerical models can be used to determine parameters or state variables which cannot be measured directly. For example, the pore velocity can be deduced from the measured pressure data using a model.

Principally, a lot of processes can be simulated numerically today. However, the reliability, the probability and the bandwidth of the results gained from the model must be evaluated critically in the context of the assumptions and abstractions made, the quality of the model's calibration and validation as well as the quality of the data used. When doing this, one should take into account that the understanding of the processes, the spatial and temporal resolution as well as available data are often limited; this is discussed in more detail in the course of this chapter (see KOBUS, DE HAAR (1995 [149]), HELMIG (1997 [98]), ZIELKE et al. (1999 [267])). In the next step, the treatment of the processes in the different model concepts is explained, modeling systems are introduced and the need for laboratory and field experiments is emphasized.

1.3 Model concepts and modeling systems

1.3.1 Model concepts

In the course of treating the different *model concepts* or *models* which are explained in the following, a number of assumptions, abstractions and linearizations are made and finally the *numerical simulations* are compared with the measurements from *nature* or *laboratory* (see fig. 1.5). In the first step, the *conceptual model* reduces the physical, chemical and biological processes that occur to those which are of interest and importance for the problems and questions to be investigated. If, for example, groynes are to be designed to reduce the amount of sediment erosion in a river, the conceptual model consists of (vertically-integrated) flow and sediment transport with high-quality turbulence, for example the $k - \epsilon$ model. For such a problem, turbulence has an important influence and must therefore be taken into account accurately (see sec. 2.6, ZIELKE et al. (1999 [267])). If, for example, the infiltration of a NAPL into the unsaturated zone is considered, the conceptual model of the infiltration zone must take the three mobile phases NAPL, air and water into account. If the NAPL is highly volatile and is to be removed with thermally enhanced soil-vapor extraction, the NAPL also occurs to a larger extent as a component in the gas phase and consequently, the conceptual model must provide three phases with three components each under non-isothermal conditions (see sec. 2.5, HELMIG (1997 [98])). If the NAPL pools on the water table, dissolves slowly in the water phase and the large-scale consequences of the NAPL infiltration are of interest, the conceptual model can be reduced to groundwater flow and transport (see sec. 2.3).

The *mathematical model* transfers the conceptual model to a mathematical formulation in a *deterministic* or *stochastic* way. Generally, the deterministic mathematical model is based on a continuum-mechanical consideration (see sec. 2.1.1), and it formulates the balance equations for mass, momentum and energy (see sec. 2.2) as well as equations of state. Additionally, it accounts for the initial and boundary conditions. A stochastic mathematical model takes the statistical behavior of model parameters or state variables into account. It is not treated further in this context, see DAGAN (1989 [68]). Mathematical models can only be solved analytically for a limited number of simple cases, for example simple domains and boundary conditions. More flexibility is given by *numerical models* because they enable the treatment of general geometries, parameter distributions as well as initial and boundary conditions. The numerical model is transfered to a computer program and *simulates* the specified processes.

The term *verification* means proving that the numerical result is correct. It is done by comparisons with analytical solutions which are only available for simple systems and single processes, by plausibility tests, for example, checking the global mass conservation, and by ensuring that programming errors have been removed. During the *calibration*, the numerical results are compared with experimental or field data. Calibration parameters are varied in physically reasonable ranges to obtain a best approximation with the data. Such parameters are, for example, the friction coefficient in a free-surface flow simulation (see sec. 2.6) or the hydraulic conductivity in a groundwater-flow simulation (see sec. 2.3). The calibration is often performed 'by hand' on the basis of the engineer's experience; however, automatic procedures also exist. The calibration offers a first indication of the model's quality. The *validation* is a further proof of the calibration under similar conditions with, however, an independent data set, i.e. the data set was not used for the verification or the calibration. 'Similar conditions' means that a free-surface flow model is calibrated for mean-water conditions in a river and validated for high water or another river. The degree of a model's validation and its reliability are again determined by experimental or field data. Experiments usually concentrate on a single phenomenon under controlled conditions and have to confirm the validation of the model corresponding to this phenomenon. Field data reflect the natural variability of the geometry and all the coupled processes. Therefore, a validation of the model by field data is desirable. However, there is often a lack of reliable field-data sets. As every new data set will give new insights into the model as well as its parameters, a model is never finally validated. It is recommended to determine the probability and the bandwidth of predicted results depending on uncertainties in the processes, the model and the data, for example, with sensitivity analyses or stochastic methods. The numerical models are the major part of the processing units in modeling systems which are introduced in the following.

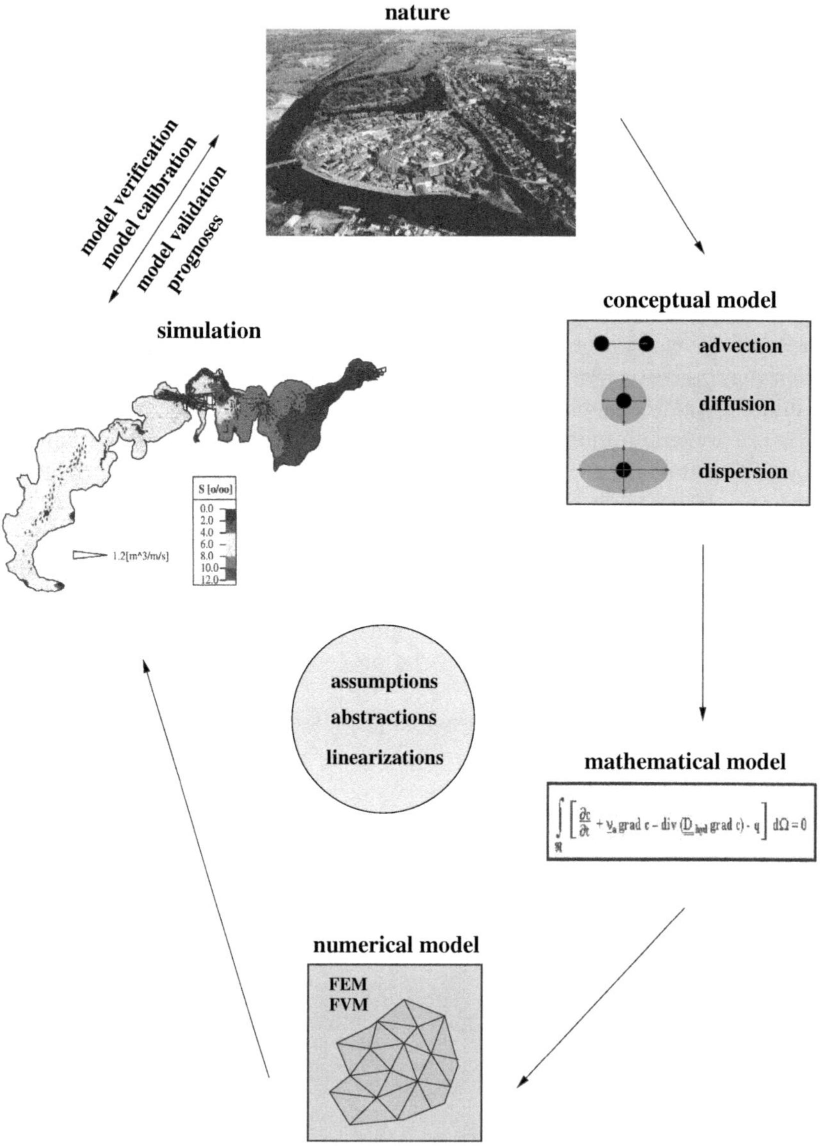

Fig. 1.5. Model concepts, after HINKELMANN et al. (2001 [110])

1.3.2 Modeling systems

Generally, a modeling system is divided into a *preprocessing, processing* and *postprocessing unit* (see fig. 1.6). In the following, a number of requirements for *state-of-the-art* modeling systems which result from current developments in numerical modeling and information processing are listed. Many existing modeling systems fulfill only some of these requirements.

During the preprocessing, the model is set up with all necessary data (see sec. 4.1). It should be possible to determine the *geometry* of the system, i.e. the computational domain and the boundaries, with the help of *CAD, GIS, remote sensing, digital water-level measurements etc. Databases* should be available to store the spatially varying *physical parameters*, for example the properties of the fluids (density, viscosity) or the porous medium (porosity, hydraulic conductivity). It should be possible to take *uncertainties* in the parameters, for example small-scale heterogeneities of a porous medium, into account by *geostatistical methods.* Finally, powerful *mesh generators* should be available. Then, all the data are passed onto the processing unit where the *numerical simulation* and, possibly, an *optimization* are carried out. Beside different *discretization methods*, the numerical simulator should offer the opportunity to use efficient solution methods, such as *parallel, adaptive Multigrid Methods* (see chap. 3, sec. 4.2). Optimization can be included by loops over the numerical simulator. Here, simulation results or a system answer can be optimized by an automatic variation of, for example, boundary conditions or other control parameters. *Visualization tools* and *statistical evalutation tools* for the numerical results should be available during the postprocessing. Furthermore, the *software quality* of a modeling system should be ensured, for example by a documentation of the functions as well as the limitations, verification and validation examples and by a user- and maintenance-friendly application environment (see FORKEL (2001 [82])).

Numerical simulation can be seen as being supported by two sets of three pillars, one set representing *data, methods* and *models* and the other representing *physics, mathematics* and *computer sciences* (see fig. 1.7). The sets strongly interact with one another.

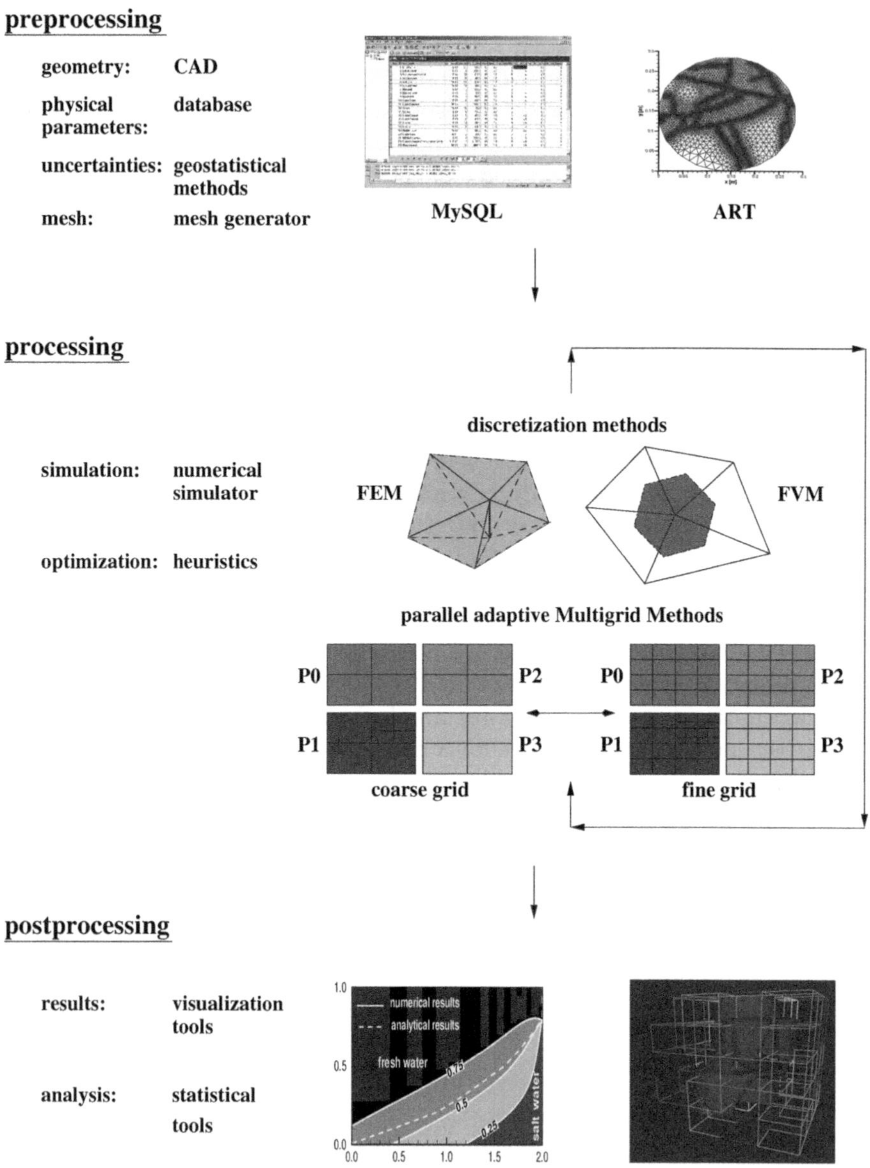

Fig. 1.6. Modeling system, after HINKELMANN et al. (2001 [110])

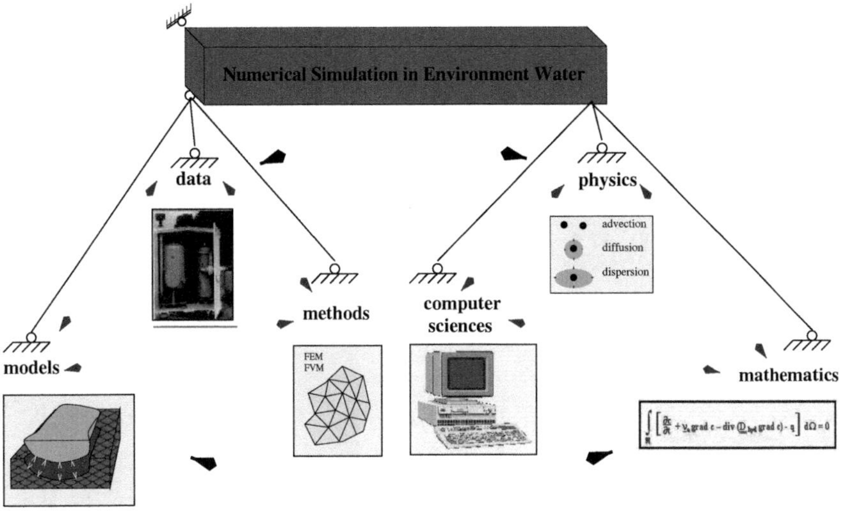

Fig. 1.7. Pillars supporting numerical simulation, after HELMIG et al. (1999 [102])

1.3.3 Need for laboratory and field experiments

Generally, single processes are considered under controlled conditions in *laboratory experiments* ranging from the column to the technical scale, while *field experiments* represent coupled processes in nature under much less controlled conditions ranging from the technical to the field scale. Both laboratory and field experiments are essential for the (further) improvement of the process understanding as well as for the (further) development and application of numerical models for all hydro- and environmental systems (see secs. 5.1 - 5.4), especially for multiphase systems in the subsurface. Together, numerical models and experiments form the basis, for example, for the development and application of subsurface-remediation technologies. Such experiments are, for example, carried out up to the technical scale in the VEGAS facility (see KOBUS et al. (1996 [148])) at the Institute of Hydraulic Engineering, University of Stuttgart, Germany. Figure 1.8, left, shows the large VEGAS tank viewed from above, which can be filled with different subsurface materials, and measurement devices.

In hydro- and environmental sciences and engineering, numerical models are continuously gaining importance and, more and more, they supplement and replace *hydraulic models*. In a hydraulic model, the area under investigation is reproduced on a reduced scale in a laboratory. This is very time-consuming and costly. In particular, several variants can be carried out comparatively easily with a numerical model, while a hydraulic model must be reconstructed.

Fig. 1.8. Laboratory experiments: subsurface remediation, left, after KOBUS et al. (1996 [148]); erosion-stability experiment, right, after WESTRICH et al. (2002 [260])

However, hydraulic models are necessary even today for processes and problems which cannot be simulated well numerically. This is the case, for example, if the erosion stability and slope-protection of an overflowing dike are investigated (see fig. 1.8, right, WESTRICH et al. (2002 [260])). The flow processes which occur can be characterized as follows: three-dimensional, non-hydrostatic, highly turbulent, sub- and supercritical, complex domain, non-flooding areas; all together, this is beyond the capabilities of current numerical process simulation based on the solution of the physical conservation laws of mass and momentum as shown in this work. However, as large experimental data sets exist, data driven modeling techniques such as *artificial neural networks* may be very suitable (see MASE et al. (1995 [174]), GENT, BOOGAARD (1998 [91])).

1.4 Deficits and objectives of the work

Deficits of the numerical simulation in hydro- and environmental sciences and engineering are presented. They are the basis for the motivation of this work.

1.4.1 Deficits

process understanding

The extent to which processes are understood is different for different problem classes and systems. While, for example, flow processes in a river are relatively well understood, there are a number of gaps regarding multiphase / multicomponent flow and transport in the subsurface. Sometimes, the complexity of a process is known, but its simplification and transformation into a numerical model is hardly possible or practicable, for example capillary-pressure hysteresis. Consequently, the corresponding numerical models have only been tested within a limited range. While, for example, river-flow models are applied up to the field scale, subsurface multiphase / multicomponent flow and transport models are generally only used up to the technical scale.

scales, couplings

The model concepts are generally only valid and practicable in a specified system on one or a limited number of space and time scales. Even in this range, there are often strong computational limitations with respect to the spatial and temporal resolution of the problem caused by complex and strongly coupled processes. For many of today's problems, the effects of small-scale processes must be taken into account on the large scale. With the exception of groundwater flow and transport processes, there are only very few *upscaling methods* which couple processes over several scales. Furthermore, the *coupling* of different systems must be improved by better consideration of the interaction processes to achieve *integrated modeling.*

data, uncertainties

For many model developments and applications, *data* are *lacking.* As a model prediction can only be as good as its data set, the probabilities as well as the bandwidths of the results should be estimated, taking uncertainties of the model as well as of the data into account, for example with stochastic methods.

information processing

From the point of view of information processing, the *standardization of interfaces* between different tools and models must be improved. Modeling systems must be extended to *decision-support systems*, taking economic, social and political aspects into account. In the context of the implementation of the *European Union's Water Framework Directive*, there is an urgent need for setting up *information systems* which form the umbrella for storing, managing and analyzing large data sets, applying modeling and decision-support systems and providing all further problem-related information to the actors as well as the public.

Overall, there are enough deficits to motivate the objectives of the work.

1.4.2 Objectives of the work

The objectives of the work are manifold. First, the state of the art of efficient numerical methods is described, an overview of information processing techniques is given and their applications in a wide range of water-related and environmental problems and systems are demonstrated. Moreover, the work contains several innovations to which the author has contributed significantly. These innovations deal with further developments of numerical methods, several extensions to the modeling system MUFTE-UG, new application fields for existing simulation methods as well as improvements in process understanding.

The work is divided into 6 chapters.

- Chapter 1 presents an introduction to numerical modeling of hydro- and environmental systems.
- Chapter 2 deals with physical model concepts for the systems subsurface and surface water as well as with mathematical model concepts based on the general form of the balance equation. The basic equations for groundwater flow and transport processes, two-phase flow and two-phase / multicomponent flow and transport processes in subsurface systems as well as flow and transport processes in surface-water systems are explained briefly.
- In chapter 3, efficient numerical methods are discussed in detail. Discretization and stabilization methods based on the Finite-Difference, Finite-Element and Finite-Volume Methods are described and applied to the basic equations of chapter 2. To treat large-scale problems, sophisticated solution methods, such as parallel and adaptive methods and fast solvers, are required as well as their interaction. The development of High-Performance Computing, parallelization strategies, the parallelization of basic tasks as well as load-balancing methods are presented. Various methods of adaptation, error estimators and indicators as well as refinement and coarsening strategies are explained. Different single-grid solvers, such as Conjugate Gradient Methods, and Multigrid solvers as well as preconditioners and non-linear solvers are introduced.
- A number of different efficient information-processing techniques associated with the efficient numerical methods are described briefly in chapter 4. Among the preprocessing tools, CAD, database-management systems,

GIS, tomography and scanning, mesh generators and geostatistical methods are dealt with. The processing part concentrates on High-Performance Computers and the extension of the modeling system MUFTE-UG. Finally, aspects of postprocessing, i.e. visualization, and WWW-based Collaborative Engineering are discussed.

- Chapter 5 demonstrates the interaction of the efficient numerical methods from chapter 3 with the efficient information-processing techniques from chapter 4. A wide range of applications in hydro- and environmental sciences and engineering is considered. A new, so-called equidimensional modeling approach for groundwater flow and transport processes in fracture-matrix systems is presented. The modeling system MUFTE-UG is used and extended to the simulation of methane-migration processes in coal mining areas, gas-water flow processes in dike systems and a multistep outflow experiment as well as to two-phase / three-component flow and transport in a coastal aquifer in the interaction area to the sea. Finally, free-surface flow and transport processes in two estuaries are investigated.
- In chapter 6, the contents is summarized and an outlook on future work is presented.

2

Physical and mathematical model concepts

This chapter starts with an introduction to physical model concepts to take different types of complex geological structures into account. For the description of flow and transport processes in such structures, all balance equations which are required in (environmental) fluid mechanics can be formulated in a generalized way, the general form of the balance equation. This equation is then applied to the mathematical modeling of different flow and transport processes in hydro- and environmental systems, i.e. groundwater flow and transport processes, two-phase flow and two-phase / multicomponent flow and transport processes in subsurface systems as well as flow and transport processes in surface-water systems.

2.1 Physical model concepts

2.1.1 Continuum-mechanical consideration

Usually, the model concepts for the *hydrosystems* subsurface and surface water are based on a continuum-mechanical consideration. Therefore, the processes occurring on the microscale must be averaged in space and / or time in order to serve as physical quantities, i.e. effective parameters and processes, on the mesoscale (see fig. 2.1).

It should be mentioned that microscale model concepts have been developed in recent years, e.g. in the context of upscaling in porous media (see sec. 4.1.4) or turbulence modeling (Direct Numerical Simulation, DNS, see sec. 2.6.1). Up to now, the application range of these model concepts has been rather limited. However, some of these microscale model concepts may open new horizons in the future.

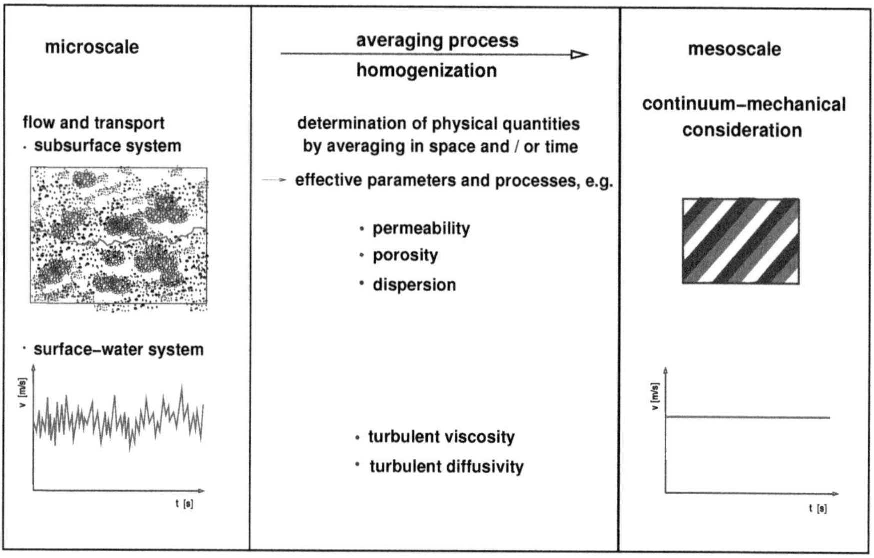

Fig. 2.1. Continuum-mechanical consideration, after HELMIG, CUNNINGHAM (2002 [101])

2.1.2 Subsurface systems

In the subsurface, the processes occurring on the microscale between single pores (see length scale d in fig. 2.2) are upscaled and averaged over certain volumes which are called representative elementary volumes or REVs. The length scale of a REV l (see fig. 2.2) is chosen so that it leads to a representative average of the property under consideration , e.g. the porosity ϕ (see fig. 2.2 and sec. 2.3.1). On the one hand, the REV must be big enough to avoid inadmissible fluctuations of the property. On the other hand, it must be small enough for spatial variations of the property under consideration to be detected (see length scale L in fig. 2.2). The REV idea is the most common model concept for porous media, subsurface systems respectively. Detailed information is given in BEAR (1972 [25]), EHLERS (1996 [74]) or HELMIG (1997 [98]).

Depending on the geological structures pore-water, karstic and fractured aquifers are distinguished in the subsurface (see fig. 1.1, KOLDITZ (1997 [152])). Generally, two main model concepts with different model approaches exist for such multiscale problems. First, these model concepts are explained for fractured systems:

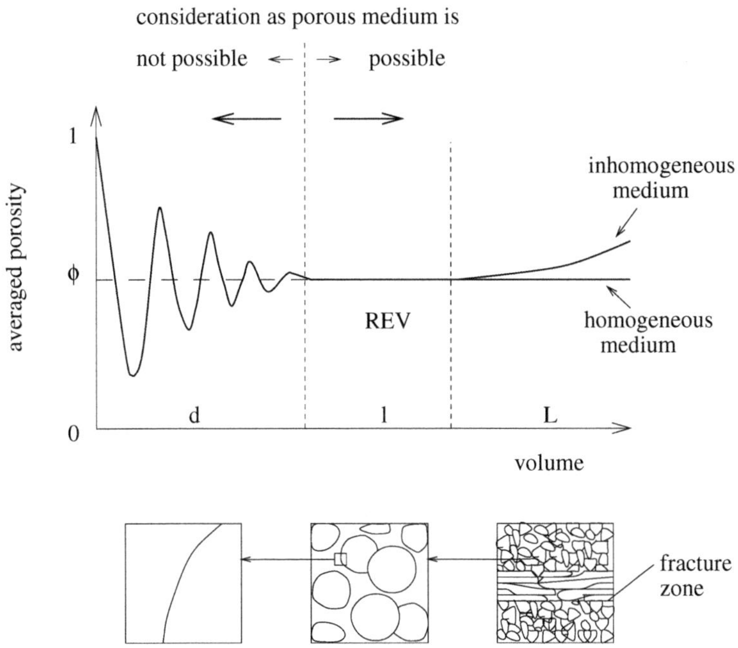

Fig. 2.2. Definition of the REV, after BEAR (1972 [25]) and HELMIG (1997 [98])

- equivalent model concept:
  single-continuum approach, double-continuum approach, multi-continua approach
- discrete model concept:
  single-fracture approach, fracture-network approach, combined approach

Equivalent model concepts are based on the REV idea and the assumption that the inhomogeneous domain can be homogenized bit by bit by shifting the observation scale. Single-continuum, double-continuum and multi-continua approaches are distinguished (see fig. 2.3). The choice of the model concept depends on the spatial scale of the problem and possibly on the data available. Single-continuum approaches are suitable if the length scales of the occurring fractures do not differ very much. A double-continuum approach is applied if the difference in length scales of the fissures and coarse fractures is significant. Then, the physical behavior is so different that this system must be described by two interacting continua consisting of coarse fractures with high permeabilities and low storage capacities and fissures with high storage capacities and low permeabilities (see LANG (1995 [163])). Additionally, a double-continuum approach should be chosen if 'exact' information of the coarse fractures is not available. Multi-continua approaches are used, if, for example, the influence of the rock matrix cannot be neglected and if this matrix cannot be reasonably homogenized together with the fissures (see NARASIMHAN, PRUESS (1987 [185]), BIRKHÖLZER (1994 [31])). The different continua are coupled by sink and source terms. Generally, double-continuum and multi-continua approaches are suitable for medium or regional scale simulations.

If the flow and transport processes in the computational domain are dominated by fractures, they must be taken into account explicitly by discrete model concepts which are subdivided into single-fracture, fracture network and combined approaches (see fig. 2.3). In single-fracture and fracture-network models, the rock matrix is considered impermeable. However, experimental (see HIMMELSBACH (1993 [107]), MALOSZEWSKI, ZUBER (1993 [171])) and numerical investigations (see PFINGSTEN (1990 [207]), KRÖHN (1991 [159])) have shown that the rock-matrix and the fracture-matrix interaction cannot be neglected for transport simulations. Fracture-network models consist of one or several possibly crossing-fracture groups without the rock matrix. Often, the fracture network approach is applied to determine equivalent continuum parameters which are integrated into models on larger scales. When the combined approach is applied, fractures or fracture networks as well as the rock matrix are taken into account. The matrix can be an equivalent continuum of a fracture system on a smaller scale. Up to now, the fractures and the matrix have generally been modeled by elements of different dimensions (see HELMIG (1993 [97]), BARLAG (1997 [13]), SONNENBERG et al. (1997 [241])). The coupling of fracture and matrix is carried out by adding up parts of the local system matrices in the Finite-Element Method or adding up fluxes in the Finite-Volume Method. However, at the fracture-matrix interface, the physical process is smoothed and the local flux conservation may be not fulfilled. This circumstance has a major influence on the transport simulation, as the flow field is its most important input parameter. New approaches are based on equidimensional considerations of fracture and matrix and require special discretization methods and solvers (see ROSEN (2000 [224]), NEUNHÄUSERER et al. (2001 [188]), OCHS et al. (2002 [193])), which are addressed in section 5.1.

In the hybrid model concept, a discrete fracture-network is coupled with a single-continuum, double-continuum or multi-continua approach (see fig. 2.3, BIRKHÖLZER (1994 [31])). The equivalent continua are not located between the fractures but are coupled with the fracture network by sink and source terms. This model concept can be applied if the occurring processes, the observed scales as well as the available data do not clearly indicate a discrete or an equivalent model concept.

Generally, a single-continuum approach is applied for pore-water aquifers, as the geological structures and the simulated processes can be homogenized bit by bit in a very reasonable way. Karstic aquifers are characterized by a double-porous structure which consists of coarse fractures like channels or tubes with

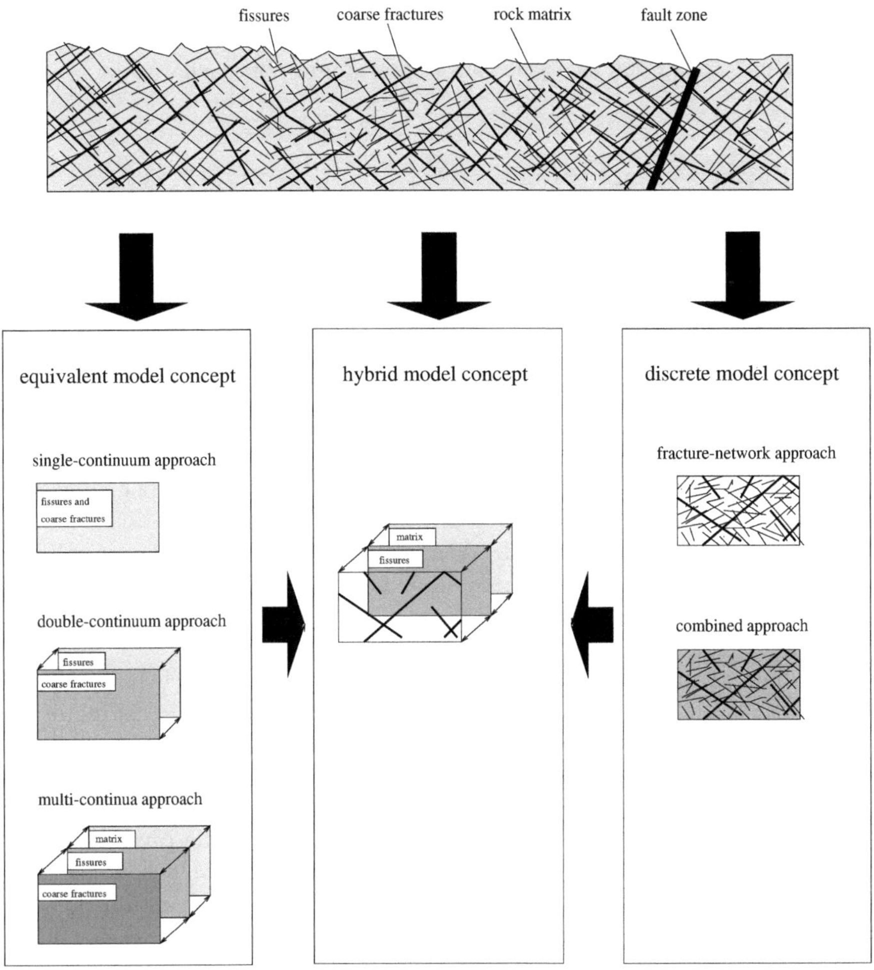

Fig. 2.3. Model concepts for subsurface systems, after NEUNHÄUSERER et al. (2000 [190])

high permeabilities and small fractures or fissures with high storage capacities. Due to this system behavior, one of two different model concepts can be chosen, the combined or the double-continuum approach (see LANG (1995 [163])). The first approach is recommended if the location, dimensions and physical properties of the coarse fractures can be estimated with reasonable accuracy. Otherwise, the double-continuum approach should be preferred.

In this work, fractured-porous aquifers are dealt with in sections 2.3, 2.4, 5.1, 5.2 using the combined approach, and pore-water aquifers are treated in sections 2.3 - 2.5, 5.2, 5.3 using the single-continuum approach.

2.1.3 Surface-water systems

In surface waters, different model concepts are chosen depending on the dominant processes (see sec. 1.1). First, one can distinguish between model concepts for *near-field* and *far-field problems.* The near-field is understood as the close surroundings of buildings, e.g. a pillar in a river, or of inlets, e.g. a pipe for waste water in a lake. The dominant processes are characterized by great local details, a three-dimensional flow field, a strong influence of turbulence and, often, erosion, for example around a pillar in a river or in a groyne field on the coastline. For such problems, the model concepts must be based on *integral methods*, *zonal methods* or *field methods.* The last-mentioned solve the three-dimensional continuity equation, the *Navier-Stokes* equations (see sec. 2.6) with a sophisticated turbulence model (see sec. 2.6.1) and, possibly, the transport equations for contaminants, sediments or heat. This is not discussed here (see JIRKA et al. (1999 [134])).

In far-field problems, the dominant processes occur on much larger spatial scales. Here, model concepts for *long* and *short waves* are distinguished. They are called *flow models* in the case of long waves (see fig. 2.4) and *wave models* in the case of short waves (see fig. 2.5). A wave with a wave length L 20 times greater than the water depth h is called a long wave, otherwise it is called a short wave. If long waves are dealt with, the pressure is generally assumed to be hydrostatic and the horizontal flow velocities are much greater than the vertical ones. Therefore, the vertical momentum equation can be considerably simplified. Further, *shallow* and *deep waters* are distinguished, causing different simplifications in the corresponding model concepts.

Water systems which are dominated by long waves are, for example, inland waters such as rivers and estuaries (see fig. 1.1, sec. 5.4). A typical long wave in an estuary is a tidal wave which has a wave length L in the order of thousand(s) of kilometers and a period T of about 12 hours (see figs. 2.4, 1.2). A typical long wave in a river is a flood wave with a length scale in the order of hundred(s) of kilometers and a time scale in the order of a few days (see fig. 1.2). As the horizontal dimensions are generally much larger than the vertical one, these systems are classified as shallow. Here, long waves in shallow waters, i.e. flow models, are dealt with in section 2.6 and applied to rivers and estuaries in section 5.4.

Short waves play an important role in coastal waters. They are generated by wind, and when moving from the sea towards the coast, they are transformed

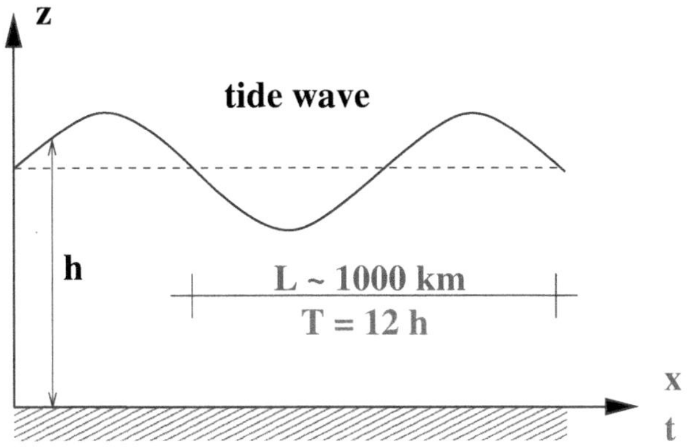

Fig. 2.4. Typical wave length and period in a flow model

because of the decreasing water level; this can lead to breaking waves (see fig. 1.1). The pressure distribution is not hydrostatic and the vertical flow velocities are in the same order of magnitude as the horizontal ones. Wave lengths are in the order of tens or hundreds of meters, and periods in the orders of seconds and minutes (see figs. 2.5, 1.2). Again, the horizontal dimensions are much larger than the vertical ones and, therefore, such systems are shallow. Depending on the problems, a superposition of long and short waves must be considered in coastal waters and marine waters. The waters mentioned last are classified as deep. Further information about wave models is found in ZIELKE, MAYERLE (1999 [268]).

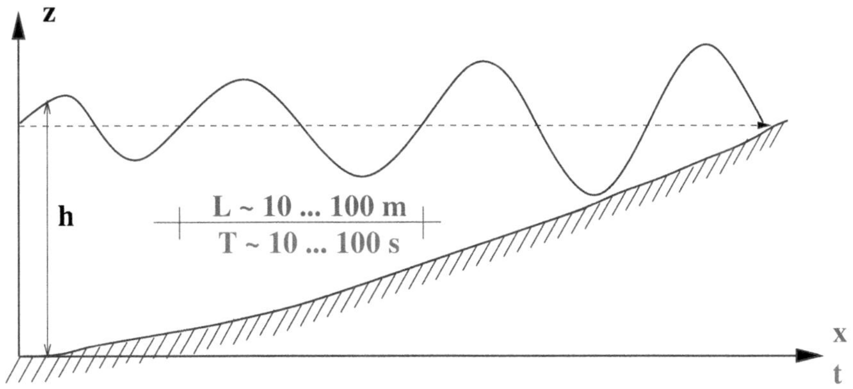

Fig. 2.5. Typical wave length and period in a wave model

The surface waters treated up to now belong to the category of *flowing* waters, while lakes and reservoirs are *standing* inland waters (see fig. 1.1). In this context, standing means that the flow velocities are very small and in equal orders of magnitude in the horizontal and vertical directions. For flowing waters, a typical flow velocity is $1m/s$, while it is $1mm/s$ for a standing water. Depending on the problems, long, short or *internal waves* or *density-induced flows* are important. When the water depth is great compared to the horizontal dimensions, a standing water can be considered deep. Although the pressure distribution is nearly hydrostatic, the simplifications which are deduced from this assumption for long waves must be examined critically before being used for standing waters. Further information about lake and reservoir modeling is given in BERGEN, FORKEL (1999 [28]).

2.2 General form of the balance equation

The balance equations for mass, momentum and energy which are required in (environmental) fluid mechanics can be formulated in a generalized manner by the so-called general form of the balance equation. A fixed *Eulerian* control volume Ω is considered and an extensive state variable $E(t)$ which is the volume integral of a scalar or vector entity $e[\underline{x}(x,y,z),t]$. $\underline{x}$ denotes a spatial vector with the components x, y, z and t stands for the time:

$$E(t) = \int_{\Omega} e(\underline{x},t)dV \tag{2.1}$$

The general form of the balance equation states that the volume integral of the temporal change of the entity e plus the fluxes $\underline{F}$ multiplied by the normal vector $\underline{n}$ and integrated over the surface $\Gamma = \partial\Omega$ minus the volume integral of sink or source terms $r(\underline{x},t)$ equals zero (see fig. 2.6). The fluxes consist of an advective part $\underline{v}(\underline{x},t)e(\underline{x},t)$ and a diffusive part $\underline{w}(\underline{x},t)$.

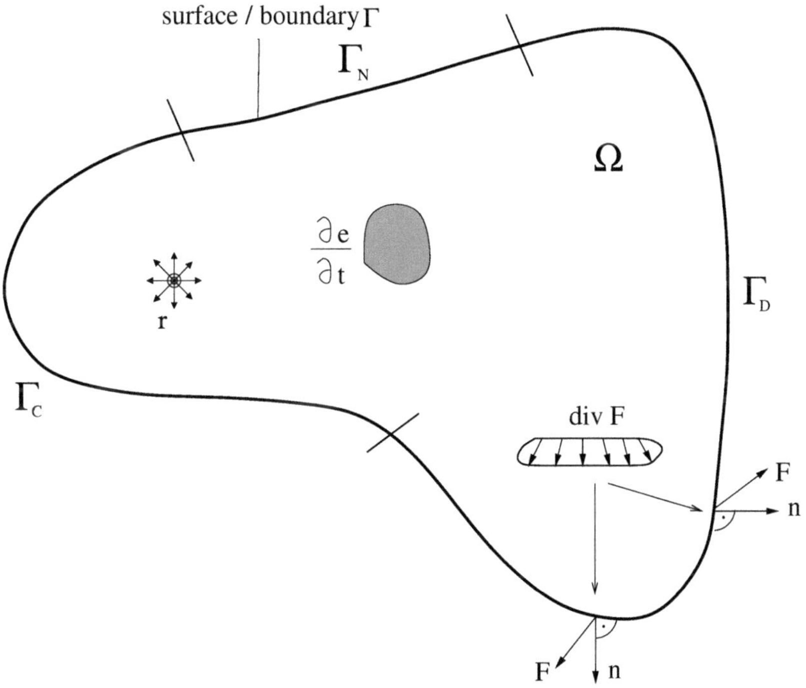

Fig. 2.6. Control volume Ω, after HELMIG (1997 [98])

In integral form, this leads to:

$$\int_{\Omega} \frac{\partial e}{\partial t} dV + \int_{\Gamma} (\underline{v}e + \underline{w})\underline{n}dO - \int_{\Omega} rdV = 0 \tag{2.2}$$

Applying the *Green-Gauss Integral Theorem*, the surface integral is transfered into a volume integral:

$$\int_{\Omega} [\frac{\partial e}{\partial t} + div\ (\underline{v}e + \underline{w}) - r]dV = 0 \tag{2.3}$$

As this equation can be applied to arbitrary control volumes, the integral form can be replaced by a differential form assuming a continuous integrand:

$$\frac{\partial e}{\partial t} + div\ (\underline{v}e + \underline{w}) - r = 0 \tag{2.4}$$

The continuity of functions $\partial e/\partial t$, $div\ (\underline{v}e + \underline{w})$ and r can be violated in so-called discontinuity areas, when jumps occur in the material properties, e.g. hydraulic conductivity, or in the state variables, e.g. sharp concentration fronts. In such cases, the control volume must be subdivided into subcontrol volumes, where the continuity condition is fulfilled, and certain jump conditions must be taken into account at the discontinuity.

For a unique solution of the initial boundary-value problem, initial conditions must be specified in the entire domain Ω as well as boundary conditions along the whole boundary Γ. The initial conditions describe the connection of the solution with the previous time and the initial state of the solution. They are given in a form:

$$e(x, y, z, t = 0) = e_0(x, y, z) \tag{2.5}$$

The boundary conditions, which may have different forms, represent the interaction with the surrounding domain. If a *Dirichlet-boundary condition* e_D is chosen, the solution function is prescribed along this boundary Γ_D. At boundaries Γ_N with *Neumann-boundary conditions* e_N, the derivative of the solution function in the direction of the exterior normal vector must be given. A *Cauchy-boundary condition* e_C is a linear combination of a *Dirichlet-* and *Neumann-*boundary condition. For further reading, see GÄRTNER (1987 [87]) or HELMIG (1997 [98]):

$$\begin{aligned} e(x, y, z, t) &= e_D(x, y, z, t) \quad on \quad \Gamma_D \\ \frac{\partial e(x, y, z, t)}{\partial \underline{n}} &= e_N(x, y, z, t) \quad on \quad \Gamma_N \\ \frac{\partial e(x, y, z, t)}{\partial \underline{n}} + \alpha e(x, y, z, t) &= e_C(x, y, z, t) \quad on \quad \Gamma_C \end{aligned} \tag{2.6}$$

2.3 Mathematical model concept for groundwater flow and transport processes

This section deals with flow processes of the (single) fluid phase water and transport processes of a single or several components in porous media (see figs. 2.7, 1.1, 1.3).

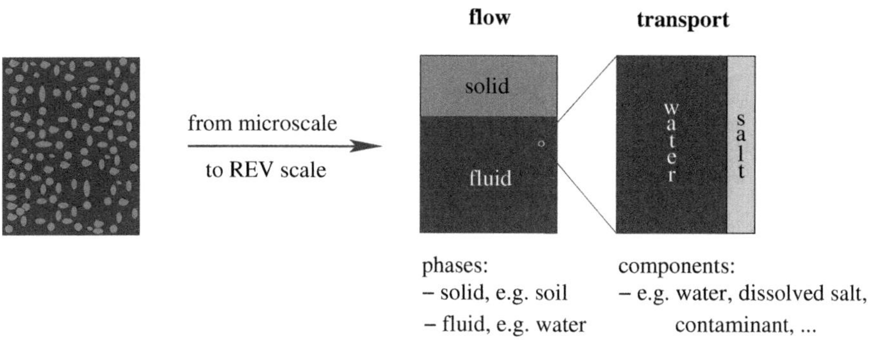

Fig. 2.7. Model concept, definition for phases and components

2.3.1 Flow processes

Continuity equation

If the general form of the balance equation (eq. 2.4) is applied to the mass balance, this means $e = \phi\rho_w$, $\underline{v} = \underline{v}_a$, $\underline{w} = 0$ and $r = q_w$. As the mean flux is much larger than its deviations, the diffusion/dispersion term is neglected. Generally, ϕ denotes the porosity of the porous medium which is the ratio of the void space to the whole volume. Here, ϕ stands for the effective porosity which only takes mobile water into account. When compared to the porosity, the effective porosity is always smaller, as it does not take water in dead-end pores or bounded water at the grain surfaces into account. For small-scale considerations, there are some empirical relationships for the effective porosity; for larger scales, this parameter is unknown a priori and must be estimated (see KINZELBACH (1992 [156])). ρ_w denotes the density of water and q_w a sink or source term of water. The pore velocity $\underline{v}_a$ in the void space is related to the *Darcy* or filter velocity by:

$$\underline{v}_a = \frac{\underline{v}_f}{\phi} \tag{2.7}$$

Thus, the mass-balance equation is:

$$\frac{\partial(\phi\rho_w)}{\partial t} + div\,(\rho_w \underline{v}_f) = q_w \tag{2.8}$$

The density and porosity depend slightly on the pressure p. Therefore, the first term in equation 2.8 is generally replaced by:

$$\frac{\partial(\phi\rho_w)}{\partial t} = \frac{\partial(\phi\rho_w)}{\partial p}\frac{\partial p}{\partial t} = \frac{\rho_w S_0}{\rho_{w0} g}\frac{\partial p}{\partial t} \tag{2.9}$$

S_0 stands for the specific storage coefficient (see LEGE et al. (1996 [164])), ρ_{w0} for a reference water density and g for gravity. Thus, equation 2.8 reads:

$$\frac{\rho_w S_0}{\rho_{w0} g}\frac{\partial p}{\partial t} + div\,(\rho_w \underline{v}_f) = q_w \tag{2.10}$$

If storage effects can be neglected, the last equations results in:

$$div\,(\rho_w \underline{v}_f) = q_w \tag{2.11}$$

If the density is a function of a tracer (see sec. 2.3.2) as salinity or temperature, an equation of state must be given. As an example, OLDENBURG, PRUESS (1995 [199]) determine the density of saline water ρ_w by the salinity S, the freshwater density ρ_f and the density of a concentrated brine ρ_b:

$$\frac{1}{\rho_w} = \frac{1-S}{\rho_f} + \frac{S}{\rho_b} \tag{2.12}$$

In such a case, density-induced flow can occur and flow and transport processes (see sec. 2.3.2) are coupled in both directions.

If the density is constant, equation 2.9 is transformed as follows:

$$\frac{S_0}{g\rho_w}\frac{\partial p}{\partial t} + div\,\underline{v}_f = \frac{q_w}{\rho_w} \tag{2.13}$$

Now, the piezometric head h is introduced together with the reference geodetic head z which corresponds to the vertical spatial coordinate:

$$h = \frac{p}{\rho_w g} + z \tag{2.14}$$

Inserting equation 2.14 into equation 2.13 results in:

$$S_0 \frac{\partial h}{\partial t} + div\,\underline{v}_f = \frac{q_w}{\rho_w} \tag{2.15}$$

If, in addition, stationary conditions are considered, equation 2.15 is simplified to:

$$div\,\underline{v}_f = \frac{q_w}{\rho_w} \tag{2.16}$$

Momentum equation

Generally, the momentum equation is replaced by the *Darcy law* in subsurface systems:

$$\underline{v_f} = -\underline{\underline{K}}_f \, grad\, h \tag{2.17}$$

In this equation, $\underline{\underline{K}}_f$ is the tensor for the hydraulic conductivity, which is symmetric:

$$\underline{\underline{K}}_f = \begin{bmatrix} K_{f,xx} & K_{f,xy} & K_{f,xz} \\ K_{f,yx} & K_{f,yy} & K_{f,yz} \\ K_{f,zx} & K_{f,zy} & K_{f,zz} \end{bmatrix} \tag{2.18}$$

The *Darcy* law has been determined experimentally. With some assumptions, it can also be derived from the *Navier-Stokes equations.* The most important simplification consists of neglecting inertia terms. Therefore, the *Darcy* law is only valid for slow laminar flows limited to *Reynolds numbers* smaller than 1.

In this context, the *Reynolds* number is defined as the ratio of inertia to viscous forces. It is given by:

$$Re = \frac{\underline{v}_f d}{\nu_w} \tag{2.19}$$

In this equation d denotes a characteristic length scale on the microscale, e.g. a grain diameter, and ν_w the kinematic viscosity of water. For larger *Reynolds* numbers, *Forchheimer's law*, which represents a non-linear relationship of $\underline{v}_f$ and $grad\, h$, can be applied (see BEAR (1972 [25])).

For large-scale simulations, the groundwater systems are often inhomogeneous and anisotropic. This is caused by different layers, geological deposition or by the shape of the material. A material consisting of long flat grains has a higher hydraulic conductivity in the longitudinal direction of the grains than in the perpendicular directions. If the coordinate directions coincide with the main directions of the hydraulic conductivity, the tensor simplifies to:

$$\underline{\underline{K}}_f = \begin{bmatrix} K_{f,x} & 0 & 0 \\ 0 & K_{f,y} & 0 \\ 0 & 0 & K_{f,z} \end{bmatrix} \tag{2.20}$$

If, additionally, the subsurface material is isotropic, this tensor reads as:

$$\underline{\underline{K}}_f = \begin{bmatrix} K_f & 0 & 0 \\ 0 & K_f & 0 \\ 0 & 0 & K_f \end{bmatrix} \tag{2.21}$$

The hydraulic conductivity tensor can be determined by experiments in the laboratory or in the field. If the main flow directions do not coincide with the global coordinate directions, the experiments are carried out in the main flow directions, and then the parameters are transformed into the global coordinate system.

The hydraulic conductivity tensor can be written as follows:

$$\underline{\underline{K}}_f = \underline{\underline{K}} \; \frac{\rho_w g}{\mu_w} \tag{2.22}$$

Here, $\underline{\underline{K}}$ represents the intrinsic permeability tensor, which is only a characteristic of the porous medium, and μ_w the dynamic viscosity of water.

For flow through fractures which are discretely taken into account (see sec. 2.1.2), a special law for the permeability can be chosen based on the *Navier-Stokes* equations, a constant aperture b and smooth walls (see WOLLRATH (1990 [264])):

$$K = \frac{b^2}{12} \tag{2.23}$$

Usually, the *Darcy* law 2.17 is put into the continuity equation 2.16. This leads to one partial differential equation for the piezometric head:

$$div\,(-\underline{\underline{K}}_f \; grad\, h) = \frac{q_w}{\rho_w} \tag{2.24}$$

For about two decades, there have been controversial and ongoing discussions concerning the *Darcy* law for single and multiphase flow, its validity as well as extensions of it; this is explained briefly in HELMIG (1997 [98]). Nevertheless, there is still almost no practicable alternative.

2.3.2 Transport processes

The main mass-transport mechanisms in porous media are advection / convection, diffusion and dispersion. Other processes such as reaction, adsorption or decay are not considered here, see KINZELBACH (1992 [156]). Advection or convection describes the movement of a tracer with the flow field in horizontal or vertical direction without changing the shape of the concentration isoareas (see fig. 2.8). Diffusion processes are a result of *Brown's* molecular movement which leads to a compensation of concentration differences, and thus causes net transport processes in the direction of lower concentrations. Diffusion is a purely physical process - in contrast to dispersion. Therefore, a diffusive spreading of a substance is independent of the direction (see fig. 2.8). Dispersion represents all transport effects which are caused by inhomogeneities of the flow field below the REV scale.

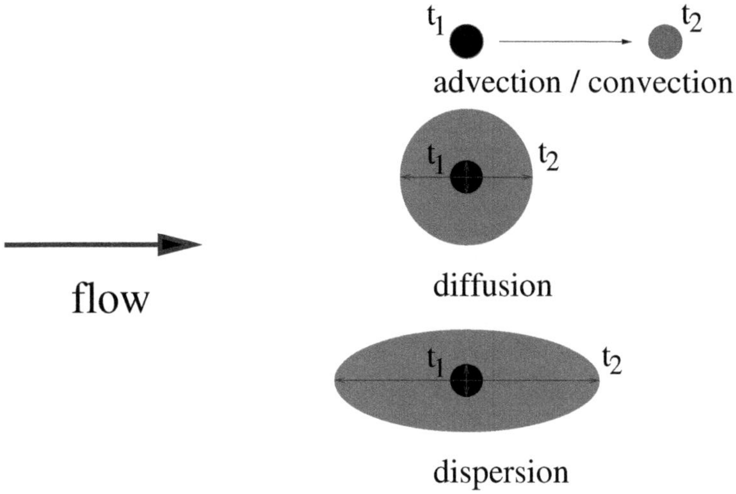

Fig. 2.8. Advection / convection, diffusion and dispersion processes, after BARLAG (1997 [13])

On the microscale, dispersion is a result of the velocity profile within a pore, the variable pore sizes as well as the flow processes around the grains (see fig. 2.9, left). For bigger REVs, fluctuating velocities due to inhomogeneities of the aquifer lead to macrodispersion (see fig. 2.9, middle and right). It is obvious that dispersion is scale-dependent. Both diffusion and dispersion are based on a *Fickian* law.

The transport equation for a tracer c is obtained from the general form of the balance equation (eq. 2.4) if $e = \phi \rho_w c$, $\underline{v} = \underline{v}_a$, $\underline{w} = -\phi \underline{\underline{D}}_{hyd} grad\,(\rho_w c)$ and $r = q_c$. Here, $\underline{\underline{D}}_{hyd}$ denotes the hydrodynamical dispersion tensor and q_c a tracer sink or source term which can take tracer sinks or sources as well as reaction, adsorption or decay processes into account. The transport equation is determined as follows:

$$\frac{\partial(\phi \rho_w c)}{\partial t} + div\,[\underline{v}_a \phi \rho_w c - \phi \underline{\underline{D}}_{hyd} grad\,(\rho_w c)] = q_c \tag{2.25}$$

The hydrodynamical dispersion tensor $\underline{\underline{D}}_{hyd}$ consists of the sum of the molecular diffusion tensor $\underline{\underline{D}}_{mol}$, which is the product of the molecular diffusion coefficient D_{mol} and the unit tensor $\underline{\underline{I}}$, and the mechanical dispersion tensor $\underline{\underline{D}}_{mech}$:

$$\underline{\underline{D}}_{hyd} = \underline{\underline{D}}_{mol} + \underline{\underline{D}}_{mech} = D_{mol}\underline{\underline{I}} + \underline{\underline{D}}_{mech} = \begin{bmatrix} D_{xx} & D_{xy} & D_{xz} \\ D_{yx} & D_{yy} & 0 \\ D_{zx} & 0 & D_{zz} \end{bmatrix} \tag{2.26}$$

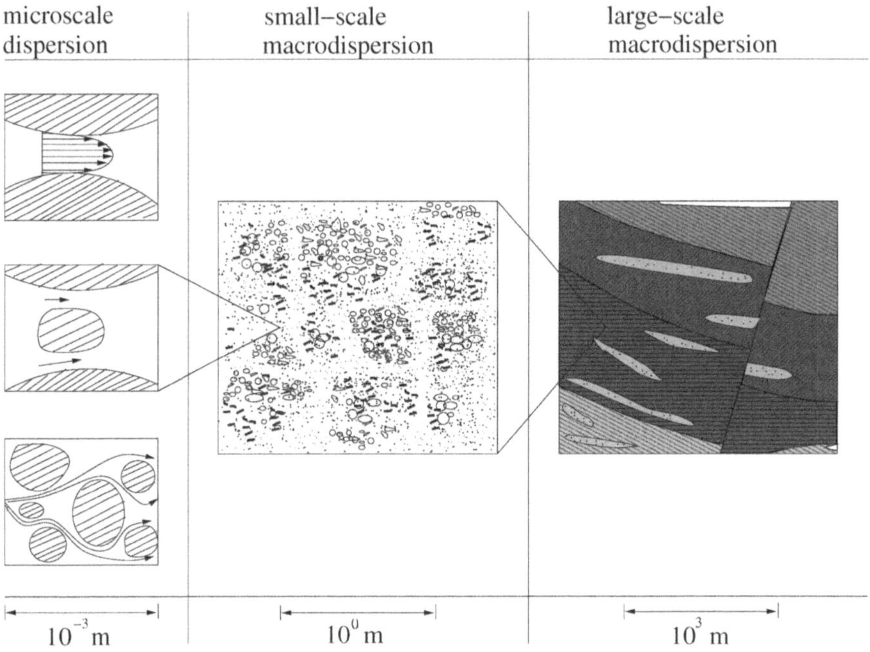

Fig. 2.9. Reasons for the variability of the flow field on different spatial scales, after KINZELBACH (1992 [156])

The hampering of the spreading by the grains for the molecular diffusion tensor is taken into account here by the multiplication with the effective porosity. Alternative approaches are given in BEAR, BACHMAT (1990 [26]) and LEGE et al. (1996 [164]).

The dispersion tensor given in equation 2.26 is anisotropic, even in an isotropic medium. It has a diagonal shape if one coordinate direction coincides with the flow direction:

$$\underline{\underline{D}}_{mech} = \begin{bmatrix} D_L & 0 & 0 \\ 0 & D_T & 0 \\ 0 & 0 & D_T \end{bmatrix} \tag{2.27}$$

Generally, the dispersion in flow direction expressed by the longitudinal dispersion coefficient D_L is about one order of magnitude greater than the dispersion in the orthogonal directions expressed by the transversal dispersion coefficient D_T. The mechanical dispersion coefficients contain information about the aquifer and the flow field:

$$D_L = \alpha_L \, |\underline{v}_a| \quad , \quad D_T = \alpha_T \, |\underline{v}_a| \tag{2.28}$$

In this equation, α_L and α_T stand for the longitudinal and transversal dispersion lengths. The most common description for the dispersion tensor is given by *Scheidegger*:

$$
\begin{aligned}
D_{xx} &= \alpha_L \frac{v_{ax}^2}{|\underline{v}_a|} + \alpha_T \frac{v_{ay}^2}{|\underline{v}_a|} + \alpha_T \frac{v_{az}^2}{|\underline{v}_a|} + D_{mol} \\
D_{xy} &= D_{yx} = (\alpha_L - \alpha_T) \frac{v_{ax}\, v_{ay}}{|\underline{v}_a|} \\
D_{yy} &= \alpha_T \frac{v_{ax}^2}{|\underline{v}_a|} + \alpha_L \frac{v_{ay}^2}{|\underline{v}_a|} + \alpha_T \frac{v_{az}^2}{|\underline{v}_a|} + D_{mol} \\
D_{xz} &= D_{zx} = (\alpha_L - \alpha_T) \frac{v_{ax}\, v_{az}}{|\underline{v}_a|} \\
D_{zz} &= \alpha_T \frac{v_{ax}^2}{|\underline{v}_a|} + \alpha_T \frac{v_{ay}^2}{|\underline{v}_a|} + \alpha_L \frac{v_{az}^2}{|\underline{v}_a|} + D_{mol}
\end{aligned}
\qquad (2.29)
$$

As already mentioned before, dispersion is highly scale-dependent. In the laboratory, $0.0001 < \alpha_L < 0.01m$ was measured for homogeneous sands and $0.07 < \alpha_L < 0.7$ was found for natural subsurface material, e.g. gravel. On the field scale, the dispersion lengths are about 4 to 5 orders of magnitude larger. This is a consequence of the macrodispersion which results from the inhomogeneities of the aquifer and does not occur in the laboratory. Macrodispersion increases with the length of the transport way (see fig. 2.10). This is caused by the influence of bigger inhomogeneities, but this effect is limited as soon as the inhomogeneities are represented by the computation of the flow field. On the field scale, the molecular diffusion is often negligible compared to mechanical dispersion, i.e. $D_{mol} << D_L$ and $D_{mol} << D_T$. For practical applications on scales larger than in the laboratory, the dispersion lengths are calibration parameters, and thus contain all other unknowns , i.e. missing information about the geological structures or effects not included in the model concept. Further information is given in KINZELBACH (1992 [156]).

If the density and porosity are constant, equation 2.25 can be simplified by applying the product rule and using the continuity equation 2.16:

$$
\frac{\partial c}{\partial t} + \underline{v}_a grad\, c - div\,(\underline{\underline{D}}_{hyd} grad\, c) = \frac{q_c - q_w c}{\rho_w \phi} \qquad (2.30)
$$

If the density is dependent on the tracer concentration, the flow and transport processes are coupled in both directions. Then equation 2.10 or 2.11 is coupled with 2.25 taking an equation of state, e.g. equation 2.12, and the *Darcy* law (eq. 2.17) into account. If the density and the porosity are constant, the flow field can be determined directly by equation 2.24 and the *Darcy* law (eq. 2.17), independently of the tracer simulation. Afterwards, the transport is determined with equation 2.30, i.e. flow and transport are coupled only in one direction. It should be mentioned that the (REV) scales for the flow and transport computations must match as well as the horizontal and vertical

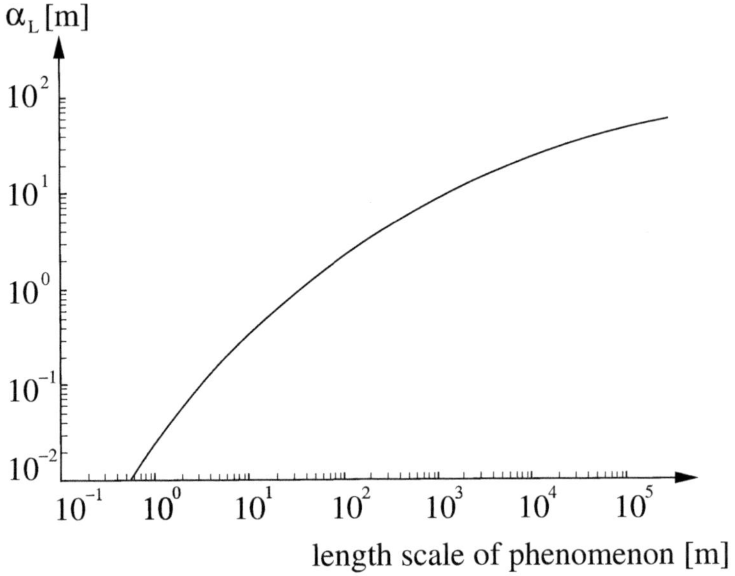

Fig. 2.10. Scale dependency of the dispersion, after KINZELBACH (1992 [156])

averaging length scales, and that the way which is passed by a tracer plume must be large compared to the REV length scale.

The flow and transport equation are both linear, the flow equation is of elliptic or parabolic type and the transport equation is of mixed parabolic / hyperbolic type. For further reading, see BEAR (1972 [25]), HÄFNER (1992 [95]), LEGE et al. (1996 [164]), KOLDITZ (1997 [152]), or SCHOTTING (1998 [232]).

2.4 Mathematical model concept for two-phase flow processes in the subsurface

If two fluid phases are not or only slightly miscible into each other, a two-phase flow model concept for a porous medium must be applied (see figs. 2.11, 1.1, 1.3). Generally, the phases water, gas (e.g. air, methane) and NAPL (Non-Aqueous Phase Liquid) are treated. A NAPL lighter than water is called LNAPL, one denser than water is called DNAPL.

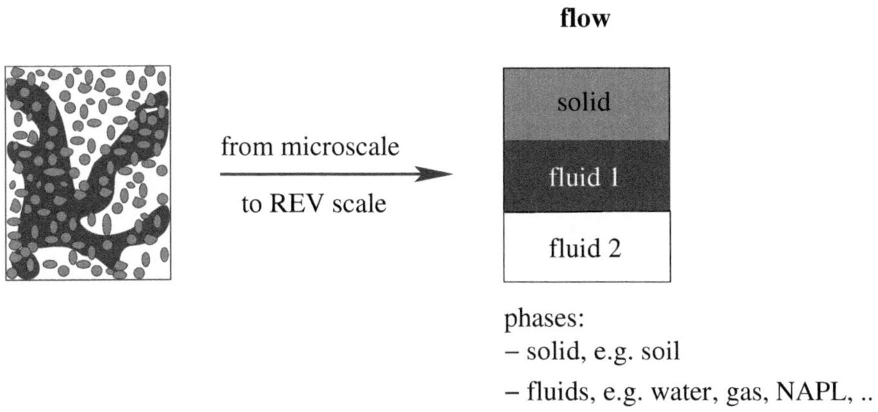

Fig. 2.11. Model concept, definition for phases

2.4.1 Continuity equation

In a two-phase system, the continuity equation must be fulfilled for each phase. For the general form of the balance equation (eq. 2.4), the following relations are valid: $e = \phi_\alpha \rho_\alpha$, $\underline{v} = \underline{v}_{f\alpha}$ and $\underline{w} = 0, r = q_\alpha$. Here, α is a subscript for the phases, the wetting phase w (e.g. water) and the non-wetting phase n (e.g. NAPL, gas) respectively. As in section 2.3.1, the continuity equation for phase α is obtained:

$$\frac{\partial(\phi_\alpha \rho_\alpha)}{\partial t} + div\,(\rho_\alpha \underline{v}_{f\alpha}) = q_\alpha \tag{2.31}$$

ϕ_α takes the fact into account that the void space is only partially filled with phase α. The saturation S_α is introduced as the ratio of the pore space filled with phase S_α to the whole pore space:

$$S_\alpha = \frac{\phi_\alpha}{\phi} \quad \Leftrightarrow \quad \phi_\alpha = S_\alpha \phi \tag{2.32}$$

If equation 2.32 is inserted into equation 2.31, this leads to:

$$\frac{\partial(S_\alpha \phi \rho_\alpha)}{\partial t} + div\,(\rho_\alpha \underline{v}_{f\alpha}) = q_\alpha \tag{2.33}$$

2.4.2 Momentum equation

First, equation 2.17 is rearranged with equations 2.22 and 2.14:

$$\underline{v_f} = -\underline{\underline{K}}\frac{\rho_w g}{\mu_w} grad\,(\frac{p}{\rho_w g} + z) = -\frac{1}{\mu_w}\underline{\underline{K}}(grad\,p - \rho_w \underline{g}) \tag{2.34}$$

The vector of the gravitational acceleration is given as:

$$\underline{g} = \begin{bmatrix} 0 \\ 0 \\ -g \end{bmatrix} \tag{2.35}$$

A number of experiments have shown that a generalized form of the *Darcy* law can describe the so-called *Darcy velocity* of each phase in a multiphase system:

$$\underline{v}_{f\alpha} = -\frac{1}{\mu_\alpha}\underline{\underline{K}}_\alpha (grad\,p_\alpha - \rho_\alpha \underline{g}) \tag{2.36}$$

The effective permeability $\underline{\underline{K}}_\alpha$ is defined as the product of the relative permeability $k_{r\alpha}$ and the intrinsic permeability $\underline{\underline{K}}$ (eq. 2.22):

$$\underline{\underline{K}}_\alpha = k_{r\alpha}\underline{\underline{K}} \tag{2.37}$$

The generalized *Darcy* law can be formulated as follows:

$$\underline{v}_{f\alpha} = -\frac{k_{r\alpha}}{\mu_\alpha}\underline{\underline{K}}(grad\,p_\alpha - \rho_\alpha \underline{g}) \tag{2.38}$$

In the last equation, the term $\frac{k_{r\alpha}}{\mu_\alpha}$ represents the mobility λ_α.

Although the relative permeability takes the mutual hampering of the two phases in the pore space into account, the generalized *Darcy* law does not account for a momentum exchange between the two phases via the interface. This seems to be one of the major shortcomings of the generalized *Darcy* law.

2.4.3 Constitutive relations

Capillary pressure-saturation relationship

In a two-phase system, there is a fundamental relationship between the wetting and non-wetting phase saturations S_w, S_n and the capillary pressure p_c. In the following, the two most common models of BROOKS, COREY (1964 [46], BC) and VAN GENUCHTEN (1980 [92], VG) which formulate the capillary pressure as a function of the saturation are introduced:

$$BC: \quad p_c(S_w) = p_d S_e^{-\frac{1}{\lambda}} \qquad \text{for} \quad p_c \geq p_d \tag{2.39}$$

$$VG: \quad p_c(S_w) = \frac{1}{\alpha}(S_e^{-\frac{1}{m}} - 1)^{\frac{1}{n}} \qquad \text{for} \quad p_c > 0 \tag{2.40}$$

$$S_e(p_c) = \frac{S_w - S_{wr}}{1 - S_{wr}} \tag{2.41}$$

In the BC model, p_d stands for the entry pressure, which is the capillary pressure required to displace the wetting phase from the largest pores. The BC parameter λ characterizes the grain-size distribution. A small value describes a single grain-size material, while a large value indicates highly non-uniform material. S_e denotes the effective (water) saturation and S_{wr} the residual water saturation. In the VG model, $n, m = 1 - 1/n, \alpha$ are form parameters characterizing the pore-space geometry. Generally, the BC and VG parameters are determined experimentally. However, it is also possible to estimate these parameters from grain-sum curves (see ARYA, PARIS (1981 [6]), JONASSON (1989 [138])). The BC parameters can be converted into the VG parameters and vice versa (see SHETA (1999 [230])). In figure 2.12, the capillary pressure-saturation relationship is shown for the BC and VG models on equal physical conditions.

As there is often a lack of experimental data, a scaling factor $p_c/\sqrt{k\phi}$ was developed by LEVERETT (1941 [167]) to transfer known parameters, for example, to another medium with a different permeability or porosity or to another two-phase system (from water-gas to water-NAPL) with a different capillary pressure. The assumption is that the scaling factor equals the one in the other porous medium or in another two-phase system.

In fractures which are discretely taken into account (see sec. 2.1.2), two different possibilities exist for the capillarity. On the one hand, the fracture can be treated as a porous medium as already described. On the other hand, the capillarity can be formulated as a function of the fracture-aperture distribution by a geostatistical model (see PRUESS, TSANG (1990 [211])).

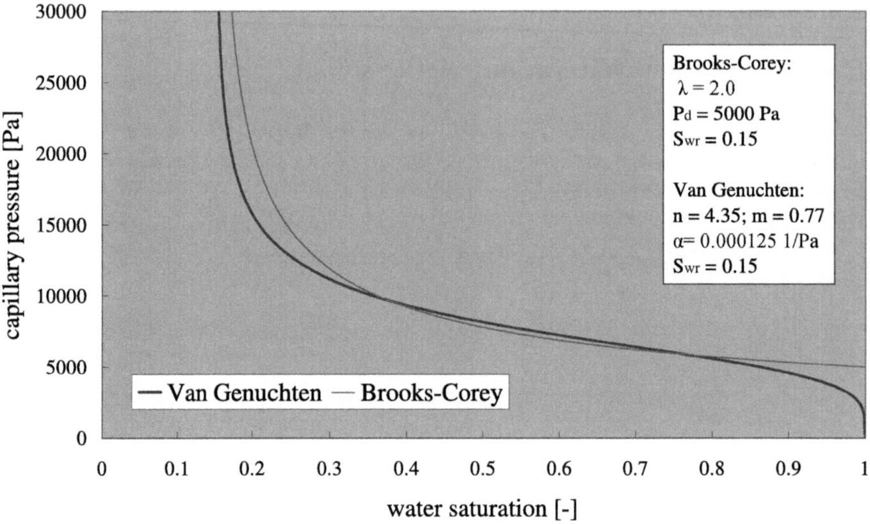

Fig. 2.12. $p_c - S_w$ relationship afer BC and VG, on equal physical conditions

Finally, it must be mentioned that the capillary pressure-saturation relationship shows a hysteresis in transient cases. Whether this phenomenon is significant or not depends on a number of things which are not discussed here (see SHETA (1999 [230])).

Relative permeability-saturation relationship

Based on pore-network models, the relative permeability-saturation relationships are determined by integrating over the capillary pressure-saturation relationship. The BROOKS, COREY (1964 [46], BC) model stems from the pore-network model of BURDINE (1953 [52]) and is as follows:

$$BC: \quad k_{rw} = S_e^{\frac{2+3\lambda}{\lambda}} \tag{2.42}$$

$$BC: \quad k_{rn} = (1 - S_e)^2 \left(1 - S_e^{\frac{2+\lambda}{\lambda}}\right) \tag{2.43}$$

The VAN GENUCHTEN model (1980 [92], VG) is determined by the pore-network model of MUALEM (1976 [182]) and has the form:

$$VG: \quad k_{rw} = S_e^{1/2} \left[1 - \left(1 - S_e^{\frac{1}{m}}\right)^m\right]^2 \tag{2.44}$$

$$VG: \quad k_{rn} = (1 - S_e)^{1/3} \left[1 - S_e^{\frac{1}{m}}\right]^{2m} \tag{2.45}$$

The parameters are the same as in the context of the capillary pressure-saturation relationship. In figure 2.13, the relative permeability-saturation relationship is given for the BC and VG models on equal physical conditions.

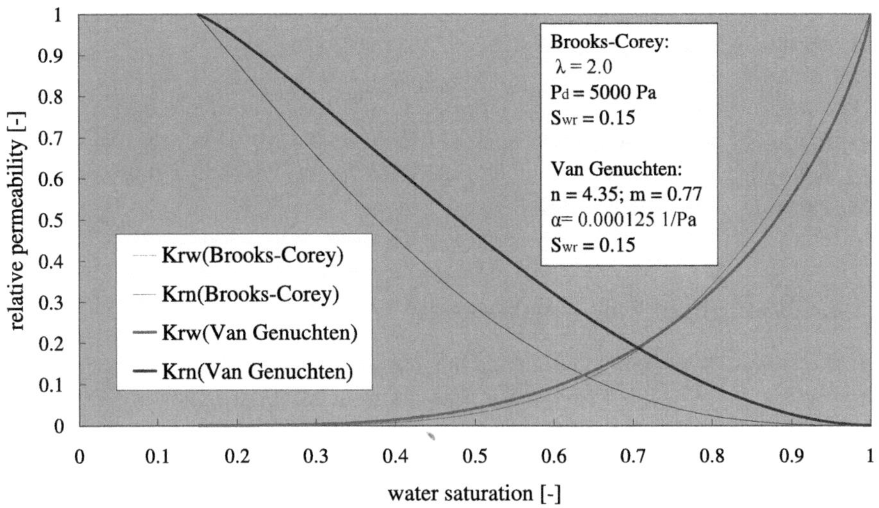

Fig. 2.13. $k_r - S_w$ relationship afer BC and VG, on equal physical conditions

In discrete fractured systems, linear relative permeability-saturation relationships can be used (see ROMM (1966 [223])) or other relations taking into account the fracture roughness, fracture aperture and the fracture contact area (see PRUESS, TSANG (1990 [211])).

Density, viscosity and porosity

Generally, the density of fluids depends on the pressure and temperature or even other parameters, see equation 2.12. For liquids, pressure dependence is often neglected and, consequently, liquids are assumed to be incompressible. Gas density is highly pressure-dependent, while the influence of the temperature is minor. Gas density is determined by the *real gas law*:

$$\rho = \frac{p}{ZRT} \tag{2.46}$$

In this equation, Z stands for the real gas factor, R for the universal gas constant and T for the temperature. If $Z = 1$, the real gas law simplifies to the *ideal gas law*.

The dynamic viscosity of liquids and gases is primarily determined by the temperature. It is important to note that the viscosity of gases is about two orders of magnitude lower than that of fluids, i.e. the mobility λ of gases (see sec. 2.4.2) is about two orders of magnitude higher.

Functional relationships for the density and viscosity of liquids and gases are given in the INTERNATIONAL FORMULATION COMMITTEE (1967 [128]) and REID et al. (1987 [216]).

The constitutive relation for the porosity as a function of pressure and temperature is given, for example, in HELMIG (1997 [98]). If effects like swelling or shrinking are considered, special constitutive relationships must be determined.

2.4.4 Two-phase flow equations

If the generalized *Darcy* law 2.38 is inserted into the continuity equation 2.33, the following system of two-phase flow differential equations is obtained:

$$\frac{\partial(S_\alpha \phi \rho_\alpha)}{\partial t} - div \left[\rho_\alpha \frac{k_{r\alpha}}{\mu_\alpha} \underline{\underline{K}} (grad\, p_\alpha - \rho_\alpha \underline{g}) \right] - q_\alpha = 0 \tag{2.47}$$

This system is completed by two further algebraic conditions. The pore volume is completely filled with the wetting and non-wetting phases:

$$S_w + S_n = 1 \tag{2.48}$$

At the interface between the two phases, a jump in the pressure which is given by the capillary pressure (see sec. 2.4.3) occurs:

$$p_n - p_w = p_c \tag{2.49}$$

Thus, two of the four unknowns p_w, p_n, S_w and S_n can be eliminated. Due to the non-linear dependencies of the capillary pressure and the relative permeability, the completed system is highly non-linear. This can even be reinforced if heterogeneous structures with strongly varying properties are investigated. Alternative formulations or choices of the primary variables have been developed, mainly depending on the choice of boundary conditions. They are briefly explained here; further reading is giving in HELMIG (1997 [98]).

If a *pressure formulation* is used, the pressures of the wetting and the non-wetting phases p_w and p_n are the primary variables and the resulting system is of mixed parabolic / hyperbolic type. This formulation is based on an inverse function which describes the saturation depending on the capillary pressure. The inverse function only exists if the capillary-pressure gradient is greater than zero ($\frac{\partial p_c}{\partial S_w} > 0$). With such a function, the saturations can be eliminated.

However, in many practical examples, for example, in the case of discrete fractures or transitions between heterogeneities, the capillary pressure gradient is very small or even zero, so that the pressure formulation cannot be chosen. Consequently, the application range of the pressure formulation is rather limited.

By introducing the total velocity of the two-phase system $v_t = v_w + v_n$ and several transformations, the *saturation formulation* is determined. The system is reduced to one equation with the primary variables S_w or S_n. One major problem results of the effect that the total velocity must be known in advance and another of the hyperbolic character for small capillary pressure gradients.

For many cases, the *pressure-saturation formulation* is the most suitable, offering the possibilities p_w, S_n or p_n, S_w or p_w, S_w or p_n, S_n as the primary variables. With the relations

$$\begin{aligned} grad\, p_n &= grad\,(p_c + p_w) = grad\, p_c + grad\, p_w \\ \frac{\partial S_w}{\partial t} &= \frac{\partial(1 - S_n)}{\partial t} = -\frac{\partial S_n}{\partial t} \end{aligned} \tag{2.50}$$

the differential equations of the wetting and non-wetting phase read as follows:

wetting phase:

$$-\frac{\partial(S_n \phi \rho_w)}{\partial t} - div\left[\rho_w \frac{k_{rw}}{\mu_w} \underline{\underline{K}}(grad\, p_w - \rho_w \underline{g})\right] - q_w = 0 \tag{2.51}$$

non-wetting phase:

$$\frac{\partial(S_n \phi \rho_n)}{\partial t} - div\left[\rho_n \frac{k_{rn}}{\mu_n} \underline{\underline{K}}(grad\, p_c + grad\, p_w - \rho_n \underline{g})\right] - q_n = 0 \tag{2.52}$$

These two equations are strongly coupled, highly non-linear and of mixed parabolic / hyperbolic type. The major advantage lies in the fact that they are not limited to small capillary-pressure gradients, i.e. the pressure-saturation formulation can be applied to discrete fractured systems and heterogeneous media.

For further reading on two-phase flow in porous media, see PRUESS (1991 [210]), HELMIG (1997 [98]) or WHITE, OOSTROM (2000 [261]).

2.5 Mathematical model concept for two-phase / multicomponent flow and transport processes in the subsurface

If the solubility of components in fluid phases as well as the transition of components into other phases is considered, a multiphase / multicomponent flow and transport model concept for a porous medium must be chosen (see figs. 1.1, 1.3). This is described here, using the example of the two phases water (subscript w) and gas (subscript g). The water phase consists of the (main) component water (superscript w), dissolved salt (superscript s) and dissolved methane (CH_4, superscript m). The gas phase contains methane (superscript m) and water vapor (superscript w). Local thermal and chemical equilibrium between the phases is assumed. However, mechanical equilibrium in a porous medium is generally violated between the phase boundaries due to the pressure jump which is caused by the capillary pressure (eq. 2.49). Methane dissolved in water degasses if the solubility is exceeded which can be caused, for example, by methane production or by decreasing pressure. Evaporation and condensation are of minor importance if exchanges of thermal energy are not considered, as is the case here. Generally, dissolution and vaporization processes are determined with the *Raoult law* and the *Henry law*. The salt component is restricted to the water phase. The phases, components and transfer processes are shown in figure 2.14.

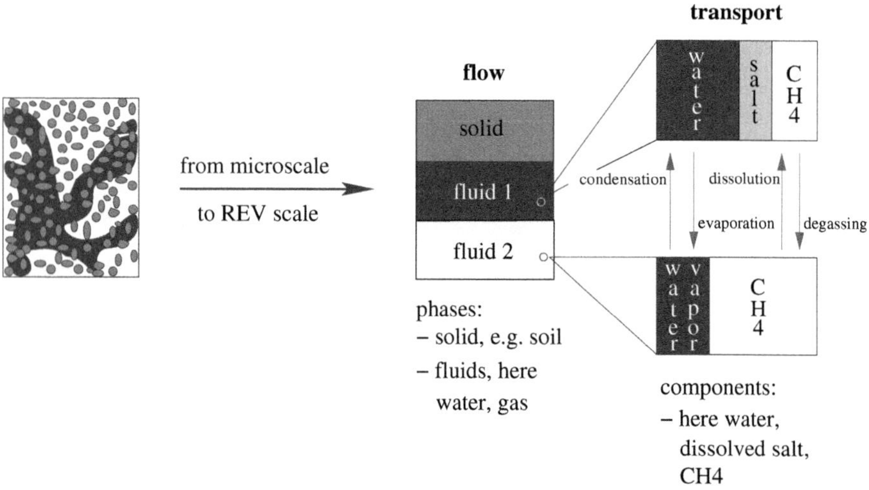

Fig. 2.14. Model concept, definition for phases, components and transfer processes

2.5.1 Continuity and momentum equation

In a two-phase / three-component model, the continuity equation is applied to each component by summing over the phases. For the general form of the balance equation (eq. 2.4) this means: $e = \phi \rho_\alpha S_\alpha x_\alpha^\kappa$, $\underline{v} = \underline{v}_{a\alpha}/S_\alpha$, $\underline{w} = -\phi \underline{\underline{D}}_{hyd\alpha}^\kappa grad(\rho_\alpha x_\alpha^\kappa)$ and $r = q_\alpha^\kappa$. Here, α is a subscript for the phases and κ is a superscript for the components. As in sections 2.3.1 and 2.4.1, the continuity equation for component κ is obtained:

$$\frac{\partial(\phi\, \rho_\alpha\, S_\alpha\, x_\alpha^\kappa)}{\partial t} + div\, \left[\rho_\alpha\, \underline{v}_{f\alpha}\, x_\alpha^\kappa - \phi\, D_{hyd\alpha}^\kappa grad\, (\rho_\alpha\, x_\alpha^\kappa)\right] = q_\alpha^\kappa \quad , \quad \alpha = w, g \tag{2.53}$$

In this equation, x_α^κ denotes the mole fraction. The other variables are explained in sections 2.3 and 2.4.

The generalized *Darcy* law (eq. 2.38) serves as the momentum equation in the same way as described in section 2.4.2.

2.5.2 Constitutive relationships

The same capillary pressure-saturation and relative permeability-saturation relationships as those explained in section 2.4.3 are used here.

Functional relationships for the density and viscosity are explained in section 2.4.3. If, additionally, the density or viscosity is a function of the composition of the water or gas phase, constitutive relations must be given, see INTERNATIONAL FORMULATION COMMITTEE (1967 [128]) or REID et al. (1987 [216]).

The mole fraction of water vapor in the gas phase is obtained by:

$$x_g^w = \frac{p_g^w}{p_g} \tag{2.54}$$

Here, p_g^w stands for the partial pressure of water vapor in the gas phase, and p_g represents the gas-phase pressure (total pressure). p_g^w is equal to the saturation pressure p_{sat}^w, presuming that water is present as a liquid phase at the same time. For both components in the gas phase, the validity of the *ideal gas law* is assumed:

$$p = \frac{n\, R T}{V} \tag{2.55}$$

In this equation, n represents the number of molecules (in mol), R the universal gas constant, T the temperature, and V the volume of the gas. The equation of state for an ideal gas is also valid for a mixture of ideal gases. Then, the total gas pressure is related to the partial pressures of the components by *Dalton's law*:

$$p_g = p_g^w + p_g^m \tag{2.56}$$

For the computation of the mole fractions in the water phase, *Henry's law* is used for the methane component:

$$x_w^m = \frac{p_g^m}{H_w^m} \tag{2.57}$$

H_w^m denotes the *Henry coefficient.*

2.5.3 Two-phase / three-component flow and transport equations

All together, there are 10 degrees of freedom consisting of 3 unknown mole fractions per phase, 2 unknown saturations and 2 unknown pressures. On the other hand, there is the same number of equations:

- 3 continuity equations (eq. 2.53) together with the generalized *Darcy* law (eq. 2.38)
- 2 supplementary conditions for the mole fractions (eq. 2.58)
- 1 supplementary condition for the saturation (eq. 2.48)
- 1 supplementary condition for the capillary pressure (eq. 2.49)
- 3 chemical equilibria (*Henry's* law (eq. 2.57), saturation-pressure condition (eq. 2.54), salt only in the water phase)

The mole fractions x_α^κ of components κ in phase α obey the constraint:

$$x_\alpha^w + x_\alpha^s + x_\alpha^m = 1 \tag{2.58}$$

The additional unknown partial pressures are determined with *Dalton's* law (eq. 2.56).

If the aforementioned equations are put together, the system is reduced to three primary variables. The choice of the primary variables depends on the phase state and can vary in space and time as a result of the appearance or disappearance of phases. Three different phase states which are listed in table 2.1 are possible. For the numerical implementation, a special variable-substitution algorithm has been developed by CLASS (2000 [62]).
The equations are highly non-linear due to the non-linear constitutive relationships and of mixed parabolic / hyperbolic type. If heterogeneous structures with strongly varying properties are investigated, the numerical complexity is even reinforced.

phase state	present phases	primary variables	appearance of phase: water	appearance of phase: gas
1	w, g	$S_w,\ x_w^s,\ p_g$	-	-
2	w	$x_w^m,\ x_w^s,\ p_w$	-	$p_{sat}^w + H_w^m x_w^m > p_g$
3	g	$x_g^w,\ x_g^m,\ p_g$	$x_g^w p_g > p_{sat}^w$	-

Table 2.1. Phase states, primary variables, and criteria for the appearance of a phase

Further reading on multiphase / multicomponent flow and transport processes in porous media can be found in PRUESS (1991 [210]), FALTA et al. (1995 [78]), HELMIG (1997 [98]), WHITE, OOSTROM (2000 [261]), or CLASS (2000 [62]).

2.6 Mathematical model concept for flow and transport processes in surface water

In this section, flow processes of the (single) phase water and transport processes of a single or several components in surface-water systems are dealt with (see figs. 2.15, 1.1, 1.2). It is restricted here to long waves (see sec. 2.1.3) and further to the *shallow-water equations* which are a special case of the *Navier-Stokes equations.* Water is regarded to be incompressible and viscous. A free water surface, a hydrostatic pressure distribution and small bottom inclinations are assumed.

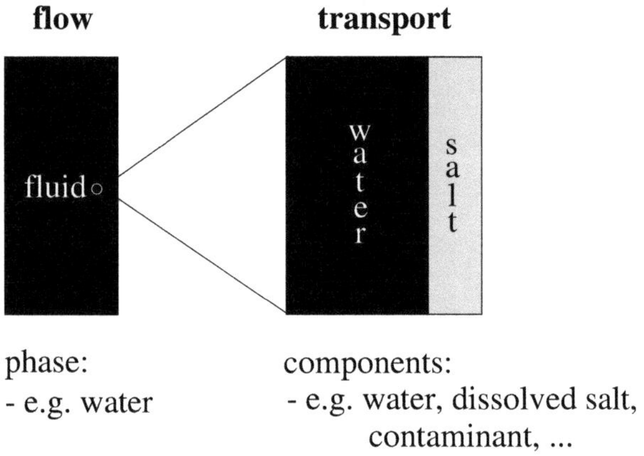

Fig. 2.15. Model concept, definition for phase and components

2.6.1 Flow processes

Continuity equation

If the general form of the balance equation (eq. 2.4) is applied to the mass balance, this originally means $e = \rho_w$, $\underline{w} = 0$ and $r = q_w$:

$$\frac{\partial(\rho_w)}{\partial t} + div\ (\rho_w \underline{v}) = q_w \tag{2.59}$$

Here, ρ_w denotes the density of water, q_w a sink or source term of water and $\underline{v}$ the velocity vector of the free-surface flow with the components v_x, v_y, v_z.

Due to incompressibility pressure-dependent density variations are negligible when compared to density variations caused by salinity or temperature influences. Usually, the *Boussinesq approximation* is applied, i.e. that density variations only have to be taken into account in the gravity term of the momentum equation 2.62 or 2.65. Therefore, equation 2.59 is simplified:

$$div\, \underline{v} = q_w / \rho_w \tag{2.60}$$

Momentum equation

In free-surface water systems, a hydrostatic pressure distribution may be assumed, if the wave length is at least twenty times larger than the water depth, because then the total vertical accelerations can be neglected. This is the case here. It should be mentioned, that a hydrostatic pressure distribution may not be given e.g. for short wave motions, deep water or flow around obstacles. In the vertical momentum equation, viscous terms can be neglected. If the general form of the momentum equation (eq. 2.4) is regarded, e is a vector entity and the following holds: $e = \rho_w h \underline{v}$, $\underline{w} = -\rho_w h \underline{\underline{\nu}}_w grad\, \underline{v}$ and $r = h(\underline{f} - grad p)$. Thus, the so-called *conservative form* of the momentum equation is obtained:

$$\frac{\partial(\rho_w h \underline{v})}{\partial t} + div\,(\rho_w h \underline{v}\underline{v} - \rho_w h \underline{\underline{\nu}}_w grad\, \underline{v}) = h(\underline{f} - grad p) \tag{2.61}$$

In this equation, h stands for the water depth, $\underline{\underline{\nu}}_w$ for the turbulent viscosity, p for the pressure and $\underline{f}$ for a momentum source term, for example the *Coriolis force* (eqs. 2.72, 2.73). Some transformations, where among other things the density of water is cancelled out of some terms because of the *Boussinesq* approximation and the continuity equation is eliminated out of the momentum equation, lead to the so-called *non-conservative form* of the momentum equation:

$$\frac{\partial \underline{v}}{\partial t} + \underline{v} grad\, \underline{v} - div\,(\underline{\underline{\nu}}_w grad\, \underline{v}) = \frac{1}{\rho_w}(\underline{f} - grad\, p) \tag{2.62}$$

To illustrate the shallow water equations a little more, especially for the vertical direction, they are given for each direction:

$$\begin{aligned} \mathbf{x-dir.:}\quad & \frac{\partial v_x}{\partial t} + v_x \frac{\partial v_x}{\partial x} + v_y \frac{\partial v_x}{\partial y} + v_z \frac{\partial v_x}{\partial z} - \frac{\partial}{\partial x}(\nu_{wh}\frac{\partial v_x}{\partial x}) \\ & -\frac{\partial}{\partial y}(\nu_{wh}\frac{\partial v_x}{\partial y}) - \frac{\partial}{\partial z}(\nu_{wv}\frac{\partial v_x}{\partial z}) = \frac{1}{\rho_{w0}}(f_x - \frac{\partial p}{\partial x}) \end{aligned} \tag{2.63}$$

$$\begin{aligned} \mathbf{y-dir.:}\quad & \frac{\partial v_y}{\partial t} + v_x \frac{\partial v_y}{\partial x} + v_y \frac{\partial v_y}{\partial y} + v_z \frac{\partial v_y}{\partial z} - \frac{\partial}{\partial x}(\nu_{wh}\frac{\partial v_y}{\partial x}) \\ & -\frac{\partial}{\partial y}(\nu_{wh}\frac{\partial v_y}{\partial y}) - \frac{\partial}{\partial z}(\nu_{wv}\frac{\partial v_y}{\partial z}) = \frac{1}{\rho_{w0}}(f_y - \frac{\partial p}{\partial y}) \end{aligned} \tag{2.64}$$

$$\mathbf{z-dir.}: \quad \frac{1}{\rho_w}\frac{\partial p}{\partial z} + g = 0 \qquad \Leftrightarrow$$

$$p = \rho_{w0} g (z_s - z) + \rho_{w0} g \int_z^{z_s} \frac{\Delta\rho_w}{\rho_{w0}} dz \tag{2.65}$$

In these equations $\rho_w, \rho_{w0}, \Delta\rho_w$ denote the density, a reference density and a density difference of water, ν_{wh}, ν_{wv} the horizontal and vertical viscosities of water, and z_s the vertical component of the water surface related to a reference height (see fig. 2.16). The viscosity consists of a molecular and a turbulent part. Generally, the turbulent part is some orders of magnitude higher when compared to the molecular one. The turbulence modeling is briefly discussed in the next section.

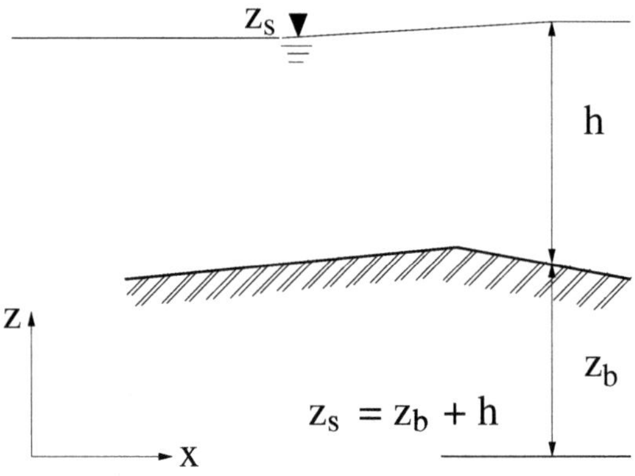

Fig. 2.16. Definitions concerning the vertical direction, after HINKELMANN (1997 [108])

Turbulence modeling

For modeling of turbulence (see fig. 1.2) there exist three main categories: *statistical turbulence models*, *Large Eddy Simulation* (LES) and *Direct Numerical Simulation* (DNS).

The statistical turbulence models are based on the *Reynolds* equations or on a temporal averaging and lead to the *eddy viscosity principle.* The simplest possibility consists of giving a constant value. Algebraic eddy viscosity models relate the eddy viscosity to a characteristic length scale for the problem considered and enable a spatial variation of the turbulence. For rivers and coastal systems, the turbulence is often determined anisotropic. Constant values are chosen for the horizontal, and a mixing length model for the vertical direction. The mixing length model of LEHFELDT (1991 [165]) looks as follows:

$$\nu_{wv} = l_m^2 g(Ri)\sqrt{(\frac{\partial v_x}{\partial z})^2 + (\frac{\partial v_y}{\partial z})^2}$$
$$\nu_{tv} = l_m^2 f(Ri)\sqrt{(\frac{\partial v_x}{\partial z})^2 + (\frac{\partial v_y}{\partial z})^2} \quad (2.66)$$

$$l_m = \begin{cases} \kappa z \; if \; z \leq 0.2h \\ 0.2\kappa h \; if \; z > 0.2h \end{cases} ,$$
$$Ri = \frac{g}{\rho_{w0}} \frac{\partial \rho_w / \partial z}{(\partial v_x/\partial z)^2 + (\partial v_y/\partial z)^2} \quad (2.67)$$

$$f(Ri) = (1 + 3Ri)^{-3} \quad , \qquad g(Ri) = (1 + 3Ri)^{-1} \quad (2.68)$$

In these equations l_m denotes the mixing length, $\kappa = 0.4$ the *Karman* constant, h the water level, ν_{tv} the vertical turbulent diffusivity (see sec. 2.6.2) and g, f damping functions which take into account layering effects and which can be determined by measurements. The *Richardson number Ri* is a measure for the stability of a layered water system. The layering is stable for $Ri > 0$ and unstable for $Ri < 0$. Further statistical models solve additional transport equations, e.g. in the $k - \epsilon$ model one transport equation for the turbulent kinetic energy and one transport equation for the turbulent kinetic dissipation (see RODI (1984 [222])). The computational effort and the accuracy are comparatively low for statistical turbulence models, and they are widely used today.

The Large Eddy Simulation (LES) is based on the assumption that eddies larger than the grid size are directly resolved by the mesh. A turbulence model is required for eddies smaller than the mesh size. Turbulence modeling is carried out by a spatial and / or a temporal averaging procedure. The computational effort and the accuracy are higher for LES when compared to the statistical models. However, LES is little used today in practical applications.

In the Direct Numerical Simulation (DNS) the meshes are refined more and more up to the size of the smallest eddies. Therefore, the computational effort is huge and the accuracy very good. The applicational range of DNS is limited to small *Reynolds* numbers.

Further information about turbulence is given in RODI (1984 [222]), LEHFELDT (1991 [165]), JIRKA et al. (1999 [134]), MALCHEREK (2000 [170]), or FORKEL (2001 [82]).

Saint-Venant equations

If the three-dimensional continuity and momentum equations are integrated over the water depth, the *Saint-Venant equations* are obtained. They are shown here in the non-conservative form:

$$\textbf{continuity eq.}: \quad \frac{\partial h}{\partial t} + v_x \frac{\partial h}{\partial x} + v_y \frac{\partial h}{\partial y} + h \frac{\partial v_x}{\partial x} + h \frac{\partial v_y}{\partial y} = q_w/\rho_w \tag{2.69}$$

$$\begin{aligned} \textbf{x}-\textbf{dir.}: \quad & \frac{\partial v_x}{\partial t} + v_x \frac{\partial v_x}{\partial x} + v_y \frac{\partial v_x}{\partial y} - \frac{\partial}{\partial x}(\nu_{wh} \frac{\partial v_x}{\partial x}) - \frac{\partial}{\partial y}(\nu_{wh} \frac{\partial v_x}{\partial y}) \\ & = \frac{f_x}{\rho_{w0}} - g \frac{\partial (h + z_b)}{\partial x} \end{aligned} \tag{2.70}$$

$$\begin{aligned} \textbf{y}-\textbf{dir.}: \quad & \frac{\partial v_y}{\partial t} + v_x \frac{\partial v_y}{\partial x} + v_y \frac{\partial v_y}{\partial y} - \frac{\partial}{\partial x}(\nu_{wh} \frac{\partial v_y}{\partial x}) - \frac{\partial}{\partial y}(\nu_{wh} \frac{\partial v_y}{\partial y}) \\ & = \frac{f_y}{\rho_{w0}} - g \frac{\partial (h + z_b)}{\partial y} \end{aligned} \tag{2.71}$$

Here, z_b stands for the vertical coordinate of the bottom related to a reference height (see fig. 2.16). The momentum source terms f_x, f_y contain the influences of bottom friction, wind stress, air-pressure gradient and *Coriolis* force:

$$f_x = -\frac{\lambda}{h} \rho_w v_x \sqrt{v_x^2 + v_y^2} \quad + \frac{c_D \rho_a}{h} v_{ax} \sqrt{v_{ax}^2 + v_{ay}^2} \quad - \frac{\partial P}{\partial x} \quad + 2 \rho_w \omega \, sin\phi \, v_y \tag{2.72}$$

$$f_y = \underbrace{-\frac{\lambda}{h} \rho_w v_y \sqrt{v_x^2 + v_y^2}}_{friction} \quad \underbrace{+\frac{c_D \rho_a}{h} v_{ay} \sqrt{v_{ax}^2 + v_{ay}^2}}_{wind} \quad \underbrace{-\frac{\partial P}{\partial y}}_{air} \quad \underbrace{-2 \rho_w \omega \, sin\phi \, v_x}_{Coriolis} \tag{2.73}$$

Here, λ stands for the *Taylor*-friction coefficient, c_D for the wind-stress coefficient, ρ_a for the density of air, v_{ax}, v_{ay} for the horizontal components of the wind velocity, P for the air pressure, ω for the rotation of the earth and ϕ for the geographical latitude. Concerning the bottom friction other laws can be chosen, and concerning the wind-stress coefficient different empirical laws depending on the wind velocity are used, see MALCHEREK (2000 [170]). Which momentum source terms have to be taken into account, is a question of the problems and the scales considered, see HINKELMANN (1997 [108]).

2.6.2 Transport processes

The main transport mechanism in surface-water systems are advection / convection and diffusion. Other processes, e.g. reaction, adsorption or decay, are not considered here, see HUNZE (1996 [125]). Similar to the viscosity (see sec. 2.6.1), the diffusion consists of a molecular and a turbulent part and generally, the turbulent part is some orders of magnitude higher when compared to the molecular part. The size of the turbulent diffusivity is in the range of the turbulent viscosity. The principle behavior of advection and diffusion was already shown for a groundwater system in figure 2.8 and is valid for a free-surface water system, too, with the exception that dispersion must be replaced by the turbulent diffusivity.

The transport equation for a tracer c is obtained from the general form of the balance equation (eq. 2.4), if $e = \rho_w c, \underline{w} = -\underline{\underline{\nu}}_t grad\,(\rho_w c), r = q_c$. Here, $\underline{\underline{\nu}}_t$ denotes the turbulent diffusivity tensor and q_c a tracer sink or source term which can take tracer sinks or sources as well as reaction, adsorption or decay processes into account. The transport equation is determined as follows:

$$\frac{\partial(\rho_w c)}{\partial t} + div\,[(\underline{v}\rho_w c - \underline{\underline{\nu}}_t grad\,(\rho_w c)] = q_c \tag{2.74}$$

If the density is constant or the concentrations are small (see MARKOFSKY (1980 [173])), the last equation can be simplified applying the product rule and the continuity equation 2.60:

$$\frac{\partial c}{\partial t} + \underline{v} grad\, c - div\,(\underline{\underline{\nu}}_t grad\, c) = \frac{q_c - q_w c}{\rho_w} \tag{2.75}$$

If the density of water depends on the tracer concentration, flow and transport is coupled in both directions. Consequently, an equation of state is required. LEHFELDT (1991 [165]) determines the density of water as a function of the salinity $c = S$ by the following formula:

$$\rho_w(S) = \rho_{w0}(1 + 0.00075S) \tag{2.76}$$

Other formulas considering the temperature or sediment concentration are given by the UNESCO (1987 [251]) or MALCHEREK (2000 [170]). For flowing surface-water systems like rivers or estuaries, the coupling between flow and transport is often weak. An exception is given e.g., if cooling water of a power plant is dumped into such a system. The coupling can be strong in standing surface-water systems like lakes or reservoirs.

The shallow-water equations are non-linear, the transport equation is linear, and both have a mixed parabolic / hyperbolic type. For further reading about the mathematical modeling of flow and transport processes in surface-water systems see VAN RIJN (1990 [219]), JIRKA (1994 [134]), HINKELMANN (1997 [108]), ZIELKE et al. (1999 [267]), or MALCHEREK (2000 [170]).

3

Efficient numerical methods

In this chapter, an introduction to different efficient numerical methods for water-related and environmental problems is given. Discretization and stabilization methods, parallel methods, adaptive methods, and fast solvers are discussed. The discretization and stabilization methods which are applied to the partial differential equations of the sections 2.3 - 2.6 lead to *algebraic equations*. As often, large-scale hydro- and environmental systems are considered for nowadays tasks in research as well as in practical applications, sophisticated methods, such as parallel and adaptive ones, must be chosen to obtain reliable solutions in acceptable time. As the major part of the numerical schemes is *implicit*, fast solvers are urgently required. The fundamentals of the most important methods are explained together with their interaction. It is beyond the objectives of this work to give a complete description of all existing methods.

3.1 Discretization and stabilization methods

3.1.1 General requirements

In most cases, the physical and mathematical model concepts dealt with in chapter 2 lead to *coupled, (non)-linear partial differential equations*, apart from some exceptions in section 2.3. As analytical solutions are rather limited for such problems and are subject to a number of constraints concerning physical parameters and boundary conditions, *numerical methods*, i.e. here *discretization techniques*, offer a more general approach. They transform the partial differential equations to *algebraic equations*, and they approximate the solution function or entity e at discrete points or nodes. In the mathematical sense, the partial differential equations are of a mixed hyperbolic / parabolic type which has a significant influence on the discretization methods.

A discretization method must be *consistent*, i.e. the discretized equation $\overline{W}$ turns into the partial differential equation W or the local *discretization error* δ tends to zero for vanishing temporal and spatial discretization sizes:

$$\delta = \lim_{\Delta t, \Delta x \to 0} = |W - \overline{W}| = 0 \tag{3.1}$$

The *order of consistency* is the minimal power of the space or time step occurring in the discretization error, i.e. if the time step size is halved in a method of second order, then the discretization error is quartered (see sec. 3.1.3). Therefore, the order of consistency is a measure of accuracy. Generally, a high order of consistency is desirable.

Apart from discretization errors, *round-off errors* r also occur; they are the difference between the exact solution of the discretized equation gx and the numerical solution of the discretized equation $\overline{g}$:

$$r = |gx - \overline{g}| \tag{3.2}$$

Round-off errors are caused by iterative solvers (see sec. 3.4) (stopping criteria) or by the fact that computers operate with a limited number of bytes for representing numbers. Generally, round-off errors should be of minor importance if suitable stopping criteria are chosen. The total error E is the sum of the discretization and round-off errors:

$$E = \delta + r \tag{3.3}$$

Additionally, *stability* is required for a discretization method; this means that small pertubations are continuously damped in the course of the simulation and do not lead to increasing oscillations. There are different methods for the *stability analysis* (see sec. 3.1.3). Stability can be achieved by different stabilization techniques, e.g. by adding artificial diffusion. However, the stabilization measures lead to *numerical diffusion / dispersion* (smearing of the front, see fig. 3.1), also not representing the exact solution. This highlights one major dilemma of numerical methods for advection / diffusion problems. Stabilization techniques are explained in sections 3.1.3 - 3.1.5.

Furthermore, a discretization method must be *convergent*, i.e. the numerical solution of the discretized equation $\overline{g}$ converges toward the exact solution of the partial differential equation g, or the total error E vanishes, for space and time steps tending to zero:

$$E = \lim_{\Delta t, \Delta x \to 0} |g - \overline{g}| = 0 \tag{3.4}$$

As the exact solution is often not known, convergence or convergence speed can be checked by observing the results at different discrete points for decreasing space and time steps. It should be mentioned that the convergence need not to be one-sided or asymptotical; the limes or the 'exact' solution can also be obtained from both sides (see fig. 3.2). If a method is consistent and stable, convergence is mathematically proven.

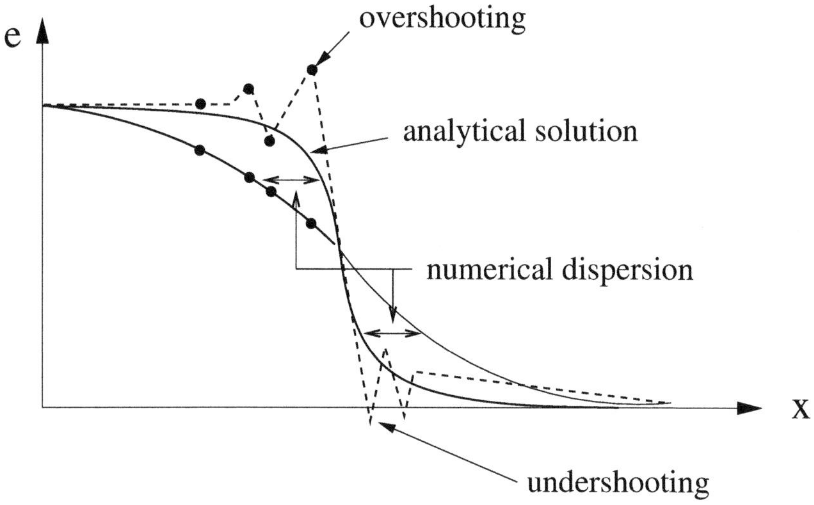

Fig. 3.1. Principle behavior of oscillations and numerical dispersion, after HINKELMANN, HELMIG (2002 [111])

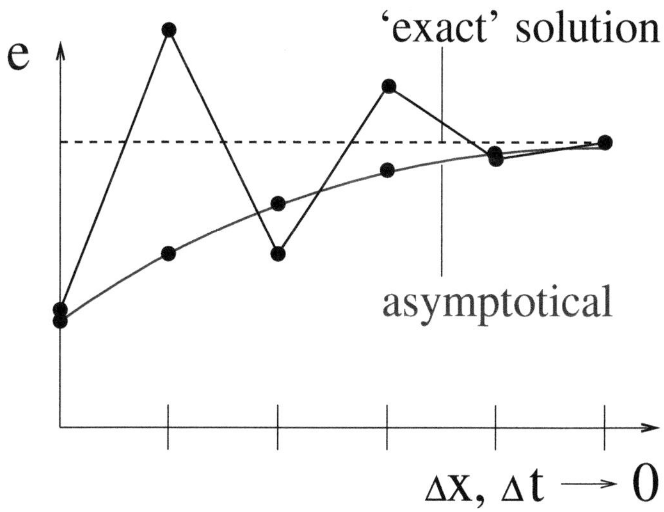

Fig. 3.2. Asymptotical and two-sided convergence behavior

Beside the characteristics already mentioned, methods should fulfill further requirements. If no sink or source terms occur in the balance equations, a method should be *monotonous*, i.e., if three nodes with increasing x-coordinate are considered, the nodal value of the middle node should vertically ly between the values of its neighbors. If monotonicity is given, no *over-* or *undershooting* occurs (see fig. 3.1). Additionally, a method should be *conservative*, i.e. the temporal change of an entity considered in a domain should be in equilibrium with the advective and diffusive fluxes across the boundaries plus source or sink terms. Futhermore, a discretization method should reproduce typical flow or transport phenomena, such as, for example, sharp fronts.

A good fulfillment of all criteria is often hardly possible. This depends on a number of interacting things: the complexity of the geometry (inner structures and boundaries), the geology (e.g. heterogeneities) and the physical problem (e.g. high non-linearities), the chosen mesh structure (structured or unstructured meshes; refinements at transition zones of inhomogeneities) as well as the possibility of choosing suitable boundary conditions, especially for multiphase flow, etc. Overall, these circumstances require different problem-dependent discretization methods. This is illustrated in the following. On structured grids, methods of a high order of accuracy have been obtained. However, in most cases a direct or reasonable transfer to unstructured meshes is not possible. On unstructured grids, a comparatively high accuracy can be achieved with 'simpler' methods by mesh refinement. Further information can be found in HELMIG (1997 [98]) or MALCHEREK (2000 [170]).

3.1.2 Time discretization

Generally, the time domain is subdivided into a number of constant or variable time steps Δt^n where the solution functions are determined. For the time t_0, the initial conditions for all unknowns must be given:

$$t^n = t^0 + n\Delta t^n \tag{3.5}$$

In this equation, t^n denotes the time of the nth time step and n the number of time steps.

The time-dependent equations determined in chapter 2 are generalized in the following way:

$$\frac{\partial e}{\partial t} = Ae \tag{3.6}$$

Here, e stands for a scalar or vector entity, and the operator A contains the spatial derivatives as well as the other terms. As an illustration, A in equation 2.64, for example, stands for:

$$A = -v_x \frac{\partial}{\partial x} - v_y \frac{\partial}{\partial y} - v_z \frac{\partial}{\partial z} + \frac{\partial}{\partial x}(\nu_{wh} \frac{\partial}{\partial x})$$
$$+ \frac{\partial}{\partial y}(\nu_{wh} \frac{\partial}{\partial y}) + \frac{\partial}{\partial z}(\nu_{wv} \frac{\partial}{\partial z}) + \frac{1}{\rho_{w0}}(f_y - \frac{\partial p}{\partial y}) \tag{3.7}$$

One-step methods

One-step methods determine the solution function on the new time level $n+1$ in just one step and take only the current time step n into account. The temporal derivative is approximated by *forward differencing* (see sec. 3.1.3):

$$\frac{\partial e}{\partial t} \approx \frac{e^{n+1} - e^n}{\Delta t^n} \tag{3.8}$$

In the following, it must be determined on which time level e is regarded in equation 3.6. If it is considered on time level n, the *explicit* or *Forward Euler Method* is obtained:

$$\frac{e^{n+1} - e^n}{\Delta t^n} = Ae^n \tag{3.9}$$

The unknowns on the new time level $n + 1$ can be completely computed by means of known values from the current time level n (see fig. 3.3) with simple mathematical operations, for example a matrix-vector product. Therefore, the computational effort for one time step is low. However, there is a limit to the time-step size for stability reasons (see sec. 3.1.3). The order of consistency is only $O(\Delta t)$ (see sec. 3.1.3).

If e is determined on the new time level $n + 1$, the *fully implicit* or *Backward Euler Method* is obtained:

$$\frac{e^{n+1} - e^n}{\Delta t^n} = Ae^{n+1} \tag{3.10}$$

The unknowns on the new time level $n + 1$ depend on each other, leading to a system of equations (see fig. 3.3, sec. 3.4). Consequently, the computational effort for one time step is high. However, for stability reasons, there is no limit to the time step size. The order of consistency is only $O(\Delta t)$ (see sec. 3.1.3).

If e is computed between the current time level n and the new time level $n+1$, the *Crank-Nicholson Method* is obtained with the Crank-Nicholson factor θ with $0 \le \theta \le 1$:

$$\frac{e^{n+1} - e^n}{\Delta t^n} = \theta Ae^{n+1} + (1 - \theta)Ae^n \tag{3.11}$$

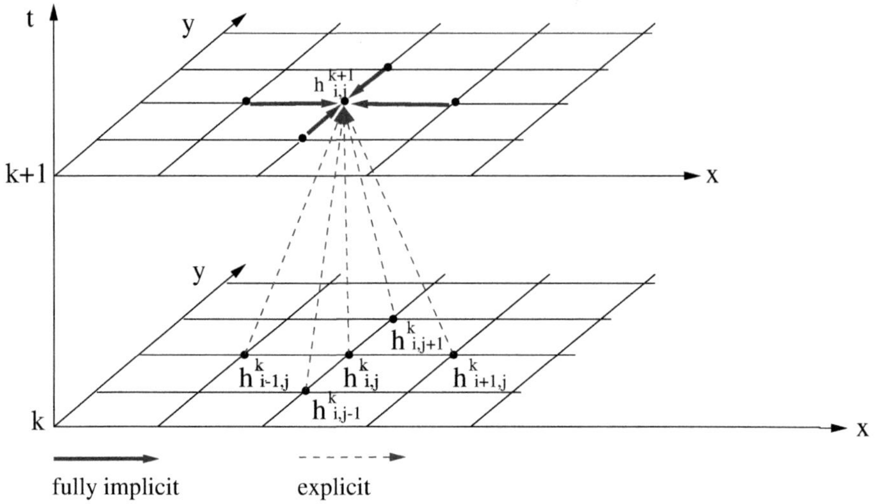

Fig. 3.3. Explicit and (fully) implicit methods, after HELMIG (2000 [99])

As this method is also implicit, the computational effort per time step is high. For $0.5 \leq \theta \leq 1$, stability is ensured; for $0 < \theta < 0.5$ this is not the case. If θ equals 0.5, the order of consistency is $O(\Delta t^2)$, otherwise only $O(\Delta t)$ (see sec. 3.1.3). For many practical applications, especially on unstructured grids, θ is chosen between 0.55 and 0.65, because values for θ closer to 0.5 may lead to stability problems.

If further terms of the *Taylor series* (see sec. 3.1.3) are taken into account, the accuracy is increased. If the term with Δt^2 is introduced, the *Lax-Wendroff Method*, which is explicit and has an order of consistency $O(\Delta t^2)$, is obtained:

$$e^{n+1} = (1 + \Delta t^n A + \frac{1}{2}(\Delta t^n)^2 A^2)e^n \quad \Leftrightarrow \quad \frac{e^{n+1} - e^n}{\Delta t^n} = Ae^n + \frac{1}{2}\Delta t^n A^2 e^n \tag{3.12}$$

If the terms with up to Δt^4 are taken into account, an explicit *Runge-Kutta Method* , which has an order of consistency $O(\Delta t^4)$, is determined:

$$e^{n+1} = (1 + \Delta t^n A + \frac{1}{2}(\Delta t^n)^2 A^2 + \frac{1}{6}(\Delta t^n)^3 A^3 + \frac{1}{24}(\Delta t^n)^4 A^4)e^n \tag{3.13}$$

This method can be easily coded (see BRONSTEIN, SEMENDJAJEW (1979 [45])).

Multi-step methods

Multi-step methods determine the solution function on the new time level $n+1$ in more than one step and / or take more time levels into account than only the current one n.

The *Leap-Frog Method* applies central differencing (see sec. 3.1.3) in time by regarding the previous time level $n-1$:

$$\frac{e^{n+1} - e^{n-1}}{2\Delta t^n} = Ae^n \tag{3.14}$$

It is an explicit method and has the order of consistency $O(\Delta t^2)$ (see sec. 3.1.3). Two time levels must be stored.

A further alternative is given by *Operator-Splitting Methods.* They split the basic equations into different parts, e.g. into an advective and a diffusive part:

$$\begin{aligned}\frac{\partial e}{\partial t} &= A_1 e + A_2 e \\ \frac{e^{ni} - e^n}{\Delta t^n} &= A_1 e \\ \frac{e^{n+1} - e^{ni}}{\Delta t^n} &= A_2 e\end{aligned} \tag{3.15}$$

In a first sub-step, for example, only the advective terms are dealt with and an intermediate result e^{ni} is determined. This is the initial condition for the second sub-step in which, for example, only the diffusive terms are treated. Thus, special methods for advection and diffusion can be applied in each sub-step which may lead to a combination of explicit and implicit methods. Stability is guaranteed for the Operator-Splitting Method if it is fulfilled in each sub-step. The order of consistency is limited by the lower order of a sub-step. If for example both methods have an order of consistency $O(\Delta t^2)$ (see sec. 3.1.3), it must be ensured that the sub-steps are commutative in order to obtain the same order of consistency for the Operator-Splitting Methods. If the sub-steps are not commutative, the Operator-Splitting Method has only the order of consistency $O(\Delta t)$.

For further reading on time discretization methods, see MALCHEREK (2000 [170]) or KOLDITZ (2000 [153]).

3.1.3 Finite-Difference Methods

The *Finite-Difference Method (FDM)* is one of the oldest methods for solving partial differential equations. The computational domain is discretized by rectangular or quadrilateral cells (see fig. 3.4). Often, the cell lengths Δx and Δz are constant or even $\Delta x = \Delta z$. The unknown variables are defined in the nodes which are placed at the centers of the cells or at the intersection points of cell boundaries (see fig. 3.4). From the geometrical point of view, it is obvious that complex boundaries or complex inner structures can only be reproduced in a very simplified way by step functions.

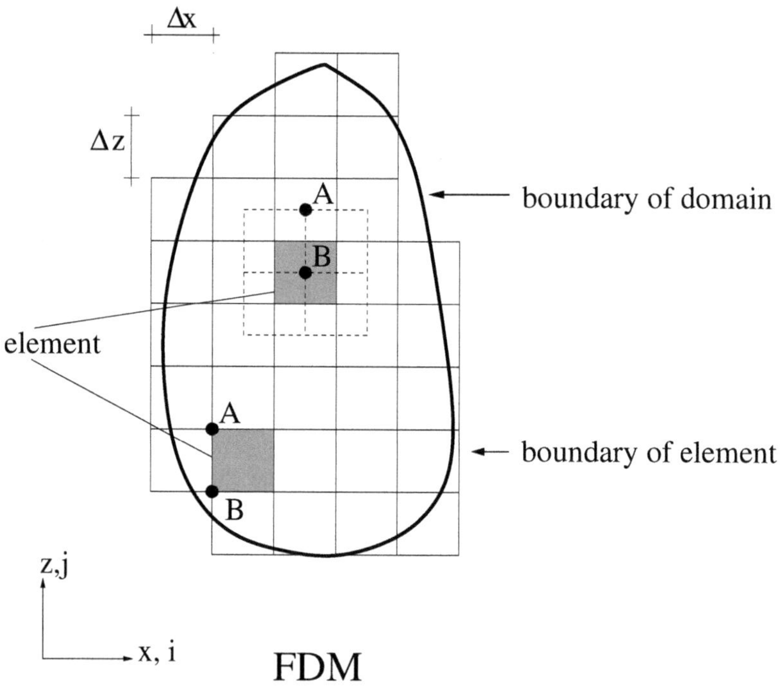

Fig. 3.4. Space discretization for the FDM, after HINKELMANN, HELMIG (2002 [111])

The basic idea of the FDM is to substitute the differential quotients by difference quotients. The equations are then put together in an explicit or implicit way (see secs. 3.1.2, 3.4). Taking into account initial and / or boundary conditions (see sec. 2.2), the solutions are obtained.

Derivatives of the unknown function e can be developed with the help of a *Taylor-series expansion*, shown here for the x-direction. For simplicity's sake, constant Δx and Δz are assumed:

$$e_{i+1,j} = e_{i,j} + \Delta x \frac{\partial e_{i,j}}{\partial x} + \frac{(\Delta x)^2}{2}\frac{\partial^2 e_{i,j}}{\partial x^2} + \frac{(\Delta x)^3}{6}\frac{\partial^3 e_{i,j}}{\partial x^3} + \frac{(\Delta x)^4}{24}\frac{\partial^4 e_{i,j}}{\partial x^4} + O(\Delta x^5) \tag{3.16}$$

The unknown function e_{i-1} is determined in a similar way:

$$e_{i-1,j} = e_{i,j} - \Delta x \frac{\partial e_{i,j}}{\partial x} + \frac{(\Delta x)^2}{2}\frac{\partial^2 e_{i,j}}{\partial x^2} - \frac{(\Delta x)^3}{6}\frac{\partial^3 e_{i,j}}{\partial x^3} + \frac{(\Delta x)^4}{24}\frac{\partial^4 e_{i,j}}{\partial x^4} - O(\Delta x^5) \tag{3.17}$$

The first derivative of a function e can be obtained in three different ways, called forward differencing (FD, from eq. 3.16), backward differencing (BD, from eq. 3.17) and central differencing (CD, eq. 3.16 minus 3.17):

$$FD: \frac{\partial e_{i,j}}{\partial x} = \frac{e_{i+1,j} - e_{i,j}}{\Delta x} - \frac{\Delta x}{2}\frac{\partial^2 e_{i,j}}{\partial x^2} - \frac{\Delta x^2}{6}\frac{\partial^3 e_{i,j}}{\partial x^3} - \frac{\Delta x^3}{24}\frac{\partial^4 e_{i,j}}{\partial x^4} - O(\Delta x^5) \tag{3.18}$$

$$\Leftrightarrow FD: \frac{\partial e_{i,j}}{\partial x} = \frac{e_{i+1,j} - e_{i,j}}{\Delta x} + O(\Delta x) \tag{3.19}$$

$$BD: \frac{\partial e_{i,j}}{\partial x} = \frac{e_{i,j} - e_{i-1,j}}{\Delta x} + O(\Delta x) \tag{3.20}$$

$$CD: \frac{\partial e_{i,j}}{\partial x} = \frac{e_{i+1,j} - e_{i-1,j}}{2\Delta x} + O(\Delta x^2) \tag{3.21}$$

The second derivative is determined by adding up equations 3.16 and 3.17:

$$\frac{\partial^2 e_{i,j}}{\partial x^2} = \frac{e_{i+1,j} - 2e_{i,j} + e_{i-1,j}}{\Delta x^2} + O(\Delta x^2) \tag{3.22}$$

The order of consistency is the minimal power of the space or time step occurring in the discretization error. Therefore, forward and backward differencing is of first order or $O(\Delta x)$, while central differencing and the second derivative is of second order or $O(\Delta x^2)$.

In the following, the principle use of Finite-Difference Methods is explained using two examples, flow processes in groundwater and in surface water.

Flow processes in groundwater

Equations 2.15 and 2.17 are recalled (see sec. 2.3.1). Furthermore, one-dimensional problems are considered with a constant hydraulic conductivity K_f and a constant storage term S_0 and without sink or source terms. This leads to the following equation for the piezometric head h:

$$S_0 \frac{\partial h}{\partial t} - div\ (K_f grad\ h) = 0 \Leftrightarrow \frac{\partial h}{\partial t} - \frac{K_f}{S_0} \frac{\partial^2 h}{\partial x^2} = 0 \tag{3.23}$$

The equation is a *diffusion equation*, and the type is parabolic.

If forward differencing (eqs. 3.9, 3.19) is applied to the time derivative and the second derivative approximation (eq. 3.22) to the spatial derivative together with the *Crank-Nicholson Method* (see 3.11), one obtains:

$$\frac{h_i^{n+1} - h_i^n}{\Delta t} - \theta \frac{K_f}{S_0} \frac{h_{i+1}^{n+1} - 2h_i^{n+1} + h_{i-1}^{n+1}}{\Delta x^2} - (1-\theta) \frac{K_f}{S_0} \frac{h_{i+1}^n - 2h_i^n + h_{i-1}^n}{\Delta x^2} = 0 \tag{3.24}$$

In the explicit case ($\theta = 0$), the unknown piezometric head on the new time level can easily be determined and the order of consistency is $O(\Delta t, \Delta x^2)$:

$$h_i^{n+1} = h_i^n + \frac{K_f \Delta t}{S_0 \Delta x^2} (h_{i+1}^n - 2h_i^n + h_{i-1}^n) \tag{3.25}$$

In the fully implicit case ($\theta = 1$), the unknown piezometric head on the new time level results from the solution of a linear and symmetric system of equations and the order of consistency is again $O(\Delta t, \Delta x^2)$:

$$-h_{i+1}^{n+1} + (\frac{S_0 \Delta x^2}{K_f \Delta t} + 2) h_i^{n+1} - h_{i-1}^{n+1} = \frac{S_0 \Delta x^2}{K_f \Delta t} h_i^n \tag{3.26}$$

If the spatial derivative is chosen between the new and the current time level, an implicit scheme is again obtained. For $\theta = 0.5$, the order of consistency is $O(\Delta t^2, \Delta x^2)$, otherwise $O(\Delta t, \Delta x^2)$:

$$-\theta h_{i+1}^{n+1} + (\frac{S_0 \Delta x^2}{K_f \Delta t} + 2\theta) h_i^{n+1} - \theta h_{i-1}^{n+1}$$

$$= (1-\theta) h_{i+1}^n + (\frac{S_0 \Delta x^2}{K_f \Delta t} + 2\theta - 2) h_i^n + (1-\theta) h_{i-1}^n \tag{3.27}$$

Generally, the flow velocities v_f are determined by the *Darcy law* (eq. 2.17), after the hydraulic heads have been computed in a 'postprocessing' step (see 3.1.6). As this is carried out by differentiation ($grad\, h$), the spatial accuracy for the *Darcy velocity* is reduced by one order of magnitude.

In the following, the stability behavior is investigated (see MALCHEREK (2000 [170]), KOLDITZ (2000 [153])) for the one-dimensional problem given in equation 3.27. When the *Matrix Method* is chosen, the discretization scheme is transformed into:

$$h^{n+1} = Th^n \tag{3.28}$$

T is a square matrix with $(N-2)$ x $(N-2)$ entries, N is the number of nodes. It is assumed that a *Dirichlet-boundary condition* is imposed at the first and the last node. The vectors h^{n+1} and h^n have $(N-2)$ entries. Stability is ensured if T is linear and if the absolute values of all *eigenvalues* λ_m are smaller than 1:

$$|\lambda_m| \leq 1\,, \qquad m = 2, 3, ..., N-1 \tag{3.29}$$

First, equation 3.27 is rewritten in the form

$$Ch^{n+1} = Dh^n \tag{3.30}$$

with the matrices C and D:

$$C = \begin{bmatrix} (\frac{S_0 \Delta x^2}{K_f \Delta t} + 2\theta) & -\theta & & & \\ -\theta & \ddots & \ddots & & \\ & \ddots & \ddots & \ddots & \\ & & \ddots & \ddots & -\theta \\ & & & -\theta & (\frac{S_0 \Delta x^2}{K_f \Delta t} + 2\theta) \end{bmatrix} \tag{3.31}$$

$$D = \begin{bmatrix} (\frac{S_0 \Delta x^2}{K_f \Delta t} + 2\theta - 2) & (1-\theta) & & & \\ (1-\theta) & \ddots & \ddots & & \\ & \ddots & \ddots & \ddots & \\ & & \ddots & \ddots & (1-\theta) \\ & & & (1-\theta) & (\frac{S_0 \Delta x^2}{K_f \Delta t} + 2\theta - 2) \end{bmatrix} \tag{3.32}$$

Second, equation 3.30 is transformed to

$$h^{n+1} = C^{-1} D h^n \tag{3.33}$$

and, with $T = C^{-1}D$, has the shape of equation 3.28.

If the explicit case $\theta = 0$ is considered, C becomes a diagonal matrix, and T can be easily computed:

$$T = \begin{bmatrix} (1-\frac{2K_f\Delta t}{S_0\Delta x^2}) & \frac{K_f\Delta t}{S_0\Delta x^2} & & & \\ \frac{K_f\Delta t}{S_0\Delta x^2} & \ddots & \ddots & & \\ & \ddots & \ddots & \ddots & \\ & & \ddots & \ddots & \frac{K_f\Delta t}{S_0\Delta x^2} \\ & & & \frac{K_f\Delta t}{S_0\Delta x^2} & (1-\frac{2K_f\Delta t}{S_0\Delta x^2}) \end{bmatrix} \qquad (3.34)$$

The eigenvalues are determined to:

$$\lambda_m = 1 - \frac{4K_f\Delta t}{S_0\Delta x^2} sin^2(\frac{m\pi}{2N}), \qquad m = 2,3,...,N-1 \qquad (3.35)$$

Stability condition 3.29 is fulfilled if the *Neumann number* is smaller than or equal to 0.5:

$$Ne = \frac{K_f\Delta t}{S_0\Delta x^2} \leq 0.5 \quad \Leftrightarrow \quad \Delta t \leq \frac{S_0\Delta x^2}{2K_f} \qquad (3.36)$$

If *Neumann-* or *Cauchy-boundary conditions* are considered, the stability is slightly reduced.

In a similar way, it can be shown that conditionally stable discretization schemes are obtained for $0 < \theta < 0.5$ if

$$\Delta t \leq \frac{S_0\Delta x^2}{2K_f(1-2\theta)} \qquad (3.37)$$

is fulfilled, and unconditionally stability is found for $0.5 \leq \theta \leq 1$.

If the boundary of the cell coincides with a jump in a coefficient - in this case the hydraulic conductivity (tensor), a *harmonic averaging* (see HELMIG (2000 [99])) should be carried out. If such jumps are within a cell, the coefficients should be *averaged arithmetically* in each cell.

In the following, the explicit method is investigated for a simple example (see SMITH (1970 [240]), HINKELMANN (2001 [110])). A one-dimensional aquifer of the length $6.4m$ with a hydraulic conductivity $K_f = 10^{-6}m/s$ and a storage term $S_0 = 10^{-4}1/m$ is considered (see fig. 3.5). The space is discretized with 64 elements of $\Delta x = 0.1m$, and a time step $\Delta t = 0.1s$ is chosen. The *Neumann-stability condition* (eq. 3.36) allows a time step up to $\Delta t \leq 0.5s$. The simulation time is $5000s = 1.4h$. The boundary conditions for the piezometric head are set to zero at the beginning and the end of the system for the whole computational time, i.e. $h_0^j = 0.0, h_{64}^j = 0.0, j = 0,1,2,...,50000$. As an initial condition, a bilinear distribution of the piezometric head with $h_{32}^0 = 1m$ in the middle, $h_0^0 = 0$ at the beginning and $h_{64}^0 = 0$ at the end is imposed (see fig. 3.5).

Equation 3.25 becomes:

$$h_i^{n+1} = 0.1h_{i-1}^n + 0.8h_i^n + 0.1h_{i+1}^n \qquad (3.38)$$

In the course of the simulation (see fig. 3.5), the initial condition is continuously reduced, until the piezometric head reaches zero in the whole domain. The results are symmetric to the axis $x = 3.2m$. They are also shown for several space and time steps in table 3.1. Here, equation 3.38 can easily be checked.

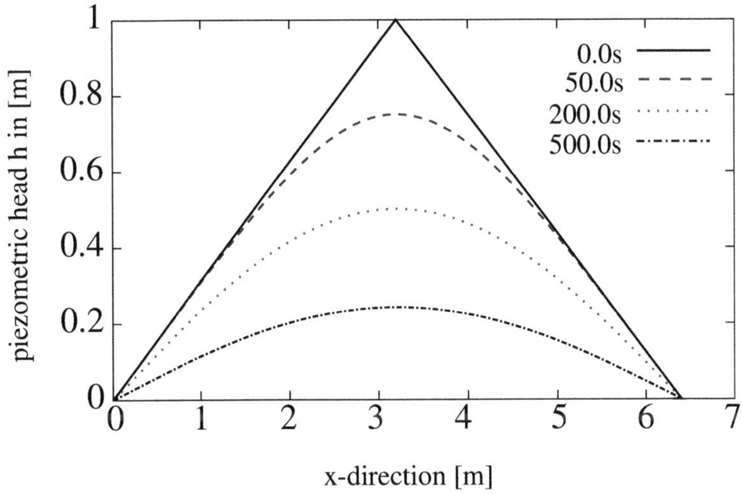

Fig. 3.5. Piezometric head distribution at different time steps, after HINKELMANN (2000 [109])

	$x = 0.0m$ $i = 1$	$x = 0.8m$ $i = 8$	$x = 1.6m$ $i = 16$	$x = 2.4m$ $i = 24$	$x = 3.2m$ $i = 32$
$t = 0.0s; n = 0$	0.000	0.250	0.500	0.750	1.000
$t = 0.1s$	0.000	0.250	0.500	0.750	0.994
$t = 0.2s$	0.000	0.250	0.500	0.750	0.989
.....					
$t = 500s$	0.000	0.093	0.172	0.225	0.243
.....					
$t = 5000s$	0.000	0.000	0.000	0.000	0.000

Table 3.1. Piezometric head at different time steps

Flow processes in surface water

The *Saint-Venant equations* in the non-conservative form 2.69 - 2.71 are considered (see sec. 2.6.1). If one-dimensional problems without viscous or turbulent terms, without sink or source terms, without friction and without bottom slope are considered, this leads to:

$$\frac{\partial h}{\partial t} + v\frac{\partial h}{\partial x} + h\frac{\partial v}{\partial x} = 0 \tag{3.39}$$

$$\frac{\partial v}{\partial t} + v\frac{\partial v}{\partial x} = -g\frac{\partial h}{\partial x} \tag{3.40}$$

In these equations h denotes the water level, v the flow velocity and g the gravity. The equations are of hyperbolic type. If only equation 3.39 is regarded and a constant v is assumed, this is a linear *advection equation.* If only equation 3.40 is considered and a constant h is assumed, this is a non-linear, inviscid *Burgers' equation.*

In the following, explicit schemes are investigated. If forward differencing is chosen in time (eq. 3.9) and forward or backward differencing in space (eqs. 3.19, 3.20), the *Upstream Difference Method* is obtained. This method is stable if a stability condition (eq. 3.45) is fulfilled; however, it is very diffusive, and it has only the order of consistency $O(\Delta t, \Delta x)$. If forward differencing is chosen in time and central differencing in space (eq. 3.21), the order of consistency is $O(\Delta t, \Delta x^2)$; however, this scheme is always unstable. A further improvement concerning stability is given by the *Diffusive Lax Method.* It uses a modified forward differencing in time by choosing an average of the neighboring nodes $i+1, i-1$ for the considered node i on the current time level and a central differencing in space:

$$\frac{h_i^{n+1} - (h_{i+1}^n + h_{i-1}^n)/2}{\Delta t} + v_i^n \frac{h_{i+1}^n - h_{i-1}^n}{2\Delta x} + h_i^n \frac{v_{i+1}^n - v_{i-1}^n}{2\Delta x} = 0 \tag{3.41}$$

$$\frac{v_i^{n+1} - (v_{i+1}^n + v_{i-1}^n)/2}{\Delta t} + v_i^n \frac{v_{i+1}^n - v_{i-1}^n}{2\Delta x} = -g\frac{h_{i+1}^n - h_{i-1}^n}{2\Delta x} \tag{3.42}$$

The unknowns are directly determined by:

$$h_i^{n+1} = \frac{h_{i+1}^n + h_{i-1}^n}{2} - \frac{\Delta t}{2\Delta x}[v_i^n(h_{i+1}^n - h_{i-1}^n) + h_i^n(v_{i+1}^n - v_{i-1}^n)] \tag{3.43}$$

$$v_i^{n+1} = \frac{v_{i+1}^n + v_{i-1}^n}{2} - \frac{\Delta t}{2\Delta x}[g(h_{i+1}^n - h_{i-1}^n) + v_i^n(v_{i+1}^n - v_{i-1}^n)] \tag{3.44}$$

This method shows a more or less diffusive behavior.

The order of consistency is increased to $O(\Delta t^2, \Delta x^2)$ if central differencing is used in space and time. This yields the *Leap-Frog Method.* However, it has an oscillatory tendency, and it should be mentioned that it requires more memory, as two previous time levels must be stored in the computer memory.

The *Lax-Wendroff* Method is a combination of the Diffusive Lax and the Leap-Frog Methods. First, the Diffusive Lax Method computes intermediate results at $i - 0.5$ and $i + 0.5$ with a time step $0.5\Delta t$. Second, the Leap-Frog Method uses these intermediate results and the result of the current time at i to determine the results at i on the new time level (see HELMIG (2000 [99])). The Lax-Wendroff Method and the Leap-Frog Method have the same order of consistency $O(\Delta t^2, \Delta x^2)$ and the same memory requirement. The computational effort of the Lax-Wendroff Method is much higher. However, this is often accepted because the numerical results of the Lax-Wendroff Method are generally superior to those of the Leap-Frog Method. Therefore, of the explicit methods up to the order of consistency $O(\Delta t^2, \Delta x^2)$, the *Lax-Wendroff* Method is recommended.

A stability analysis for all explicit methods mentioned in this section shows that the *Courant number Cr* must be smaller than one or that there is a time step limitation:

$$Cr = \frac{(|v| + c)}{\Delta x / \Delta t} \leq 1 \Leftrightarrow \Delta t \leq= \frac{\Delta x}{|v| + c} \tag{3.45}$$

In this equation $c = \sqrt{gh}$ denotes the *wave velocity.* The *Courant* number is space and time dependent. However, a maximum Cr or a minimum Δt can be estimated before a numerical simulation.

Concerning the stability behavior, it is often advantageous to choose a spatial average for the considered node (eqs. 3.43, 3.44):

$$h_i^n = \frac{h_{i+1}^n + h_{i-1}^n}{2}, \qquad v_i^n = \frac{v_{i+1}^n + v_{i-1}^n}{2} \tag{3.46}$$

If an implicit scheme is taken, non-linear and non-symmetric systems of equations occur (see sec. 3.4). If the *Crank-Nicholson Method* is chosen, the equations look as follows:

$$\begin{aligned} &\frac{h_i^{n+1} - h_i^n}{\Delta t} + \theta v_i^{n+1} \frac{h_{i+1}^{n+1} - h_{i-1}^{n+1}}{2\Delta x} + (1-\theta) v_i^n \frac{h_{i+1}^n - h_{i-1}^n}{2\Delta x} \\ &\quad + \theta h_i^{n+1} \frac{v_{i+1}^{n+1} - v_{i-1}^{n+1}}{2\Delta x} + (1-\theta) h_i^n + \frac{v_{i+1}^n - v_{i-1}^n}{2\Delta x} = 0 \end{aligned} \tag{3.47}$$

$$\frac{v_i^{n+1} - v_i^n}{\Delta t} + \theta v_i^{n+1} \frac{v_{i+1}^{n+1} - v_{i-1}^{n-1}}{2\Delta x} + (1-\theta) v_i^n \frac{v_{i+1}^n - v_{i-1}^n}{2\Delta x} = -\theta g \frac{h_{i+1}^{n+1} - h_{i-1}^{n+1}}{2\Delta x} - (1-\theta) g \frac{h_{i+1}^n - h_{i-1}^n}{2\Delta x} \quad (3.48)$$

For $\theta = 0.5$, the order of consistency is $O(\Delta t^2, \Delta x^2)$; for $0.5 < \theta \leq 1$, it is $O(\Delta t, \Delta x^2)$. The method is unstable for $0 \leq \theta < 0.5$ and unconditionally stable for $0.5 \leq \theta \leq 1$ if a linearization has been carried out.

The Diffusive *Lax* (eqs. 3.43, 3.44) and the *Lax-Wendroff* Methods are compared using a simple example (see BUSSE (2000 [54]), HINKELMANN et al. (2001 [110])). A one-dimensional channel of $3000m$ length and $10m$ width is considered (see fig. 3.6). The space is discretized with 240 elements of $\Delta x = 12.5m$, and a time step of $\Delta t = 1s$ is chosen. The *Courant* stability condition (eq. 3.46) $max\ Cr \approx 0.86 < 1$ is fulfilled. The system is open at $x = 0$ where a *Dirichlet* boundary is given for the water level. The system is closed at the end with an imposed *Dirichlet* boundary for the flow velocity $v = 0$. Constant initial conditions are prescribed for the water level $h_0(x, t_0) = 6.4m$, and the flow velocity is set to $v_0(x, t_0) = 0$. A surge wave of $\Delta h = 1.5m$ is given at the inflow boundary.

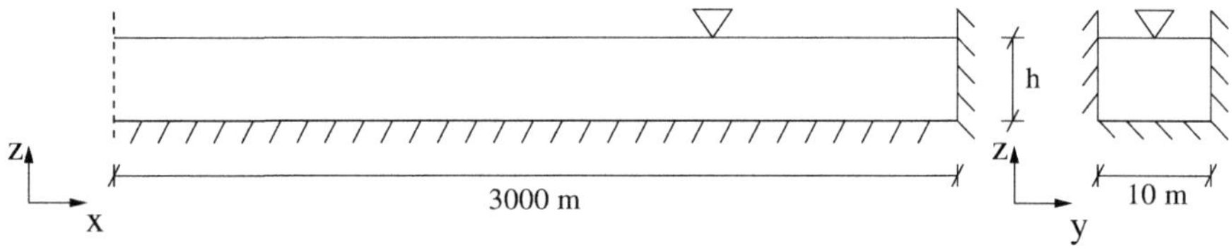

Fig. 3.6. System, after BUSSE (2000 [54])

In figure 3.7, the propagating wave is shown after $270s$. The Diffusive *Lax* Method is characterized by little numerical dispersion or smearing of the front, while the *Lax-Wendroff* Method has a sharper front, but with 'small' oscillations.

Additional information

It must be mentioned that problems arise for certain kinds of boundary conditions, e.g. when nodes which are outside the domain are required. If higher accuracy is desired, in principle more terms of the *Taylor* series can be taken into account. However, this is not very practicable, as higher derivatives which lead to further problems, e.g. with boundary conditions, must be determined.

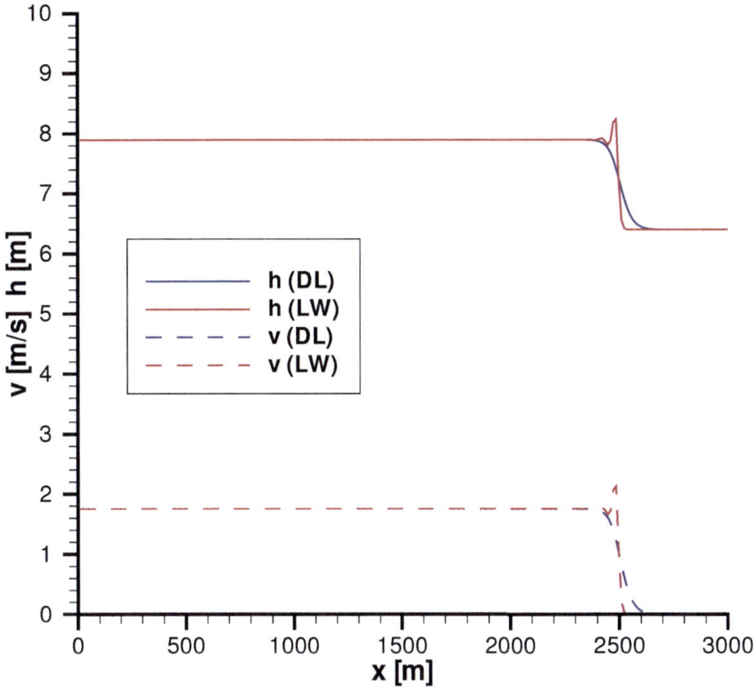

Fig. 3.7. Water level and flow-velocity distribution after 270s, after BUSSE (2000 [54])

The Finite-Difference Method is comparatively easy to understand and to 'translate' into computer programs. Rules can be mathematically proven in several cases such as consistency and stability. Therefore, it is suitable to show the principle effects and contexts. However, the Finite-Difference Method does *not necessarily* guarantee mass or momentum conservation; this is one of the greatest disadvantages of this method. The differential quotients in the basic equations are approximated by differencing, but this procedure does not imply conservativity. Due to the structured meshes, complex boundaries and inner structures can only be taken into account very roughly. Consequently, the Finite-Difference Method is no longer gaining importance.

In this section 3.1.3 only a few methods were mentioned. Further information is given in SMITH (1970 [240]), FINLAYSON (1992 [81]), HELMIG (2000 [99]), MALCHEREK (2000 [170]), KOLDITZ (2000 [153]), or HINKELMANN, HELMIG (2002 [111]).

3.1.4 Finite-Element Methods

The *Finite-Element Method (FEM)* has been widespread in hydro- and environmental engineering for several decades. The computational domain is subdivided in many small finite elements with nodes; in 2D general triangles or quadrilaterals are chosen. The unknown variables are generally defined at the nodes, sometimes at the center of the element or the center of the edges as well. The course of the unknowns over the elements is determined by *interpolation functions*. For each element, the underlying differential equation is treated by minimising an integral formulation. Then the single equations are put together, in most cases resulting in a system of equations (see sec. 3.4). Finally, the solution for the whole system is determined, taking initial and / or boundary conditions (see sec. 2.2) into account. Due to the possibly unstructured meshes, it is very suitable for complex boundaries and complex inner structures (see fig. 3.8).

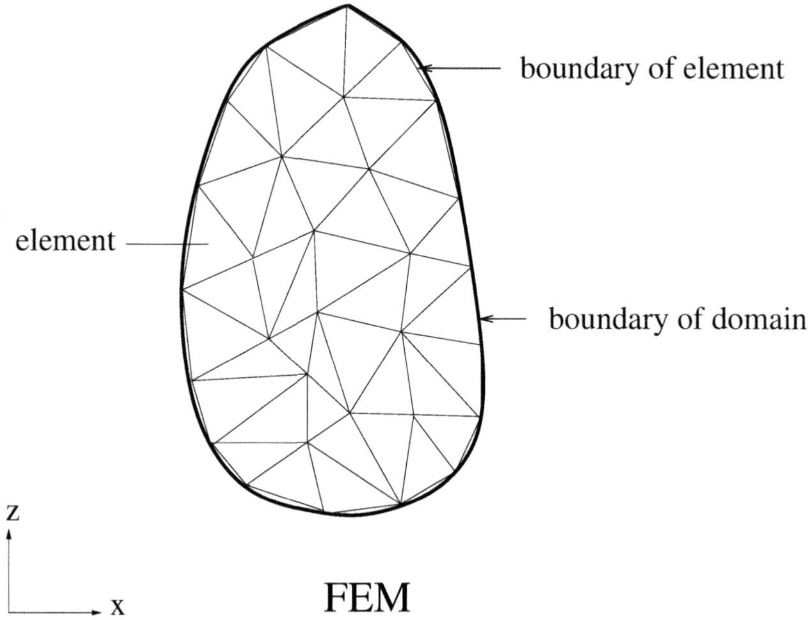

Fig. 3.8. Space discretization for the FEM, after HINKELMANN, HELMIG (2002 [111])

In the following, the principle use of the Finite-Element Method is explained using two examples, transport processes in surface water and in the subsurface.

Transport processes in surface water

The transport equation 2.75 is recalled and sink and source terms as well as advection are neglected (see sec. 2.6.2):

$$\frac{\partial c}{\partial t} - div\,(\underline{\underline{\nu}}_t grad\, c) = 0 \tag{3.49}$$

In this equation, c stands for the tracer concentration and $\underline{\underline{\nu}}_t$ for the diffusivity. As advection is not taken into account, only molecular diffusivity occurs. The type of equation is parabolic, and it is also called *diffusion equation.* The FEM is highly suitable for parabolic and elliptic problems.

First, the *semidiscrete method* which means here that the space is discretized with the FEM and the time with the FDM is introduced. This is the most common method for time-dependent problems which are treated with the FEM. For the time derivative, forward differencing (eqs. 3.9, 3.19) is applied:

$$\frac{\partial c}{\partial t} \approx \frac{c^{n+1} - c^n}{\Delta t} \tag{3.50}$$

In space, the *isoparametric concept* is used. This means that the geometry and the unknown variables are described by the same interpolation functions. Here, linear functions which lead to a second order of consistency in space are chosen. The unknown variables on the new time level c^{n+1} are approximated by $\tilde{c}^{n+1}$ or by summing up the products of the interpolation functions N and the nodal values $\hat{c}^{n+1}$, e.g. for a quadrilateral:

$$c^{n+1} \approx \tilde{c}^{n+1} = \sum_{i=1}^{4} N_i \hat{c}_i^{n+1} = N_1 \hat{c}_1^{n+1} + N_2 \hat{c}_2^{n+1} + N_3 \hat{c}_3^{n+1} + N_4 \hat{c}_4^{n+1} \tag{3.51}$$

The interpolation functions N_i, which are also called *shape functions*, are shown for a quadrilateral in figure 3.9. N_i equals 1 at node i and decreases to 0 at all other nodes of the element.

If approximation 3.51 is inserted in the diffusion equation 3.49, it no longer equals 0 exactly, but an error or *residual* ϵ occurs:

$$\frac{\partial \tilde{c}}{\partial t} - div(\underline{\underline{\nu}}_t\, grad\, \tilde{c}) = \varepsilon \tag{3.52}$$

In the *Method of Weighted Residuals*, this residual is multiplied with a weighting function W, the product is integrated over the computational domain Ω, and the integral is forced to become 0:

$$\int_{\Omega} W_j\, \varepsilon\, dV = 0 \tag{3.53}$$

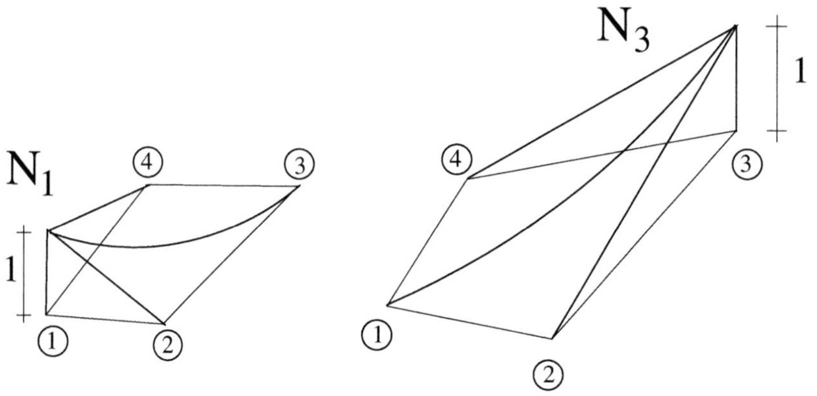

Fig. 3.9. Linear shape functions for quadrilaterals, after HINKELMANN, HELMIG (2002 [111])

Thus, the error vanishes in the average over the whole computational domain. This is important for the conservativity characteristics. When the *Standard-Galerkin Method* is applied, the weighting function equals the interpolation function:

$$W_j = N_j \tag{3.54}$$

Equations 3.52 and 3.54 are inserted into 3.53:

$$\int_{\Omega} N_j \left[\frac{\partial \tilde{c}}{\partial t} - div(\underline{\underline{\nu}}_t \, grad\, \tilde{c}) \right] dV = 0 \tag{3.55}$$

The *Green-Gauss Integral Theorem* is applied:

$$\int\limits_{\Omega} N_j \frac{\partial \tilde{c}}{\partial t} dV - \int\limits_{\Gamma} N_j \, \underline{\underline{\nu}}_t \, grad\, \tilde{c} \, \underline{n} \, dO + \int\limits_{\Omega} grad\, N_j \, \underline{\underline{\nu}}_t \, grad\, \tilde{c} \, dV = 0 \tag{3.56}$$

Γ denotes the boundary or edge of the element and $\underline{n}$ the normal vector (see fig. 2.6). The boundary integrals vanish along interior edges of the domain because they occur twice with opposite signs and thus cancel each other out. At the outer boundary of the domain, the boundary integrals are determined by *Neumann-* or *Cauchy*-boundary conditions (eq. 2.6). As a gradient of the concentration, which is often not known, must be given for this kind of boundary, this integral is often set to zero.

The concentration is determined by the *Crank-Nicholson* Method (eq. 3.11). For $\theta = 0.5$, the order of consistency is $O(\Delta t^2, \Delta x^2)$, otherwise it is $O(\Delta t, \Delta x^2)$. It should be mentioned that an implicit scheme is obtaind even for $\theta = 0$. The method is conditionally stable for $\theta < 0.5$, and unconditionally stable for $0.5 \leq \theta \leq 1$ (see HUGHES (1987 [123])). As the index notation is used in the following, the underlining for the tensors is omitted:

$$\frac{\hat{c}_i^{n+1} - \hat{c}_i^n}{\Delta t} \underbrace{\int_\Omega N_i \, N_j \, dV}_{mass\ matrix\ \ M_{ij}} +$$

$$\left[\Theta \, \hat{c}_i^{n+1} + (1-\Theta)\hat{c}_i^n\right] \underbrace{\int_\Omega grad \, N_i \, \nu_{t\,i,j} \, grad \, N_j \, dV}_{diffusion\ matrix\ \ D_{ij}} = 0 \qquad (3.57)$$

The derivatives of the interpolation functions with respect to the *Cartesian* coordinates x, z must be computed. As they are functions of the natural coordinates r, s, a transformation relation is required (see fig. 3.10) which is given by the *Jacobian matrix*:

$$\begin{bmatrix} \partial c/\partial r \\ \partial c/\partial s \end{bmatrix} = \underbrace{\begin{bmatrix} \partial x/\partial r \ \partial z/\partial r \\ \partial x/\partial s \ \partial z/\partial s \end{bmatrix}}_{Jacobian\ matrix\ \ J_{ij}} \begin{bmatrix} \partial c/\partial x \\ \partial c/\partial z \end{bmatrix} \Leftrightarrow \begin{bmatrix} \partial c/\partial x \\ \partial c/\partial z \end{bmatrix} = \underline{\underline{J}}^{-1} \begin{bmatrix} \partial c/\partial r \\ \partial c/\partial s \end{bmatrix} \qquad (3.58)$$

$$x = \sum_{i=1}^{4} N_i \, x_i \quad , \quad z = \sum_{i=1}^{4} N_i \, z_i \quad , \quad dV = dx \, dz = det J \, dr \, ds$$

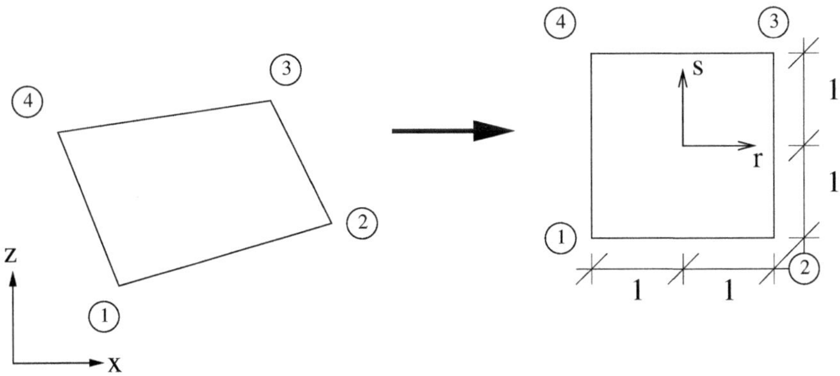

Fig. 3.10. Transformation from *Cartesian* to natural coordinates, after HINKELMANN, HELMIG (2002 [111])

The *Jacobian* matrix is inserted in equation 3.58 to obtain:

$$\underbrace{\int_\Omega N_i\, N_j\, det\, J\, dr\, ds}_{mass\ matrix\ \ M_{ij}} \frac{\hat{c}_i^{n+1} - \hat{c}_i^n}{\Delta t}$$

$$+ \underbrace{\int_\Omega grad\, N_i\, J_{ik}^{-1}\, \nu_{t\,k,l}\, J_{lj}^{-1}\, grad\, N_j\, det\, J\, dr\, ds}_{diffusion\ matrix\ \ D_{ij}} \left[\Theta\, \hat{c}_i^{n+1} + (1-\Theta)\hat{c}_i^n\right] = 0 \tag{3.59}$$

This yields the following *symmetric* system of equations for the unknown concentration c^{n+1}:

$$\left[\frac{1}{\Delta t} M_{ij} + \Theta\, D_{ij}\right] \hat{c}_i^{n+1} = \left[\frac{1}{\Delta t} M_{ij} + (\Theta - 1)\, D_{ij}\right] \hat{c}_i^n \tag{3.60}$$

It is mentioned that there is a similar meaning of the *Jacobian* matrix in the context of non-linear solvers (see sec. 3.4.5, eq. 3.142). For linear triangles, the mass and diffusion matrices can be analytically integrated. However, for general quadrilaterals, the *Jacobian* matrix consists of quotients of rational functions. Therefore, the integration is carried out numerically, generally with the *Gauss-Point integration*, as follows:

$$\int_\Omega f(r,s) dV = \sum_{i=1}^{n_{gp}} \sum_{j=1}^{n_{gp}} \alpha_i \alpha_j f(r_i, s_j) \tag{3.61}$$

The integration of the function $f(r,s)$ is substituted by a sum of values of the function at given coordinates $f(r_i, s_j)$ multiplied by weighting factors α_i and α_j. For the *Gauss-Point* integration, the special coordinates as well as weighting factors for different numbers of *Gauss-Points* n_{gp} are given in the literature (see HELMIG (2000 [99])). The *Gauss-Point* integration is superior to other methods, because it exactly integrates a polynomial of $(2n_{ng} - 1)$ degree with n_{gp} *Gauss-Points*. For linear quadrilaterals, 2 *Gauss-Points* per direction r and s are generally sufficient.

The computation of the element matrices, e.g. M_{ij}, D_{ij}, requires more CPU-time in the Finite-Element Method than in the Finite-Difference and the Finite-Volume Methods. This is the case, even though several parts of the computation must be carried out only once for all quadrilateral elements. Only the parts which are affected by the *Jacobian* matrix must be determined for every element and, of course, the numerical integration.

After the system matrices have been determined for each element, they are *assembled* to the *global system matrix* and the global right-hand-side terms

according to the neighborhood relationships. A linear symmetric system of equations must be further treated.
If the boundary of an element coincides with a jump in a coefficient - in this case the diffusivity (tensor), the FEM leads to an *arithmetic averaging.* If such jumps are within an element, the coefficients should be averaged element-wise.

In the following, a simple quasi-one-dimensional example is investigated. A *PCG Method* with a diagonal preconditioner (see sec. 3.4.1, 3.4.2) is chosen as the solver. The space is discretized with 50 elements of $\Delta x = 0.1m$ with 51 nodes, and a time step of $\Delta t = 10^6 s \approx 11.6d$ is chosen. The molecular diffusivity is set to $\nu_t = 10^{-9} m^2/s$ and $\Theta = 1$. Although there is no stability constraint, the *Neumann* number (eq. 3.36) is given , $Ne = \frac{\nu_t \Delta t}{\Delta x^2} = 0.1$, for the reader's information. The boundary conditions for the concentration c are set to zero at the beginning and the end of the system. As an initial condition, the concentration c_0 is set to 1 in the middle of the system at node 26; c_0 drops to zero at the neighboring nodes and is zero in the remaining system (see fig. 3.11). Equation 3.60 simplifies to:

$$\left[\frac{1}{\Delta t} M_{ij} + D_{ij} \right] \hat{c}_i^{n+1} = M_{ij} \hat{c}_i^n \tag{3.62}$$

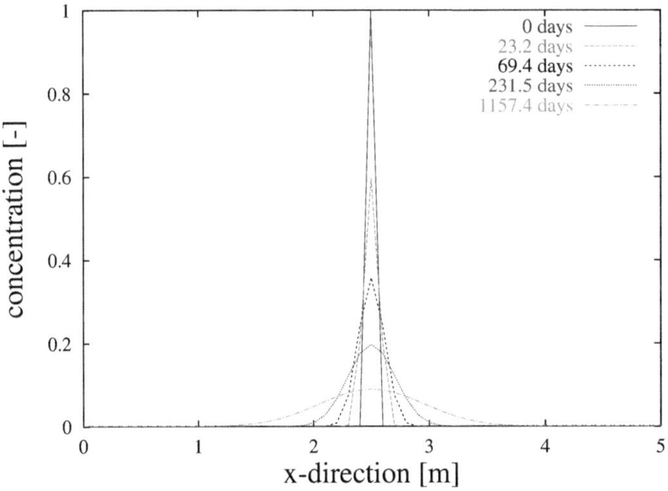

Fig. 3.11. Concentration distribution at different time steps

In the course of the simulation (see fig. 3.11), the initial peak is continuously reduced. The solution function becomes very smooth.

Transport processes in the subsurface

Now, the transport equation 2.30 is considered and only sink and source terms are neglected (see sec. 2.3.2):

$$\frac{\partial c}{\partial t} + \underline{v}_a grad\, c - div\,(\underline{\underline{D}}_{hyd} grad\, c) = 0 \tag{3.63}$$

In this equation, c denotes the tracer concentration, $\underline{v}_a$ the pore velocity and $\underline{\underline{D}}_{hyd}$ the hydrodynamic dispersion tensor. The transport equation is of mixed parabolic / hyperbolic type. If advection is neglected, this is a *diffusion equation*, and the type is parabolic. If dispersion is not taken into account, one obtains a linear *advection equation* and its type is hyperbolic.

The one-dimensional *Peclet number Pe* which is the ratio of the pore velocity multiplied by the element length Δx to the dispersion coefficient is introduced:

$$Pe = \frac{|v_a|\ \Delta x}{D_{hyd}} \tag{3.64}$$

The Peclet number is high if advection dominates. Therefore, it is an important number for indicating the tendency to instabilities.

The one-dimensional *Courant number* in a subsurface system is defined in a similar way to that shown in equation 3.45 as the ratio of the pore velocity v_a to the 'mesh velocity' $\Delta x/\Delta t$:

$$Cr = \frac{|v_a|}{\Delta x/\Delta t} \leq 1 \Leftrightarrow \Delta t \leq = \frac{\Delta x}{|v_a|} \tag{3.65}$$

For rectangular meshes, the *Peclet* and *Courant* numbers can be computed for each direction. For unstructured meshes, the element length Δx is replaced by the *characteristic discretization length l*. However, the determination of the characteristic discretization length is not unique. The diameter of the inner circle or the path line of a particle, which follows the velocity vector of the center of the element through the element, can be chosen (see fig. 3.12).

Basically, the advective part can be modeled in the same way as shown above for the *Standard-Galerkin* Method, i.e. the weighting function equals the interpolation function $W_j = N_j$ (eq. 3.54). As index notation is used in the next equation, the underlining for the vectors and tensors is omitted. If element-wise constant pore velocities are assumed, this yields:

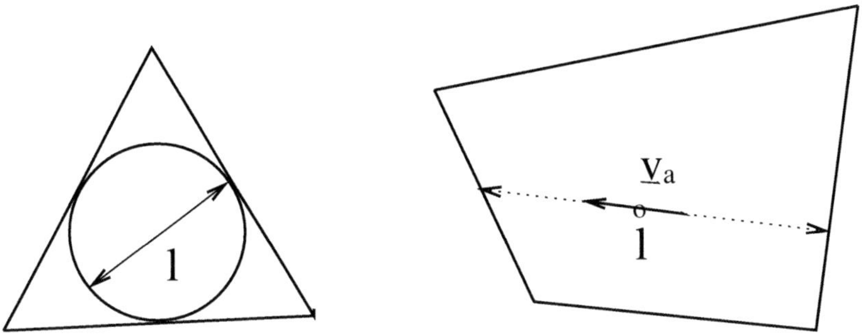

Fig. 3.12. Estimation of the characteristic element length

$$\underbrace{\int_{\Omega} N_i\, N_j\, det\, J\, dr\, ds}_{mass\ matrix\ \ M_{ij}} \frac{\hat{c}_i^{n+1} - \hat{c}_i^n}{\Delta t}$$

$$+ \underbrace{\int_{\Omega} v_a grad\, N_i\, J_{ik}^{-1} N_j\, det\, J\, dr\, ds}_{advection\ matrix\ \ A_{ij}} \left[\theta\, \hat{c}_i^{n+1} + (1-\theta)\hat{c}_i^n\right]$$

$$+ \underbrace{\int_{\Omega} grad\, N_i\, J_{ik}^{-1}\, D_{hyd\, k,l}\, J_{lj}^{-1}\, grad\, N_j\, det\, J\, dr\, ds}_{dispersion\ matrix\ \ D_{ij}} \ \left[\theta\, \hat{c}_i^{n+1} + (1-\theta)\hat{c}_i^n\right] = 0 \tag{3.66}$$

$$\left[\frac{1}{\Delta t} M_{ij} + \theta\, (A_{ij} + D_{ij})\right] \hat{c}_i^{n+1} = \left[\frac{1}{\Delta t} M_{ij} + (\theta - 1)(A_{ij} + D_{ij})\right] \hat{c}_i^n \tag{3.67}$$

The advection matrix is non-symmetric. The order of consistency in time depends on θ, while second-order consistency in space is obtained. Thus, one has $O(\Delta t^2, \Delta x^2)$ for $\theta = 0.5$ and $O(\Delta t, \Delta x^2)$ for $0 \leq \theta < 0.5$, $0.5 < \theta \leq 1$. The practical use of the *Standard-Galerkin* Method is very limited, as the following stability condition must be ensured (see PERROCHET, BEROD (1993 [206])):

$$Cr\ Pe \leq 2 \tag{3.68}$$

An improvement is given by *Petrov-Galerkin Methods.* The idea is to take more information from the upstream flow direction by choosing modified weighting functions:

$$W_j = N_j + \alpha\, T_j \tag{3.69}$$

In the classical *Petrov-Galerkin Method*, a polynomial T_j of higher degree is added, while the *Streamline-Upwind Petrov-Galerkin Method (SUPG Method)* adds a polynomial T_j of lower degree (see fig. 3.13). For the SUPG Method the weighting functions are:

$$W_j = N_j + \alpha \frac{\underline{v_a}}{|\underline{v_a}|} \cdot gradN_j \tag{3.70}$$

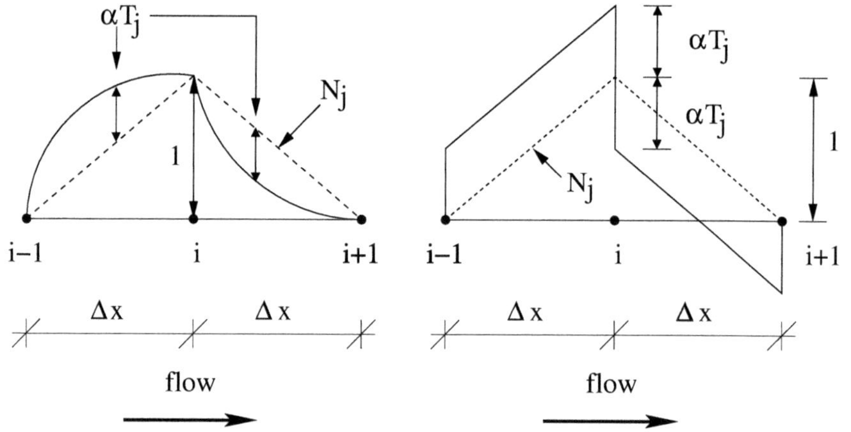

Fig. 3.13. Petrov-Galerkin (left) and Streamline-Upwind Petrov-Galerkin Methods (right), after HINKELMANN, HELMIG (2002 [111])

The modified weighting functions behave like artificial diffusion in the flow direction (not orthogonal to the flow direction, thus no cross diffusion). Therefore, the oscillations disappear, but the front is damped, and a more or less slight over- or undershooting (see fig 3.1) occurs. Generally, the SUPG Method is - more or less - superior to the classical one.

The *upwind parameter* α is free a priori. For one-dimensional stationary problems, an optimal α was determined by CHRISTIE et al. (1976 [59]):

$$\alpha_{HB} = cot\,(Pe/2) - 2/Pe \tag{3.71}$$

For multi-dimensional instationary problems, NOORISHAD et al. (1992 [192]) make some recommendations:

$$\alpha_{No} = \begin{cases} 0 & for\ Cr\,Pe \leq 2 \\ Cr - 2/Pe & for\ Cr\,Pe > 2 \end{cases} \tag{3.72}$$

Thus, no upwinding is required for $Cr\,Pe \leq 2$.

As the upwinding behaves like an artificial diffusion / dispersion, it also can be introduced into the dispersion tensor instead of choosing modified weighting functions (see PERROCHET, BEROD (1993 ([206])). Equation 2.29 is recalled

$$D_{xx} = \alpha_L \frac{v_{ax}^2}{|\underline{v}_a|} + \alpha_T \frac{v_{az}^2}{|\underline{v}_a|} + D_{mol} \tag{3.73}$$

with the longitudinal and transversal dispersion lengths α_L, α_T, the molecular diffusion D_{mol} and the component D_{xx} of the dispersion tensor. An additional longitudinal dispersion length $\overline{\alpha_L}$, which is added to α_L, is defined as:

$$\overline{\alpha_L} = \begin{cases} |\underline{v}_a| \Delta t/2 - \alpha_L - D_{mol}/|\underline{v}_a| & if \quad \alpha_L + D_{mol}/|\underline{v}_a| < |\underline{v}_a| \Delta t/2 \\ 0 & else \end{cases} \tag{3.74}$$

The upstream techniques in the FDM (see sec. 3.1.3) resemble the upwind techniques in the FEM.

For *Petrov-Galerkin* Methods described in this context, the order of consistency in time is determined by the choice of θ in the *Crank-Nicholson* Method. In space, the order of consistency depends on the *Courant* and *Peclet* numbers. Second-order consistency is approached if dispersion dominates, i.e. $Cr\,Pe < 2$. In the case of high advection, i.e. $Cr\,Pe >> 2$, only first-order consistency is approached. Furthermore, there is no stability constraint, and linear non-symmetric systems of equations must be dealt with further.

Higher orders of consistency are achieved if further terms of the *Taylor* series are taken into account; this leads to *Taylor-Galerkin Methods* (see DONEA et al. (1987 [72]), KRÖHN (1991 [159])). However, only a few *Taylor-Galerkin* Methods have been developed for unstructured grids.

Flux-corrected transport is also a very promising stabilization technique as it maintains the monotonicity of the solution (see FINLAYSON (1992 [81]), see sec. 3.1.5).

A comparison of the *Standard-Galerkin* and the *SUPG* Method (eqs. 3.66, 3.69) is shown for a one-dimensional example (see BARLAG (1997 [13])). A *BiCGSTAB Method* with a diagonal preconditioner (see sec. 3.4.1, 3.4.2) is used as the solver. The system is $100m$ long with variable element lengths $\Delta x = 0.25m, 0.5m, 1.0m$. A time step $\Delta t = 2.5 \cdot 10^5 s$ is chosen. The pore velocity is constant $v_a = 10^{-6} m/s$, and the molecular diffusivity is set to $D_{mol} = 6.25 \cdot 10^{-10} m^2/s$, while mechanical dispersion is neglected. The *Courant* number in the smallest element equals 1 and the *Peclet* number in the latter element is 400. As an initial condition, $c_0(x, t_0) = 0$ is given in the whole system. The *Dirichlet*-boundary condition $c(x = 0, t) = 1.0$ is imposed on the left side of the system (see fig. 3.14).

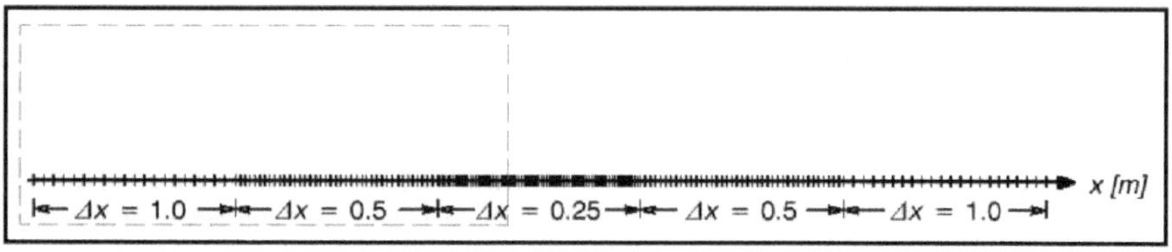

Fig. 3.14. System, after BARLAG (1997 [13])

Figure 3.15 shows the results after $t = 4 \cdot 10^7 s$, when the front has propagated approximately $40m$. The results are compared with an analytical solution of OGATA, BANKS (1961 [195]). The *Standard-Galerkin* Method oscillates, while the *SUPG* Method is smooth; however, it smears the front somewhat.

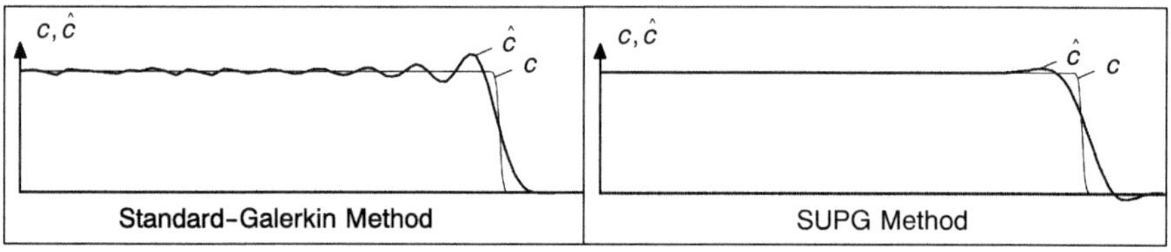

Fig. 3.15. Comparison of the Standard-Galerkin with the SUPG Method, after BARLAG (1997 [13])

Additional information

As an alternative to the semidiscrete method, space-time *Finite-Element Methods* were developed for which the time is discretized by the FEM (see NITSCHE (1985 [191]), JOHNSON (1992 [137])). They have some advantages concerning the error analysis (see sec. 3.3.2). However, implicit methods lead to very large systems of unknowns with the dimension number of unknowns in space multiplied by the number of time steps. Overall, such methods are not widespread. In this context, *Eulerian-Lagrangian Localized Adjoint Methods* or *ELLAM* Methods which are based on a space-time FEM formulation and guarantee mass conservation (see CELIA et al. (1990 [55]), HERRERA et al. (1993 [105])) are mentioned. The edges of the elements follow characteristics in the space-time domain, and the weighting functions fulfill the adjoint basic equation (see CIRPKA, HELMIG (1997 [61])). The computational effort is comparable to classical instationary FEM.

Apart from linear shape functions, quadratic and cubic functions also exist. For the special case of the shallow-water equations, the so-called *Babuska-Brezzi condition* is mathematically proven (see MALCHEREK (1990 [170])). This condition states that stability is ensured if the velocity is approximated by a higher polynomial than the water level. Therefore, often linear functions are chosen for the water level and quadratic functions for the flow velocities in the case of the shallow-water equations. *Discontinuous-Galerkin Methods* use jump functions to better approximate *sharp-front problems* or problems with large jumps in the coefficients (see JOHNSON (1992 [137]), RIVIERE et al. (2000 [221]), RIVIERE, WHEELER (2000 [220])).

As an alternative to *Galerkin* Methods, others are also used, e.g. the *Least-Square* or *Subdomain-Collocation Methods* (see HELMIG (1997 [98]), KOLDITZ (2000 [152]), sec. 3.1.5).

FORSYTH (1991 [83]) developed the *Control-Volume Finite-Element Method (CVFEM)* by combining the *Standard-Galerkin* Method for the spatial discretization with a lumped mass matrix which is multiplied by the temporal derivative (eq. 3.59). A lumped mass matrix is a diagonal matrix where the mass which is distributed over the whole domain is replaced by equivalent point masses in the nodes. FORSYTH (1991 [83]) has proven that the CVFEM guarantees local mass conservation which is a characteristic of the Finite-Volume Method (see HELMIG (1997 [98]), sec. 3.1.5).

All methods discussed in section 3.1.4 require initial and boundary conditions (see sec. 2.2) for the solution. If higher accuracy is desired, higher shape functions can be chosen in the FEM; this increases the effort of computing the system matrices and the number of unknowns. Often, it is more economical to leave linear shape functions and increase the number of elements, e.g. with the help of *adaptive methods* (see sec. 3.3).

The FEM guarantees global mass and momentum conservation which is a consequence of the formulation, because the latter states that the residual should vanish in the whole computational domain. Unstructured meshes enable an excellent approximation of complex boundaries and inner structures. Compared to the FDM, these are two important advantages for the FEM.

In this section 3.1.4, only certain methods were discussed. Further information is found in BATHE (1986 [23]), ZIENKIEWICZ, TAYLOR (1991 [269]), JOHNSON (1992 [137]), CIRPKA, HELMIG (1997 [61]), HELMIG (1997 [98]), HELMIG (2000 [99]), MALCHEREK (2000 [170]), KOLDITZ (2000 [152]), or HINKELMANN, HELMIG (2002 [111]).

3.1.5 Finite-Volume Methods

The *Finite-Volume Method (FVM)* has been established in engineering sciences for several decades. Traditionally, it has its origin in the fields of the *Reynolds equations* and technical flows, and it is often applied to structured grids. If it is used for rectangular grids (see fig. 3.4), this method is also called *Integral-Finite-Difference Method (IFDM).* Lately, the FVM has been used for unstructured meshes; in this context, it is comparable to the FEM (see fig. 3.8). A control volume is defined for every node. There are three different ways for 'constructing' a control volume. In the *cell-centered* FVM, the control volume is the element or cell, and the unknowns are determined in the center of the element (see fig. 3.16, left). If the *node-centered* FVM is chosen, two different ways of defining a control volume in the node patch are distinguished. One way is to build the polygon of the verticals on the midpoints of the element edges (see fig. 3.16, middle). However, it must be mentioned that there is a contraint concerning the mesh geometry. The mesh must fulfill the *Voronoi property*, i.e. the angles in all triangles are not allowed to differ much from 60° and in those quadrilaterals from 90° (see FUHRMANN, LANGMACH (1998 [85])). The other way is to build the control volume by the polygon of the centers of the elements and the centers of the edges (see fig. 3.16, right). As there are theoretically no limitations to the last method, it is recommended.

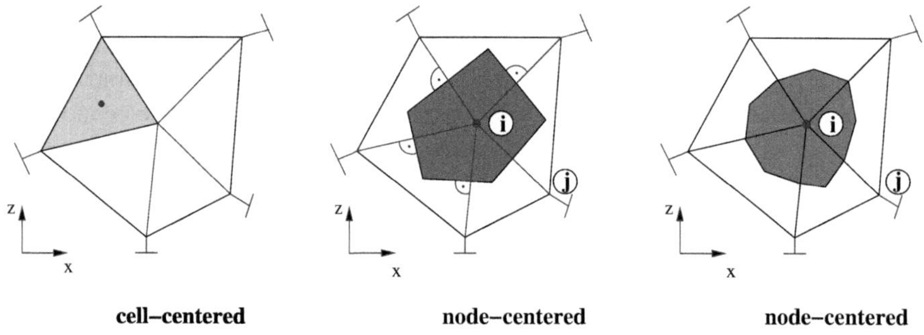

Fig. 3.16. Control volumes for the FVM

The idea of the FVM consists of integrating the differential equations in their conservative form (see sec. 2.6.1) over all control volumes. In this way, a local conservation of the considered equation is guaranteed. The basic steps of the FVM are explained for the general form of the balance equation (eq. 2.3):

$$\int_{\Omega} \frac{\partial e}{\partial t} dV + \int_{\Omega} div\, \underline{F} dV - \int_{\Omega} r dV = 0 \tag{3.75}$$

In this equation, e denotes a scalar or vector entity, $\underline{F}$ a flux term consisting of an advective and a diffusive / dispersive part $\underline{F} = \underline{v}e + \underline{w}$, $\underline{v}$ the flow velocity, r a sink / source term and Ω the domain of the control volume. If the *Green-Gauss* Integral Theorem is applied, the volume integral is transformed into a surface integral. Γ stands for the surface of the control volume and $\underline{n}$ for the normal vector (see fig. 2.6):

$$\int_{\Omega} \frac{\partial e}{\partial t} dV + \int_{\Gamma} \underline{F}\,\underline{n} dO - \int_{\Omega} r dV = 0 \tag{3.76}$$

Each term in the last equation, i.e. the temporal derivative, the flux, and the sink / source term, must be determined. Then the equations are combined, often to a system of equations which must be solved (see sec. 3.4). Taking initial and / or boundary conditions (see sec. 2.2) into account, a solution is obtained. The unstructured meshes enable the FVM to approximate complex boundaries and complex inner structures very well.

In the following, the principle use of the FVM is introduced using two examples, groundwater and two-phase flow in subsurface systems. For a better understanding, the methods are explained for a quadratic control volume, and some remarks are added concerning unstructured meshes.

Flow processes in groundwater

The stationary groundwater flow equation 2.24 is recalled and the sink / source terms are neglected (see sec. 2.3.1):

$$div\,(-\underline{\underline{K}}_f\; grad\, h) = 0 \tag{3.77}$$

In this equation, $\underline{\underline{K}}_f$ denotes the hydraulic conductivity tensor and h the piezometric head. The type of equation is elliptic, and it is also called a *Laplace equation.*

If one compares the last equation with equation 3.75, one sees $\underline{F} = -\underline{\underline{K}}_f\; gradh$. According to equation 3.76, the problem to be solved is:

$$\int_{\Gamma} (-\underline{\underline{K}}_f\; grad\, h)\underline{n} dO = 0 \tag{3.78}$$

The gradients of h at the boundaries of the control volume must be determined:

$$\underline{F} = -\underline{\underline{K}}_f \, grad \, h = -\begin{bmatrix} K_{f,xx} & K_{f,xz} \\ K_{f,zx} & K_{f,zz} \end{bmatrix} \begin{bmatrix} \partial h/\partial x \\ \partial h/\partial z \end{bmatrix}$$

$$= -\begin{bmatrix} K_{f,xx}\frac{\partial h}{\partial x} + K_{f,xz}\frac{\partial h}{\partial z} \\ K_{f,zx}\frac{\partial h}{\partial x} + K_{f,zz}\frac{\partial h}{\partial z} \end{bmatrix} \quad (3.79)$$

This is illustrated for the x-direction in figure 3.17. The gradients in x-direction are replaced by central differencing between $i-1, j\,; i, j$ and $i, j\,; i+1, j$. The gradients in z-direction are determined by central differencing between $j+1$ and $j-1$, at $i-0.5$ as an average of $i-1$ and i, and at $i+0.5$ as an average of i and $i+1$. Inflow is positive and outflow negative. One thus obtains:

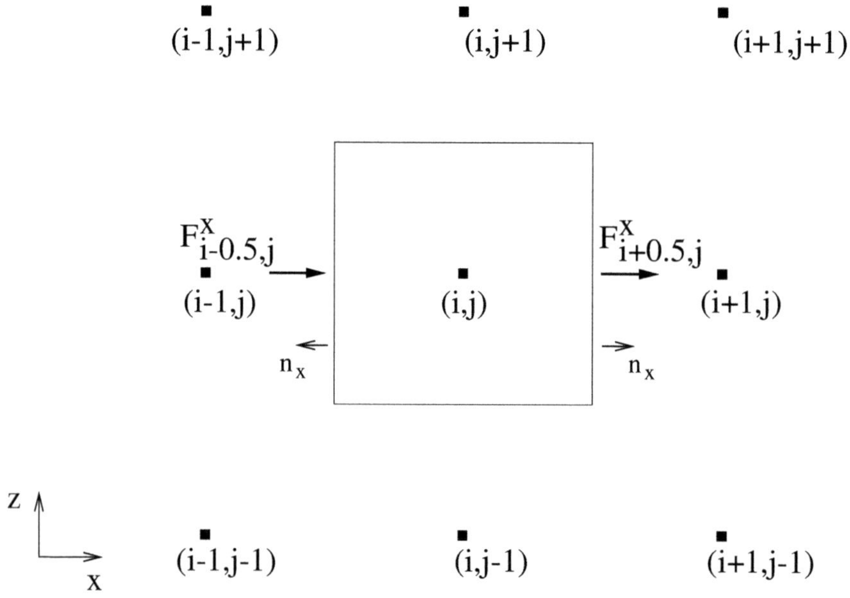

Fig. 3.17. Fluxes at the boundaries of the control volume in x-direction, after HINKELMANN, HELMIG (2002 [111])

$$\int_\Gamma F^x \, n_x dO = \int_\Gamma (F^x_{i-0.5,j} + F^x_{i+0.5,j}) n_x dO$$

$$= -K_{f,xx}\frac{h_{i,j} - h_{i-1,j}}{\Delta x}\Delta z$$

$$-K_{f,xz}\left(\frac{h_{i,j+1} - h_{i,j-1}}{2\,\Delta z} + \frac{h_{i-1,j+1} - h_{i-1,j-1}}{2\,\Delta z}\right)\frac{\Delta z}{2}$$

$$+K_{f,xx}\frac{h_{i+1,j}-h_{i,j}}{\Delta x}\Delta z$$
$$+K_{f,xz}\left(\frac{h_{i+1,j+1}-h_{i+1,j-1}}{2\,\Delta z}+\frac{h_{i,j+1}-h_{i,j-1}}{2\,\Delta z}\right)\frac{\Delta z}{2} \quad (3.80)$$

The hydraulic conductivities must be determined at the boundaries of the control volume, e.g. $K_{f,xx}$ at $i-0.5, j$ for the first term in equation 3.80 and $K_{f,xz}$ at $i+0.5, j$ or the last term in equation 3.80. The fluxes for the z-direction are obtained in the same way. Thus, a system of equations is built up.

If the coordinate directions coincide with the main directions of the hydraulic tensor (see sec. 2.3.1), the boundary integral 3.78 becomes:

$$\begin{aligned}\int_\Gamma(-\underline{\underline{K}}_f\, grad\, h)\underline{n}dO = &-K_{f,xx}\tfrac{\Delta z}{\Delta x}(h_{i+1,j}-2h_{i,j}+h_{i-1,j})\\ &-K_{f,zz}\tfrac{\Delta x}{\Delta z}(h_{i,j+1}-2h_{i,j}+h_{i,j-1})=0\end{aligned} \quad (3.81)$$

If, additionally, the subsurface material is isotropic and $\Delta x = \Delta z$, the last equation is further simplified:

$$\int_\Gamma(-\underline{\underline{K}}_f\, grad\, h)\underline{n}dO = -K_f(h_{i+1,j}+h_{i-1,j}+h_{i,j+1}+h_{i,j-1}-4h_{i,j})=0 \quad (3.82)$$

There is no stability constraint, and the order of consistency is $O(\Delta x^2)$ for the piezometric head. Linear symmetric systems of equations must be treated further. For the same reasons as those explained in section 3.1.3, the order of consistency is only $O(\Delta x)$ for the *Darcy* velocity (see sec. 3.1.6).

The similarity between the Finite-Volume and the Finite-Difference Method (see sec. 3.1.3) is shown for an isotropic subsurface material and $\Delta x = \Delta z$. If the Finite-Difference formula for the second derivative 3.22 is applied to equation 3.77, this yields:

$$-K_f(\frac{h_{i+1,j}-2h_{i,j}+h_{i-1,j}}{\Delta x^2}+\frac{h_{i,j+1}-2h_{i,j}+h_{i,j-1}}{\Delta z^2})=0 \quad (3.83)$$

If the equations 3.82 and 3.83 are compared, it is obvious that both methods lead to the same result in this specific case. Of course, this not the case in general.

If the boundary of the control volume coincides with a jump in a coefficient - in this case the permeability (tensor), a *harmonic averaging* (see HELMIG (2000 [99])) should be carried out. If such jumps are within a control volume, the coefficients should be *averaged arithmetically* in each control volume.

In the following, a quasi one-dimensional example based on equation 3.82 is investigated (see WITTE (2000 [262]), WITTE et al. (2001 [263])). A

PCG Method with a diagonal preconditioner (see sec. 3.4.1, 3.4.2) is applied as the solver. The system has a length of $1000m$, a width of $500m$ with $\Delta x = \Delta z = 20m$ and a constant hydraulic conductivity of $K_f = 10^{-4} m/s$. It is closed along $x = 0m$ and $x = 500m$. Along the open boundaries $z = 0m$ and $z = 1000m$, the piezometric head is imposed as $h = 100m$ and $h = 99m$ respectively. Figure 3.18 shows the results, a linear course of the hydraulic head and a constant parallel flow. The numerical results agree with the analytical solution; this can be proven easily.

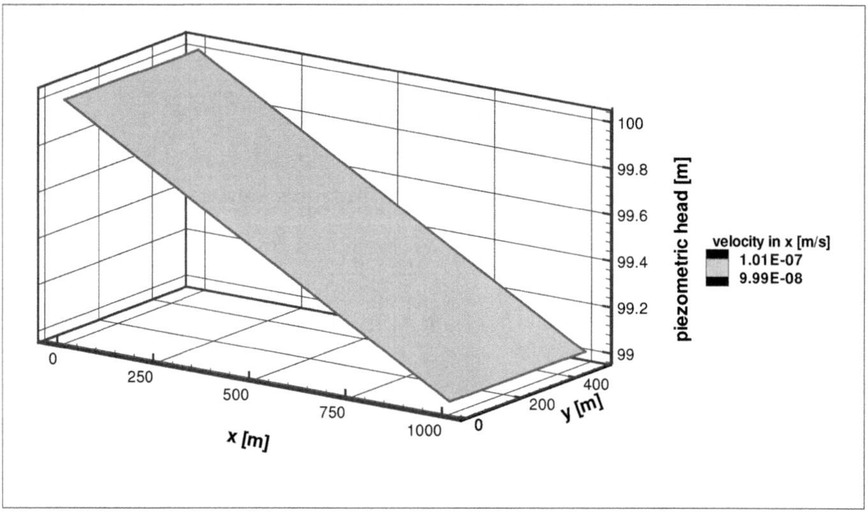

Fig. 3.18. Piezometric head and flow velocity, after WITTE et al. (2001 [263])

Two-phase flow processes in the subsurface

The two-phase flow equations in the pressure-saturation formulation 2.51 and 2.52 are considered (see sec. 2.4.4):

wetting phase:

$$-\frac{\partial(S_n \phi \rho_w)}{\partial t} - div \left[\rho_w \lambda_w \underline{\underline{K}}(grad\, p_w - \rho_w \underline{g})\right] - q_w = 0 \tag{3.84}$$

non-wetting phase:

$$\frac{\partial(S_n \phi \rho_n)}{\partial t} - div \left[\rho_n \lambda_n \underline{\underline{K}}(grad\, p_c + grad\, p_w - \rho_n \underline{g})\right] - q_n = 0 \tag{3.85}$$

In these equations, the subscripts n, w stand for the non-wetting and wetting phase, S for saturation, ϕ for porosity, ρ for density, λ for mobility, $\underline{\underline{K}}$ for the permeability tensor, p for pressure, p_c for capillary pressure and $\underline{g}$ for the vector of gravity. The equations are strongly coupled, highly non-linear and of mixed parabolic / hyperbolic type.

As shown in the context of the FEM (see sec. 3.1.4), a semidiscrete method is applied, i.e. here, that the space is discretized with the FVM and the time with the FDM. A fully implicit *Euler* scheme (see sec. 3.1.2) is chosen for the determination of the unknowns. The similarity of upwind or upstream methods in the FDM, FEM and FVM is mentioned (see secs. 3.1.3, 3.1.4). The boundary and volume integrals are solved for the control volume i according to equation 3.77 (see fig 3.16). This leads to the following non-linear non-symmetric algebraic equations for each time step (see HELMIG (1997 [98])):

wetting phase:

$$
\begin{aligned}
&-\left[(S_n\rho_w)_i^{n+1} - (S_n\rho_w)_i^n\right]\phi B_i/\Delta t \\
&-\rho_{w,ij}^{n+1}\lambda_{w,ij}^{n+1}\int_{\Gamma_i}\underline{\underline{K}}gradN_j\underline{n}dO\;\underbrace{(p_{w,j}^{n+1}-\rho_{w,j}^{n+1}g-p_{w,i}^{n+1}+\rho_{w,i}^{n+1}g)}_{discrete\ flow\ \ df_w}-q_{w,i}^{n+1}B_i=0
\end{aligned}
\tag{3.86}
$$

non-wetting phase:

$$
\begin{aligned}
&-\left[(S_n\rho_n)_i^{n+1} - (S_n\rho_n)_i^n\right]\phi B_i/\Delta t \\
&-\rho_{n,ij}^{n+1}\lambda_{n,ij}^{n+1}\int_{\Gamma_i}\underline{\underline{K}}gradN_j\underline{n}dO\;\underbrace{(p_{w,j}^{n+1}+p_{c,j}^{n+1}-\rho_{n,j}^{n+1}g-p_{w,i}^{n+1}-p_{c,i}^{n+1}+\rho_{n,i}^{n+1}g)}_{discrete\ flow\ \ df_n} \\
&-q_{n,i}^{n+1}B_i=0
\end{aligned}
\tag{3.87}
$$

In these equations, B_i denotes the control volume. The surface integral over the boundaries of the control volume can be determined according to the FEM; this is explained later. The densities are averaged along the edges (see fig. 3.16):

$$
\rho_{\alpha,ij}^{n+1} = 0.5(\rho_{w,i}^{n+1} + \rho_{w,j}^{n+1}), \qquad \alpha = w, n \tag{3.88}
$$

The mobility term can be determined with fully upwinding, i.e. the value at the upstream node of the discrete flow direction is chosen:

$$
\lambda_{\alpha,ij}^{n+1} = \begin{cases} \lambda_{\alpha,i}^{n+1}\ if\ df_\alpha \geq 0 \\ \lambda_{\alpha,j}^{n+1}\ if\ df_\alpha < 0 \end{cases}, \qquad \alpha = w, n \tag{3.89}
$$

This method is also called *Fully Upwind Box Method*. The computation of the mobilities in equation 3.89 corresponds to the *positive transmissivity condition* after FORSYTH (1991 [83]). In order to fulfill this condition, the mesh geometry must be given in such a way that negative coefficients of the permeability tensor caused by the element geometry are avoided. Other possibilities for choosing the mobility term are discussed later.

As already addressed before, the similarity in method of the Finite-Element and the Finite-Volume Methods is demonstrated. The *Subdomain-Collocation* Method can be interpreted as a *Galerkin* Method (see sec. 3.1.4) with special weighting functions (see fig. 3.19). The weighting function W_j equals 1 in a certain *area* around the considered node and 0 outside:

$$W_j(x) = \begin{cases} 1 \; if \quad x \in area \\ 0 \; otherwise \end{cases} \tag{3.90}$$

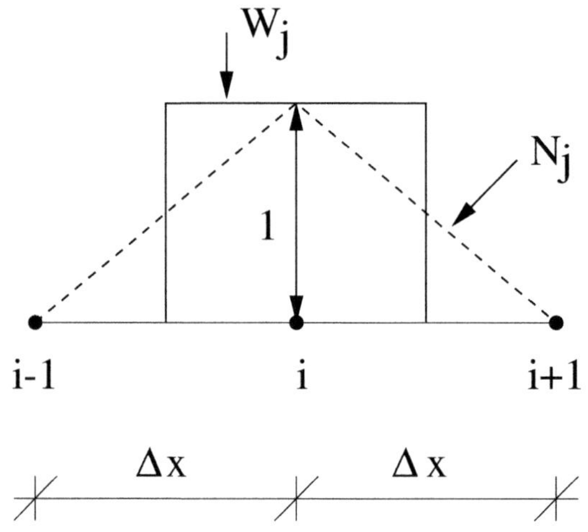

Fig. 3.19. Shape and weighting functions for Subdomain-Collocation Method, after HELMIG (1997 [98])

If the area around the considered node (see fig. 3.19) and the control volume (see fig. 3.16) coincide, the similarity between the FVM and CVFEM is given. It can be shown that the FVM and the CVFEM (see sec. 3.1.4) lead to very similar results (see FORSYTH (1991 [83]), HELMIG (1997 [98])). The major difference results from the fact that the FVM determines the flux term $\underline{F}$ by a five-point stencil on a rectangular grid and the CVFEM by a nine-point stencil, i.e. the CVFEM also takes the diagonal nodes into account. Moreover, the advantages of the two discretization methods can be combined. If, for example, the control volume is defined as shown in figure 3.16, right, it is a problem in the FVM to determine the gradients of the unknowns along the control-volume boundaries. However, this is easy in the FEM, as the gradients of unknowns are computed with the help of the gradients of the shape functions. Thus, the fluxes perpendicular to the boundaries of the control volumes in equations 3.86 and 3.87 are computed with a *one-point integration*, also called the *midpoint rule*.

The Fully Upwind Box Method is only of first-order consistency in space and time $O(\Delta x, \Delta t)$, and there is no stability constraint. In convection dominated cases, the equations are *self-sharpening* with respect to the front profile caused by the non-linearities in the flux terms (see HELMIG (1997 [98])).

As an alternative to the Fully Upwind Method, the mobility term can be determined by a linear interpolation between the mobilities at node i and j (see fig. 3.16) with an upwind parameter α_{ij} shown here for the wetting phase:

$$\lambda_{w,ij}^{n+1} = \alpha_{ij}\lambda_{w,i}^{n+1} + (1 - \alpha_{ij})\lambda_{w,j}^{n+1} \tag{3.91}$$

Again, the upwind parameter, which depends on the flow direction and the ratio of convection to diffusion, must be determined heuristically. For different reasons, which are partially given later, this method is not recommended.

A comparative study of different discretization methods regarding aspects of monotonicity, mesh geometry, convection / diffusion and heterogeneities (see HELMIG (1997 [98])) demonstrates that the Fully Upwind Box Method is monotonous, applicable to unstructured grids, locally mass-conservative and represents the correct physical behavior at the cost, however, of a certain numerical diffusion.

Using a comparatively simple, quasi one-dimensional example, the displacement of one immiscible fluid by another, for example oil by water, is simulated (see HELMIG (1997 [98]), PAUL et al. (2000 [205])). A $p_w - S_n$ formulation is chosen. The Fully Upwind Box Method is applied for spatial discretization and a fully implicit *Euler* scheme for the temporal discretization. A *BiCGSTAB Method* with a Multigrid preconditioner in combination with a *Newton-Raphson Method* for the non-linearities (see sec. 3.4) is chosen as the solver. The system is shown in figure 3.20. If gravitational and capillary effects and sink / source terms are not taken into account, if the fluids are incompressible (here $\varrho_w = \varrho_n = 1000 kg/m^3$) and have the same viscosity (here $\mu_w = \mu_n = 0.001 kg/(ms)$), and if the porosity is constant (here $\phi = 0.2$), equations 3.86 and 3.87 can be considerably simplified and solved analytically (see BUCKLEY, LEVERETT (1942 [50])). The permeability is set to $\underline{\underline{K}} = 10^{-7} m^2$, and the relative permeability-saturation relationship is determined after BROOKS, COREY (1964 [46]) with the distribution index $\lambda = 2.0$ and the residual saturations $S_{wr} = S_{nr} = 0.2$; capillarity is neglected. The system has a length of $300 m$, subdivided into 64 elements of $\Delta x = 4.6875 m$. A constant time step of $\Delta t = 5d$ is used. As initial conditions, the oil saturation is set to $S_{n0} = 0.8$ and the water pressure to $p_{n0} = 2 \cdot 10^5 Pa$. At the left boundary, the oil saturation $S_w = 0.2$ and the water pressure $p_w = 2 \cdot 10^5 Pa$ are imposed while, at the right boundary, the water flux is prevented ($q_w = 0.0$), and the flux of the oil phase $q_n = -3 \cdot 10^{-4} kg/(m^2 s)$ is given.

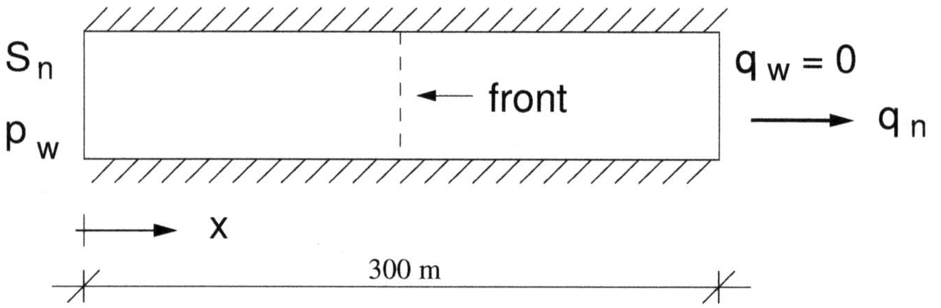

Fig. 3.20. System for the *Buckley-Leverett* problem, after HELMIG (1997 [98])

In figure 3.21, a comparison of the simulation results with the analytical solution for the water saturation after $500d$ is presented. The numerical solution is monotonous and somewhat diffusive. The overall agreement between both solutions is quite reasonable.

Additional information

For transport equations, the upwind parameters (eq. 3.91) can be computed in a similar way to the FEM (see sec. 3.1.4). For multidimensional problems, a streamline-orientated determination of the upwind parameters is advantageous for the reduction of unwanted cross diffusion (see HELMIG (1997 [98]), NEUNHÄUSERER et al. (2001 [188])). On rectangular grids, quadratic upwinding, known as the *Quick Method*, can be used (see SCHÖNUNG (1990 [231]), MALCHEREK (2000 [170])).

Problems with more than one unknown per node, e.g. shallow-water equations, can also be solved on so-called staggered grids, i.e. the different unknowns are defined at different positions, e.g. the water level in the center of the element and the flow velocity in the center of the edges. Then, two sets of control volumes must be determined (see SCHÖNUNG (1990 [231]), MALCHEREK (2000 [170])).

A very promising class of discretization methods comprises those which maintain the monotonicity of the solution, e.g. *Slope-Limiter, Flux-Limiter Methods*. The main idea of the Slope-Limiter approach is an extension of the *Go-*

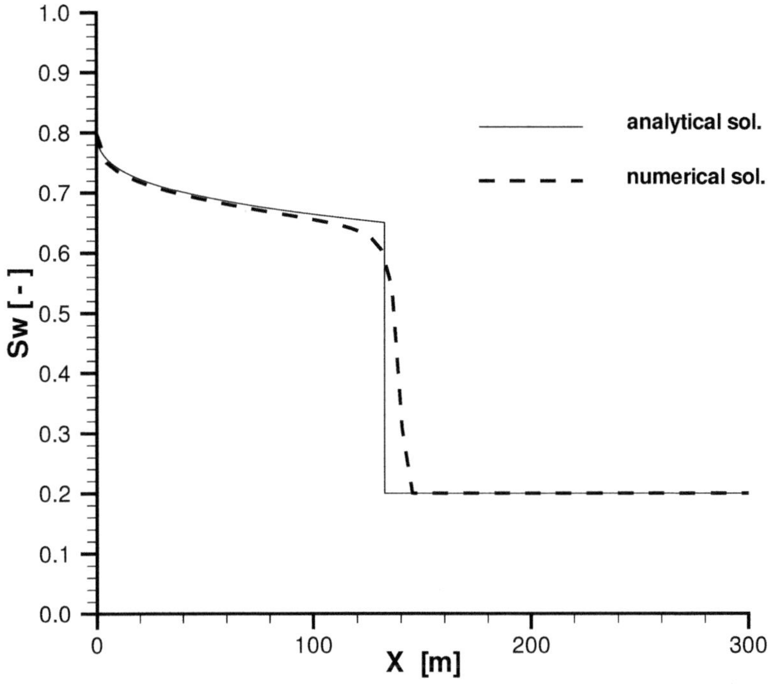

Fig. 3.21. Comparison of numerical and analytical results for the *Buckley-Leverett* problem, after PAUL et al. (2000 [205])

dunov Method for hyperbolic conservation laws. The *Godunov* Methods assume piecewise constant, cell-averaged functions in the control volume. The solutions on the new time level are obtained by shifting each cell solution of the old time level a certain distance (determined by multiplying the advective velocity with the length of the control volume) in the flow direction and then averaging the intermediate results over the control volumes. It should be mentioned that the Fully Upwind Box Method corresponds to the *Godunov* scheme. Rather than using such piecewise constant functions on the new time level, piecewise linear functions are applied in the Slope-Limiter approach. There are different ways of determining the linear functions (see FINLAYSON (1992 [81])). Depending on the slopes, the order of consistency in space lies between $O(\Delta x)$ and $O(\Delta x^2)$. For one-dimensional problems without sink or source terms, the Slope-Limiter Methods are explicit; otherwise they are implicit and even lead to non-linear algebraic equations. The Slope-Limiter approach can be applied to linear, e.g. the *advection equation*, or to non-linear partial differential equations, e.g. the *Burgers' equation.* If polynomials of higher order are chosen for contructing the slopes, this leads to *Essentially Non-Oscillatory*

(ENO) Methods (see FINLAYSON (1992 [81])). Flux-Limiter methods resemble Slope-Limiter Methods (see LE VEQUE (1992 [252])). It should be mentioned that Slope-Limiter and Flux-Limiter Methods can only be applied to the FVM, not to the FEM or FDM. The major drawback of the limiter methods is that they are generally restricted to structured grids, because it is unclear how to determine the slopes on 2D or 3D unstructured grids. Further information about limiter methods is found in FINLAYSON (1992 [81]) and LE VEQUE (1992 [252]).

Another monotonicity-preserving method, the *Flux-Corrected Transport Method (FCT)*, has been developed in the context of the transport equation (see FINLAYSON (1992 [81])). FCT solves a time step twice, once using a monotonous low-order scheme and once using an oscillatory high-order scheme. Antidiffusive fluxes are introduced by substracting low- from high-order fluxes, and they are added to obtain the solution on the new time level. However, the antidiffusive fluxes are restricted so that no new maxima, compared to the previous time step and the low order solution, occur. Although a time step must be computed twice, the advantage compared to the limiter methods results from the fact that non-linearities are avoided. FCT can also be applied to the FEM (see CIRPKA, HELMIG (1997 [61])).

The Finite-Volume Method guarantees local conservation of mass, momentum and energy which is a consequence of the formulation, because the differential equations are fulfilled in each control volume. In this case, the FVM is superior to the FEM and the FDM. When compared to the FDM, the unstructured meshes allow an excellent approximation of complex boundaries and inner structures.

This section provides only an introduction to different methods. Further information is given in SCHÖNUNG (1990 [231]), FINLAYSON (1992 [81]), CIRPKA, HELMIG (1997 [61]), HELMIG (1997 [98]), BASTIAN (1999 [18]), HELMIG (2000 [99]), MALCHEREK (2000 [170]), KOLDITZ (2000 [153]), FARTHING, MILLER (2001 [80]), TORO (2001 [249]), or HINKELMANN, HELMIG (2002 [111]).

3.1.6 Some other methods

The *Method of Characteristics* can easily be applied to hyperbolic or advection equations, e.g. inviscous shallow-water equations (see MALCHEREK (2000 [170])). The partial, space- and time-dependent, differential equations are transformed into ordinary, time-dependent, differential equations which can easily be solved along the characteristics by time-integration schemes, e.g. the explicit *Euler* or *Runge-Kutta* Methods (see sec. 3.1.2)), to determine the solution on the new time level. Although these methods are explicit, there is no stability constraint concerning the time step. However, they cannot be used alone to solve mixed hyperbolic / parabolic problems.

Another important method of solving problems which are of mixed hyperbolic / parabolic type is the *Operator-Splitting Method*, also called *Fractional-Step Method* (see GALLAND et al. (1991 [86]), EWING, WEEKES (1998 [76])). The terms in the differential equations are split up into hyperbolic and parabolic ones. Then, for each part, a numerical method specially suitable for the type of problem class is applied. The hyperbolic terms, for example, can be treated very well with the *Method of Characteristics* using a *Lagrangian* point of view, and the parabolic terms with the FEM or FVM using an *Eulerian* point of view (see HINKELMANN (1997 [108]), HINKELMANN, ZIELKE (2000 [119])). In the field of two-phase flow in porous media, the two basic equations can be decoupled if two incompressible fluids, e.g. oil and water, are considered; this is the standard case in petroleum engineering. The IMPES concept (IMplicit Pressure Explicit Saturation) is used in such cases. First, the pressure equation, which is formulated implicitly, is solved and second, the saturation equation, which is formulated explicitly, is solved, both by suitable methods (see AZIZ, SETTARI (1979 [9]), HUBER, HELMIG (1996 [122]), HELMIG (1997 [98])).

Generally, the flow velocities in groundwater flow have a spatial consistency one order lower than the hydraulic head, because they are calculated in a postprocessing step by the *Darcy* law, i.e. by a derivative of the previously determined hydraulic head (see sec. 3.1.3). *Mixed* and *Mixed-Hybrid Methods* do not insert the *Darcy* law into the continuity equation, but formulate the hydraulic head and the flow velocities as unknowns. Thus, both can be computed with the same order of spatial consistency, i.e. the order of spatial consistency for the flow velocity is increased by one (see CHAVENT, ROBERTS (1991 [58])). This is a major advantage when modeling transport processes because the flow field can strongly influence the transport simulation and the computation of an highly accurate flow field is desirable. In a similar way, *Mixed* and *Mixed-Hybrid* Methods have been applied to two-phase flow in porous media (see DURLOWSKY (1993 [73]), EWING, WEEKES (1998 [76]), HELMIG, HUBER (1998 [103])).

The methods described in sections 3.1.3 - 3.1.5 are continuum methods and are based on an *Eulerian* point of view. *Particle Methods* which are based on a *Lagrangian* point of view can also be chosen for single or multiphase transport modeling. Here, the 'fate' of each virtual particle carrying mass is followed in space and time in the computational domain (see CORDES, KINZELBACH (1992 [64]), KINZELBACH (1992 [156])). When the *Random-Walk Method* is used, the advective movement is driven by the flow field and the dispersive movement is determined by a statistical approach (see LA BOLLE et al. (1996 [33]), WOLLSCHLÄGER (1996 ([265])). After each time step, the concentrations are calculated by summing up all the particles in each element or control volume. Particle Methods have no problems with sharp fronts and no negative concentrations occur. However, the redistribution of the particles to the elements or control volumes leads to a certain numerical diffusion and, in order to get reasonable results, a large number of particles must be simulated which may cause a long calculating time.

Modeling flow and transport processes in the subsurface always involves *uncertainties.* Three different ways of dealing with uncertainties in order to make reliable predictions are briefly addressed here. One can use *conservative assumptions* which have a high measure of security, but this may be not economical. Uncertainties in certain parameters and their influences on the simulation results can be estimated by *sensitivity analyses*, i.e. by investigating what happens if these parameters vary in reasonable limits. Finally, *stochastical modeling* can be applied which leads to bandwidths of results with mean values and standard deviations. As the flow field is the most important input parameter for the transport equation, it is common to examine the influence of different permeability distributions assuming the same mean value and varying the standard deviation and the correlation lenghts. Simulations can be carried out applying the *Monte-Carlo* or *Turning-Band Methods.* Stochastical methods for subsurface modeling are an important research field which is gaining importance. For further reading, see KINZELBACH (1992 [156]), BARDOSSY (1992 [15]) or WACKERNAGEL (1998 [257]) and sec. 4.1.6.

3.2 Parallel methods

The tremendous ongoing increase in computer technology in recent years has led to a change in the methods applied in environmental and engineering sciences. In addition to theory and experiment, so-called *High-Performance Computing (HPC)*, also called *Scientific Computing*, based on mathematical models, computer simulation and optimization, has become the third pillar of research. HPC offers new ways of broadening the knowledge of complex coupled processes and phenomena in research as well as of addressing problems of large space and time scales in environmental and engineering practice, and there is no doubt that a number of tomorrow's fundamental problems will be solved with its help. HPC is dominated by *parallel computer systems* consisting of *scalar processors* or *vector units*. In order to obtain the maximum benefit from HPC architectures (see sec. 4.2), existing algorithms must be adapted to the special hardware and new algorithms must be developed.

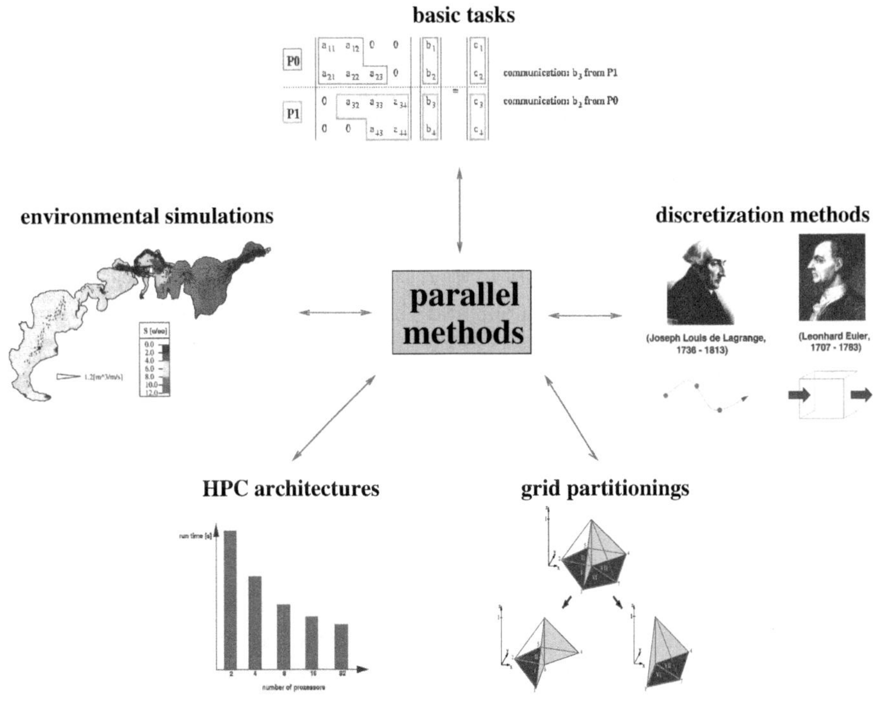

Fig. 3.22. Interactions between parallel methods and other parts of the numerical simulation, after HINKELMANN (2000 [109])

In figure 3.22, the most important interactions between parallel methods and other parts of the numerical simulation are shown, and they will be discussed in this section.

3.2.1 Development of High-Performance Computing

For about two decades, *vector* and *parallel computers* have been used for modeling hydro- and environmental systems. After a starting period, during which the HPC systems were mainly employed in research, (single-unit) vector computers were in a productional use from the late eighties. Parallel computers have gained more and more importance since the early nineties, and they have dominated HPC since the middle of the nineties. There is a small number of parallel systems with more than 512 or 1024 processors, but a large number with fewer than 128 or just a few hundred processors. Parallel processors can all share the same memory, i.e. *shared memory*, or they can each have their own memory, i.e. *distributed memory*, which is connected by a fast *communication network*. Often, smaller units of a parallel system, between 2 and about 32 processors, operate with shared memory, and several such units are connected and operate with *distributed memory*. This memory concept is called *hybrid*. Single-unit vector processors have disappeared from the market in recent years. However, parallel vector computers, equipped with shared memory for smaller numbers of units and with hybrid memory for large numbers, have established themselves in the past few years. Further information concerning architectures and information processing of HPC systems is given in sec. 4.2 and HINKELMANN (1997 [108], 2000 [109]). The USA has a leading position in the hardware development and the use of HPC systems. The market is developing very rapidly. The performance of the fastest systems worldwide approximately doubles every year (see MEUER et al. (2001 [177])). Nowadays, the fastest supercomputers achieve several Teraflops, and experts predict Petaflops computers by the end of the first decade of the 21st century (see BELL, GRAY (2001 [27])).

3.2.2 Parallelization strategies

Generally, parallel systems work with a *programming model* which is called *Single Program Multiple Data (SPMD)*. SPMD means that the same program is stored on each processor and executed with the data available for the processor. The overall aim is to achieve *scalable algorithms* which obtain a *speedup* close to the ideal one (see sec. 4.2). Below the SPMD model, two different programming models which are called *data parallel* and *message passing* are distinguished.

In the data-parallel programming model, which is particularly suitable for shared-memory machines, the parallelization and the exchange of data is carried out to a large extent by a compiler, i.e. no work need to be done by hand. A user or developer can further increase the performance by integrating directives into the code. Only one global data or addressing space exists. The data-parallel model performs well for algorithms which are based on *structured grids* and for problems which can be *synchronized* well. The major drawback of the data-parallel programming model lies in fact that, generally, it is *not portable* to distributed-memory systems. As the number of processors which can reasonably scale with shared memory is limited up to about 100, data-parallel models cannot be used for very large-scale problems. In this context, *High-Performance Fortran (HPF)* (see KOELBEL et al. (1993), SCHÜLE (1998 [233])) and *OpenMP* (see CHANDRA et al. (2000 [57])) are mentioned. HPF has been continously vanishing from the market, however, since about a few years a further variant, *JaHPF*, is under development, especially in Japan, and winning certain importance. Nevertheless, OpenMP has more or less become the standard tool for data parallelism.

The message-passing programming model has been developed for distributed-memory parallel computers, and it is also applicable to shard-memory systems, generally without loss of performance. Here, communication routines carry out the data exchange between the processors. On the one hand, these routines can be implemented into the code by hand. On the other hand, much of this 'hand work' can often be avoided, because a number of general-purpose parallel tool-boxes have been developed in recent years, for example ScaLAPACK (see BLACKFORD et al. (1997 [32])) and UG (see BASTIAN (1996 [17]), BASTIAN et al. (1997 [19])). In this context, it should be mentioned that BIRKEN (1998 [30]) has demonstrated with an advanced use of *object-oriented methods* that the amount of work done by hand can be substantially reduced, even for a complex software system like UG. The *Message-Passing Interface (MPI)* (see WALKER, DONGARRA (1996 [258])) and *Parallel Virtual Machine (PVM)* (see GEIST et al. (1994 [90])) have become standard communication interfaces, and they enable portability among nearly all the existing parallel platforms. In recent years, MPI or MPI-2 has spread much more than PVM. The message-passing programming model enables the treatment of algorithms based on unstructured grids. When compared to data-parallel models, message-passing models may scale over a much larger number of processors.

Generally, *explicit* discretization methods can be parallelized faster and more efficiently than *implicit* ones. For implicit schemes, there are two different parallelization strategies, called *Domain-Decomposition Methods* and *algebraic parallelization.*

Domain-Decomposition Methods are based on a division of the computational domain into overlapping or non-overlapping subdomains. In general, each subdomain is assigned to one processor. The problem to be solved is split up into a *local problem* in each subdomain and an *interface problem* which consists of the interprocessor nodes, i.e. the nodes belonging to more than one subdomain / processor. The local problem can be parallelized easily and efficiently without or with little communication. In most cases, the interface problem requires little CPU time compared to the local problem. A coupling of the subproblems is done by *Schwarz* or *Schurcomplement Methods* (see HACKBUSCH (1991 [94])). The major disadvantage of Domain-Decomposition Methods lies in the fact that the local problems may need very different CPU times for their solutions. This is the case when, for example, a sharp front occurs only in one or a few subdomains, and then a good *load balancing* (see sec. 3.2.4) cannot be achieved. Additionally, the convergence of iterative solvers (see sec. 3.4) depends on the number of processors. Domain-Decomposition Methods are used for elliptic or parabolic problems (see LÄMMER (1997 [161])). For the reasons mentioned above, they are not suitable for mixed hyperbolic / parabolic problems which are typical in hydro- and environmental engineering.

When the algebraic parallelization strategy is chosen, most of the serial algorithm remains the same, and every processor computes its part of the basic algebraic tasks, such as matrix-vector products or vector operations, with its assigned data (see sec. 3.3). Generally, the convergence of iterative solvers (see sec. 3.4) does not depend on the number of processors, and a good load balancing (see sec. 3.2.4) can be expected. However, algebraic parallelization requires more communication than Domain-Decomposition Methods. Overall, algebraic parallelization is superior for most environmental problems.

3.2.3 Parallelization of basic tasks

In this section, the parallelization of several *problem-independent* tasks is explained. These parallel tasks are later combined for some fluid mechanical problems. In principle, they are also suitable for other problem classes which need the same basic parallel tools. A detailed description is given in HINKELMANN (1997 [108], 2000 [109]).

Communication

When the message-passing programming model is chosen, *data exchange* or *message passing* between processors is required for certain mathematical operations, and it is performed by interprocessor communication. The time which has to be invested for a communication t_{com} is determined by equation 3.92

with the *start-up time* t_{st}, the amount of data m and the rate of data transmission r_{tr}:

$$t_{com} = t_{st} + m\, r_{tr} \tag{3.92}$$

The communication basically consists of the routines *send data* and *receive data*. Sending and receiving can be *synchronous* or *asynchronous*. When synchronous communication is chosen, the sender and receiver are blocked until the sender has received a confirmation of receipt. In the asynchronous case, program execution is continued directly after the send or receive command. Thus, an overlapping of communication and program execution is possible. Synchronous communication is very safe, asynchronous very economical. Combining asynchronous sending with synchronous receiving is recommended. Moreover, it should be mentioned that there are *blocking* and *non-blocking* communications which are similar to synchronous and asynchronous communications; for details, see the handbooks of MPI and PVM. The standard send function for exchanging data between two processors using MPI and the program language C is shown in the following example:

MPI_Ssend (buf, count, datatype, dest, tag, comm)

void	**buf*	address of data to be sent
int	*count*	number of elements of the MPI datatype which *buf* contains
MPI_Datatype	*datatype*	MPI datatype
int	*dest*	destination process for the message
int	*tag*	marker to distinguish between different messages
MPI_Comm	*comm*	communicator shared by the sending and receiving processes

This is also called *point-to-point communication* whereby point means process or processor. Other communication functions such as *broadcast* (send a message to every processor), *global sum* (sum up the partial values of a variable among all processors) and *synchronization* (stop at a predefined mark until every processor has reached it) are derived from the basic sending and receiving routines. The communication routines are available as *optimized* subroutines in the MPI and PVM libraries (see WALKER, DONGARRA (1996 [258]), GEIST et al. (1994 [90])).

One has to distinguish between *local* and *global communication*. For a local communication, only a part of all the processors exchange data. Let us assume that $K2$ in figure 3.23 is partially computed on the processors $P0$ and $P2$, a local exchange between these processors is required for the determination of the full value of $K2$. It must be mentioned that the development of a communication pattern is much more complex for unstructured grids than for structured grids. Generally on unstructured grids, each processor has to ex-

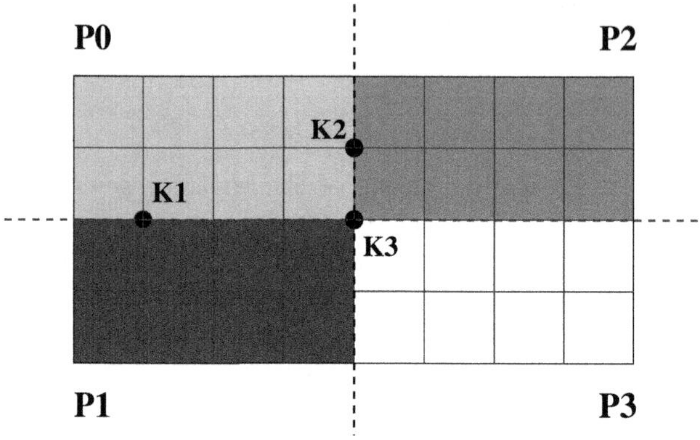

Fig. 3.23. Interprocessor nodes, local and global communication, after HINKELMANN (2000 [109])

change a different amount of data with a different number of other processors for a local communication, and this should be done in parallel with minimal communication (see HINKELMANN (2000 [108])).

When a global communication, also called *collective communication*, is carried out, a message is exchanged between all the processors. Let us assume that $K3$ is partially computed on the processors $P0, P1, P2$ and $P3$, an exchange between these processors is necessary for the determination of full value of $K3$ (see fig. 3.23).

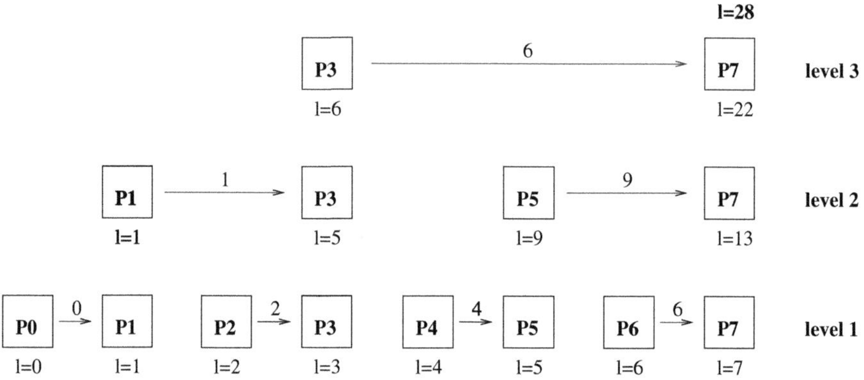

Fig. 3.24. Tree structure for global sum, after HINKELMANN (2000 [109])

To improve the understanding of *parallel programming* or *programming in networks* (see sec. 4.2), the determination of a global sum is explained. Imagine, the variable l contains the processor number on each processor, and its global sum is being looked for. It is possible, for example, that the processors $P0$ to $P6$ send their values to $P7$ which computes the result, while all other processors are waiting the whole time they are not making their single communication. Overall, $P7$ has to receive 7 messages, and the communication effort O_{com} increases linearly with the number of processors p, i.e. $O_{com}(p)$. When a *tree structure* is applied, this communication is carried out over three levels (see fig. 3.24). On the first level, 4 messages are exchanged in parallel, on the second level 2 messages and on the third level 1 message. Thus, $P7$ only receives 3 messages, and the communication effort only increases logarithmically, i.e. $O_{com}(log_2 p)$. This procedure is similar for numbers of processors not equal to 2^n with $n = 1, 2, 3, \ldots$. Finally, it is mentioned again that such routines are available in the MPI and PVM libraries.

Data structures and algebraic operations for the message-passing programming model

The numerical solution methods discussed in section 3.1 are generally characterized by *sparse matrices*, and there are many techniques for storing these matrices economically, depending on whether they are designed for *serial* or *parallel* computers and *scalar* or *vector* processors (see SCHWARZ (1991 [235]), GEIST et al. (1994 [90]), HINKELMANN (1997 [108])). Overall, storage techniques which only address non-zero terms are very efficient.

For the interprocessor nodes (see fig. 3.23), there are two different possibilities for storing the values of the corresponding variables. When the *inconsistent*, also called *geometric*, storage technique is applied, only partial values of the full value are stored on each processor. This occurs, for example, when a system matrix in the FEM (see sec. 3.1.4) is only locally assembled in its subdomain; this is done without communication. As shown later, the global system matrix is not assembled for matrix-vector operations. However, this method requires somewhat more storage space, because the interprocessor nodes are multiply stored. The storage technique is called *consistent* or *algebraic* if the full value is assigned to the variables on each processor (see fig. 3.25). Communication is needed for computing the global system matrix in the FEM; for further algebraic operations, a reorganisation of the data distribution among the processors is necessary to obtain a good load balance (see sec. 3.2.4).

Adding or substracting vectors can be done fully in parallel without communication. For inconsistent storaging, a local communication must be carried out if the full values are needed for a further operation, e.g. a division. *Scalar products* or *norms* are first fully computed in parallel on each processor with its assigned data without communication. If the inconsistent storage technique is used, the interprocessor nodes must be multiplied by a factor depending on the number of processors / subdomains to which they belong. Finally, a global communication leads to the global sum (see fig. 3.25).

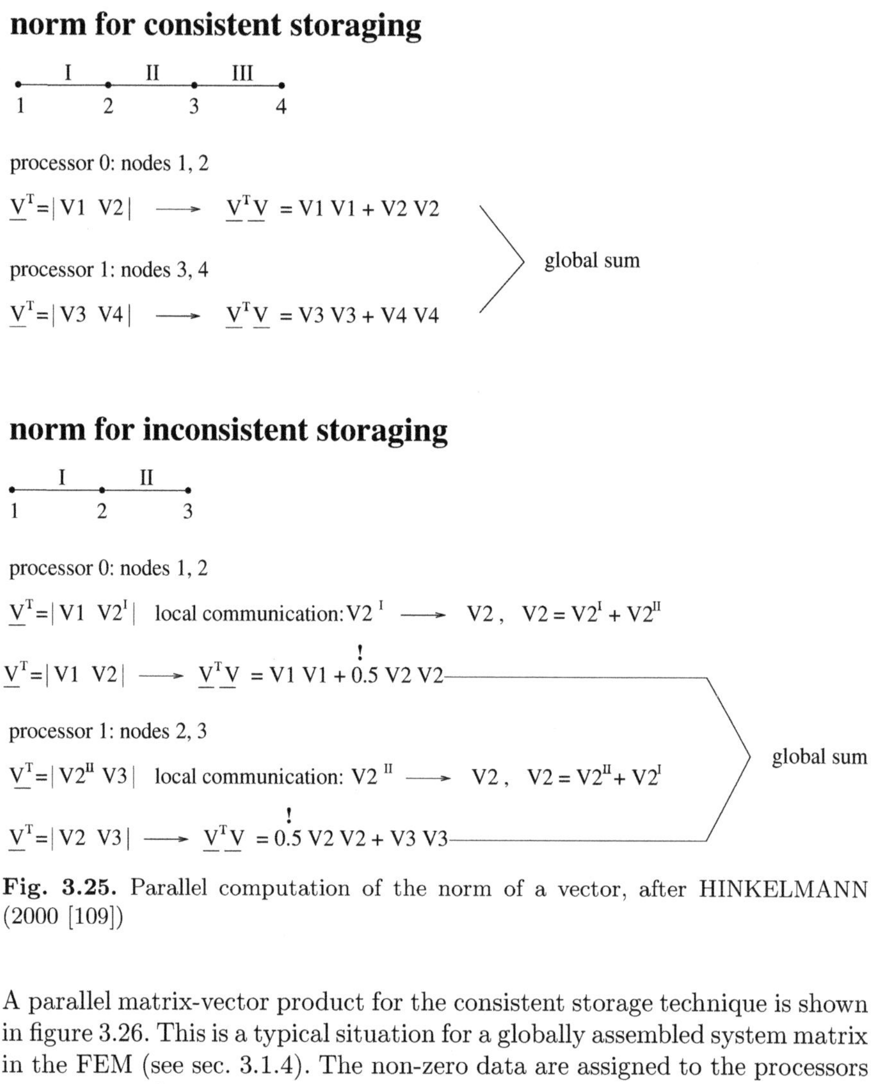

Fig. 3.25. Parallel computation of the norm of a vector, after HINKELMANN (2000 [109])

A parallel matrix-vector product for the consistent storage technique is shown in figure 3.26. This is a typical situation for a globally assembled system matrix in the FEM (see sec. 3.1.4). The non-zero data are assigned to the processors row-wise; a column-wise assignment is also possible. For large matrices, a major part of the matrix-vector product is fully computed in parallel without communication. However, for a smaller part, communication is required, for example $P0$ needs b_3 from $P1$. Further information is given, for example, in KLAAS (1996 [144]).

A parallel matrix-vector product for the inconsistent storage technique is shown in figure 3.27. This is a typical situation for a locally assembled system matrix in the FEM. Again, only the non-zero terms are assigned to the processors. For the parallel matrix-vector product, just the interprocessor nodes

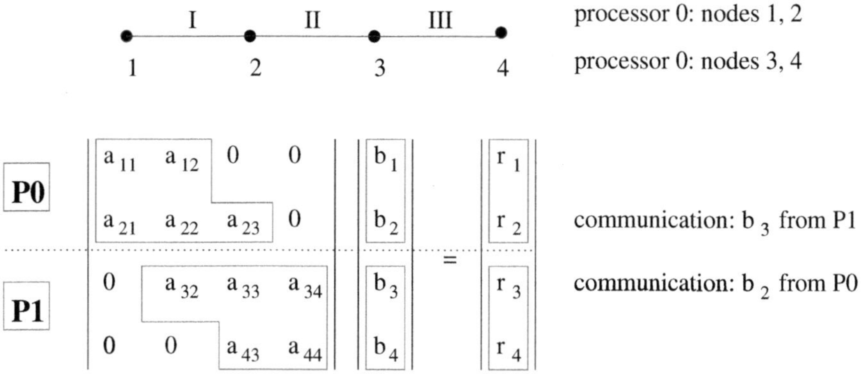

Fig. 3.26. Parallel matrix-vector product for consistent storage technique, after HINKELMANN (2000 [109])

of the vector perform a local communication to obtain their full values. If the full right-hand term is needed for further operations, a local communication must be carried out again.

As far as the efficiency of the parallel matrix-vector product is concerned, the differences between inconsistent and consistent storaging are small (see HINKELMANN (2000 [109])). The inconsistent storage techniques need less communication, but more storage than the consistent variant. In parallel programming, one often has to decide whether certain operations are performed with additional communication or additional storage and computing effort. Finally, it should be mentioned that the (parallel) matrix-vector products require the major part of CPU time of the solvers (see sec. 3.4), and also a major part of the whole CPU time for many numerical simulations.

In the context of the message-passing programming model, *parallel overhead* arises. It consists of communication time, load imbalance (e.g. wait times, see sec. 3.2.4) and additional computing time as well as additional storage requirements. The parallel overhead has to be taken into account when the message-passing programming model is chosen. This is the reason, why no *ideal speedup* (see sec. 4.2) is generally obtained. Of course, the parallel overhead should be small in order to get an efficient parallel algorithm.

If a parallel unit consists of *vector processors*, a further run-time speedup can be obtained. In a first step, an *auto-vectorization* is carried out by the compiler. Generally, the compiler can only partially detect the vectorization possibilities. Therefore, the developer has to enlarge the code by further compiler directives. Large loops can be well vectorized, as an efficient *pipelining* (see sec. 4.2) is possible. In the pipelines, no *data dependencies* or *redundancies* may occur. Generally, the aspects *direct addressing, structured grid, explicit*

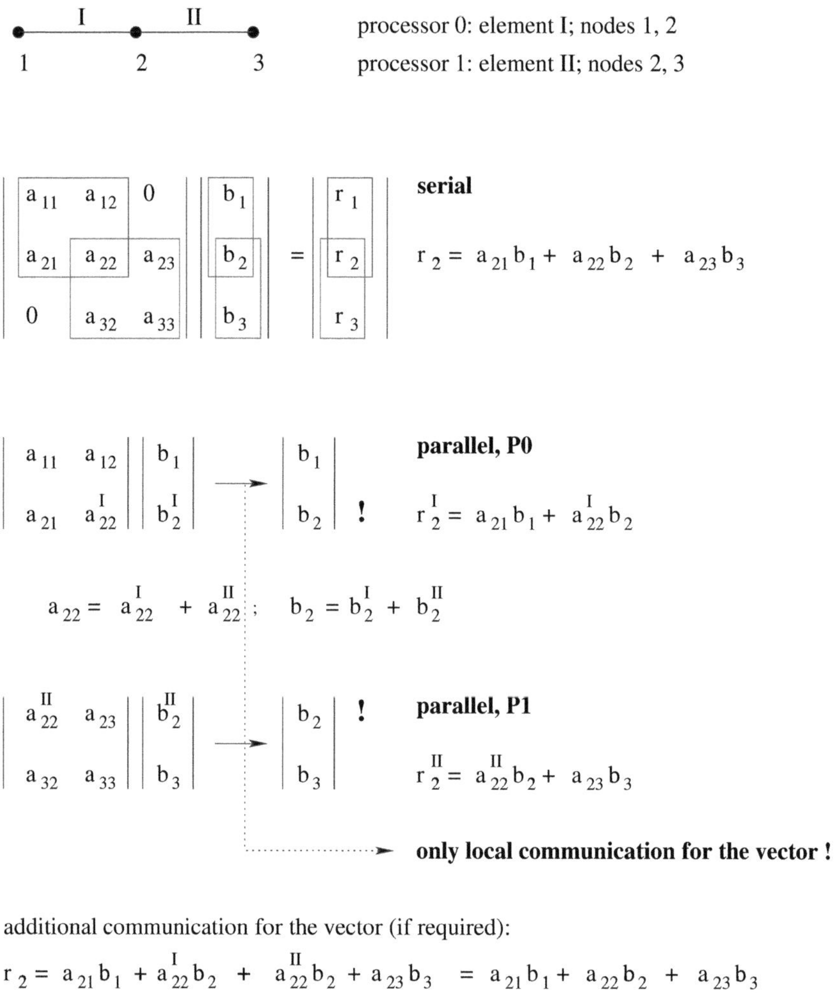

Fig. 3.27. Parallel matrix-vector product for inconsistent storage technique, after HINKELMANN (2000 [109])

algorithm and *dense matrix* are very suitable for vectorization. If they are not or only partially given, a vectorization may be reasonable depending on the special problem. Several algorithms cannot be vectorized efficiently, for example *direct solvers*, *Multigrid solvers* (see sec. 3.4) and different *Particle Methods* (see sec. 3.2.5).

Further information about the topics discussed in this section is given in HINKELMANN (2000 [109]).

Data structures and algebraic operations for the data-parallel programming model

As already mentioned before, the data-parallel programming model is suitable for parallel systems with shared memory. For *synchronous* problems and algorithms operating on *structured grids*, a large part of the parallelization including the load balancing (see sec. 3.2.4) for algebraic operations is carried out by the compiler. Futher optimizations may be possible and must be added to the source code by the developer. If the two mentioned prerequisites are not fulfilled, the data-parallel programming model should not be chosen.

3.2.4 Load balancing

Problem description

The problem of load balancing is concerned with dividing the computational load among the processors in such a way that they are all equally burdened and, for the message-passing programming model, interprocessor communication is minimized. For the partial differential equations treated here, load balancing is done by partitioning the computational domain into subdomains and assigning each subdomain together with the corresponding data to one processor (see fig. 3.28).

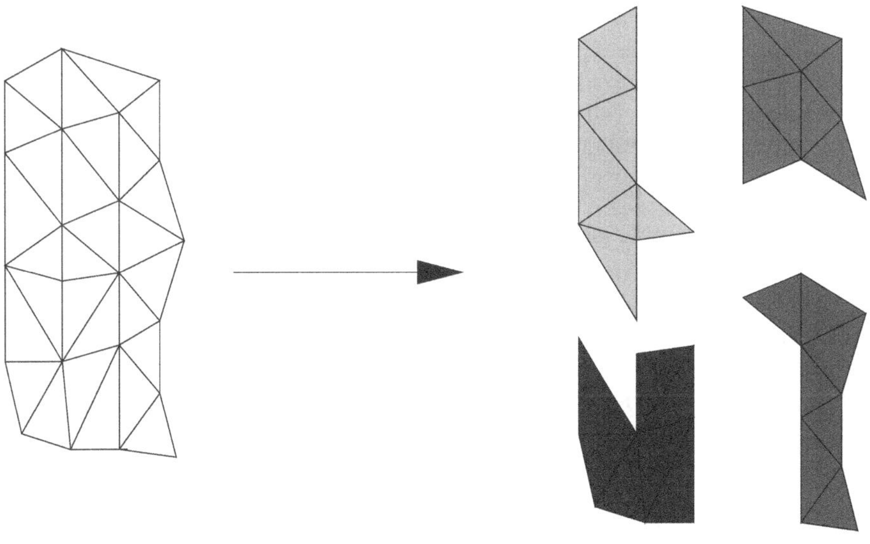

Fig. 3.28. Grid partitioning into four subdomains, after HINKELMANN (2000 [109])

It should be mentioned that *time-parallel* approaches also exist (see BURMEISTER, HACKBUSCH (1996 [53])). However, they have not gained acceptance for different reasons.

On *structured grids* the domain is divided row-wise, column-wise or block-wise. In the data-parallel programming model, this task is carried out by the compiler which splits the corresponding vectors up. Grid partitioning is much more complex for unstructured grids than for structured ones. In a general formulation, the load balancing problem leads to an *optimization task* of *NP-complete complexity* (see GAREY et al. (1976 [88])). As no suitable exact solution methods exist for such problems today, *suboptimal* methods must be chosen.

For many problems in which the CPU-time requirements in the subdomains do not vary much a *static* grid partitioning is carried out in a preprocessing stage, and it does not change during the numerical simulation. If, for example, *adaptive methods* (see sec. 3.3) are used, the CPU time in the subdomains may temporarily increase significantly caused by adaptive mesh refinement and coarsening (see sec. 3.3). Then, a *dynamic load balancing* should be applied, and the CPU load must be redistributed among the processors during the simulation.

Solution methods

A grid partitioning can be carried out in such a way that the number of nodes in the subdomains is equal. However, the number of elements then differs slightly. Within an implicit numerical simulator, the CPU time for the solvers (see sec. 3.4) is generally dominated by loops running over the number of nodes. When the number of elements is distributed equally, then a small imbalance in the number of nodes occurs in the subdomains. In the latter case, the load-balancing techniques should be applied to the *dual mesh* or the *dual graph*. A dual mesh is generated when every element is replaced by its center of gravity, and these nodes are connected according to the neighborhood conditions (see fig. 3.29). In a numerical simulator, the CPU time for evaluating the system matrices is used by loops running over the number of elements. It is also possible to introduce a *cost function* which consists of a weighted sum of the number of nodes and elements or even other parts, e.g. the number of boundary nodes, and to partition the grid to achieve an equal distribution of the costs.

Two widely used procedures for subdividing the grids are explained. In the *recursive bisection*, a domain is split up into two subdomains, and this is then repeated recursively in the subdomains in several steps up to the desired number $2^n, n = 1, 2, 3, \ldots$. With a few extensions, a subdivision into a general number of subdomains is possible. When a *clustering technique* is chosen, the grid partitioning into the desired number of subdomains is carried out in one step. The recursive bisections have the advantage that a better load balancing can be obtained. However, they need more CPU time.

Three different groups of load-balancing methods are distinguished: *geometrically orientated, graph-orientated* and *further heuristic methods.* They can be combined with the recursive bisection or the clustering technique; some of the important ones are discussed in the following.

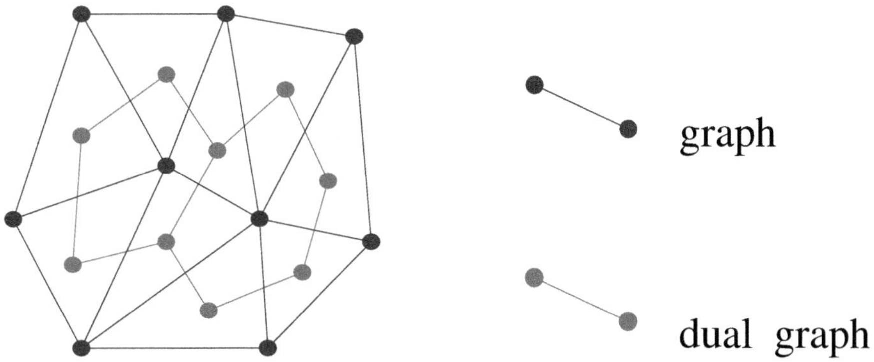

Fig. 3.29. Mesh and dual mesh, after HINKELMANN (1997 [108])

Geometrically orientated methods use geometric information of the mesh. The *Recursive-Coordinate Bisection (RCB)* determines the axis with the longest extension and subdivides the mesh along this axis (see fig. 3.30). If the investigation concerns which coordinate axis leads to an interprocessor boundary with the least nodes or edges, this variant is called modified RCB. As the grid partitioning depends on the choice of the coordinate system, further improvement is offered by *Recursive-Inertial Bisection (RIB)*, which divides the mesh along the *inertial axis.* Geometrically orientated methods are simple and require little CPU time. However, in complex meshes, subdomains which are not connected with each other can occur. Such a situation can cause a breakdown of certain parts of an algorithm. Additionally, the communication is not taken into account during the solution of the load-balance problem, and this can cause large interprocessor boundaries with many nodes and edges or much communication time.

Graph-orientated methods are based on the connectivity information of the mesh or the graph / dual graph. The *Recursive-Graph Bisection* determines the diameter of the graph or the maximal graph distance to partition the mesh. One of the most promising techniques is the *Recursive-Spectral Bisec-*

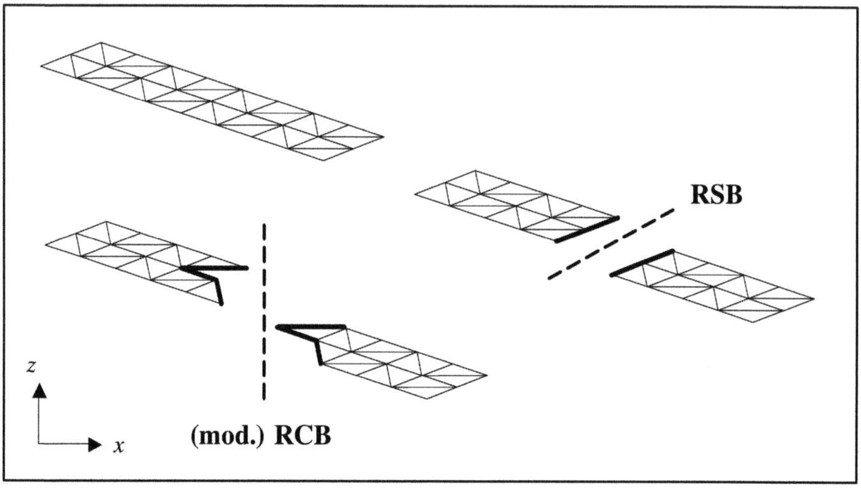

Fig. 3.30. Modified Recursive-Coordinate and Recursive-Spectral Bisection, after HINKELMANN (1997 [108])

tion (RSB) which minimizes the number of common nodes or edges on the interprocessor boundary - and thus the CPU time for local communication, when a graph is subdivided into two subgraphs with the same number of nodes or edges (see fig. 3.30). The RSB is based on computing the *Fiedler vector* which is the eigenvector corresponding to the second smallest eigenvalue of the graph-connectivity matrix. Graph-orientated methods are superior to geometrically orientated ones. If the initial mesh is connected, then it can be mathematically proven that at least one subgraph is connected when RGB or RSB are applied. However, graph-orientated methods require more CPU time, for example, because an eigenvalue solver is required for RSB.

Heuristic methods use other information. The mesh partitioning for the dual graph often results in evaluating the interprocessor boundaries in the form of a 'sawtooth pattern' (see fig. 3.31). The *Kernighan-Lin heuristic (KL)* can be used to obtain a further improvement of this solution. KL switches elements from one subdomain to the other if the number of common nodes can be decreased. Thus, KL operates like a smoother. KL needs a good starting subdivision, otherwise it requires too much CPU time. Further methods are given in the literature at the end of section 3.2.4.

If the CPU time changes significantly in a subdomain during a simulation, for example as a result of adaptive mesh refinement (see sec. 3.3) or *Local Multiplicative Multigrid* (see sec. 3.4.4), a *dynamic* load balancing should be applied. If (adaptive) Local Multigrid is applied, a load imbalance not only occurs within a level in the horizontal direction, but also between the levels

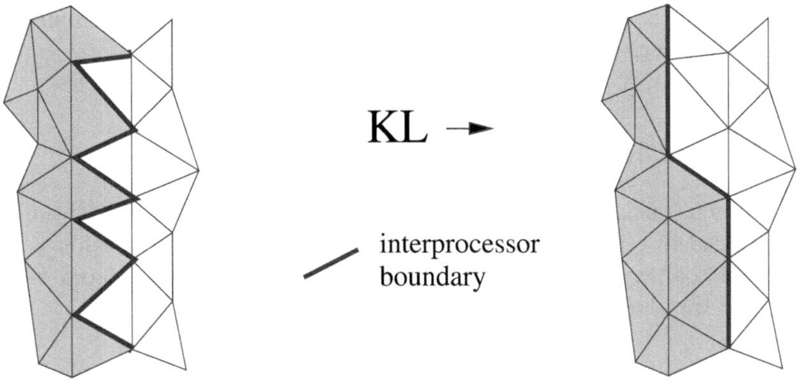

Fig. 3.31. Kernighan-Lin heuristic, after HINKELMANN (1997 [108])

in the vertical direction. In Multigrid Methods, the same nodes on different grid levels (see sec. 3.4.4) can be assigned to different processors. In practice, equally loaded clusters are built consisting of a few elements on a certain grid level together with the corresponding refined elements up to a further given grid level, and the partitioning is then carried out for these clusters (see LANG (2001 [162])). In a dynamic case, the quality of the grid partitioning is of only minor importance; what is important is the CPU time required for the grid partitioning technique and that required for the *data redistribution* and *load migration*. Therefore, the benefit of the CPU-time expensive graph-orientated methods is very questionable. Often, simple grid-partitioning techniques, e.g. RCB, and clustering techniques are chosen. A *cost function* must be determined which indicates a load redistribution during the simulation if a certain load imbalance is exceeded. The following data redistribution and load migration is very complex. Using graph-based data structures and *object-orientated methods*, it can be carried out very efficiently with methods based on the concept of *distributed objects*. For example, with distributed objects, it is comparatively easy to move not only objects or elements between processors, but whole element trees. For parallel adaptive Multigrid Methods, the CPU time for the load migration is much higher than that for the grid partitioning in general. Although dynamic load balancing may not seem to be economical with regard to the CPU time, it has been demonstrated that it requires only a small percentage of the CPU time for large-scale computations; thus, dynamic load balancing is not a serious handicap for overall scalable simulation methods.

After the mesh has been partitioned, the subdomains must be assigned to the processors; this is called *mapping*. Adjacent subdomains should be assigned to adjacent processors, because then the communication time is minimal. Often, this is not or cannot be taken into account during the grid partitioning. However, it must be mentioned that on parallel systems nowadays the mapping is no longer of such importance because of the fast routing in the communication networks (see sec. 4.2).

Additional information

Three-dimensional mesh partitioning can be significantly simplified if the mesh is structured in one coordinate direction where the mesh resolution is comparatively coarse. This mesh structure is also called 2.5 mesh, and is often chosen for the shallow-water equations (see sec. 2.6).

It should be mentioned that there are different public-domain tools for grid partitioning which offer the most standard as well as dynamic load-balancing techniques (see CHACO ([56]), ParMETIS ([202]), UG / DDD ([250]), EXDASY ([77])). Aspects of parallel input and output are given and discussed in HINKELMANN (1997 [108]), UG / DDD ([250]) and COVISE ([65]).

Further reading on load-balancing methods can be found in HENDRICKSON, LELAND (1993 [104]), FARHAT, LESOINNE (1993 [79]), SIMON et al. (1993 [239]) and HINKELMANN (1997 [108]), with special focus on dynamic load balancing in BASTIAN (1996 [17]), BIRKEN (1998 [30]), or LANG (2001 [162]).

3.2.5 Particle Methods and series

Particle Methods may be very suitable for parallel computation, as they are characterized by *coarse granularity* (see sec. 4.2). The basic idea of Particle Methods is the following: a given number of particles is traced from time step to time step through the computational domain controlled by a velocity field; after each time step, the particles in each element or control volume are counted to compute the concentration distribution. The particle methods differ in the way how the advective and diffusive / dispersive paths as well as the tracing are determined. This is not discussed here (see sec. 3.1.6).

For parallelization, the particles are distributed equally among the processors and, in principle, a very efficient parallel algorithm can be obtained. This is easy to handle on a shared-memory system. On a distributed-memory system, two problems may occur (see fig. 3.32). One is when an algorithm based on a particle-method follows path lines starting from nodes of the computational domain, e.g. the Method of Characteristics (see sec. 3.2.6, HINKELMANN (1997 [108])). If such nodes adjoin several subdomains (**!** in fig. 3.32) and are stored on the corresponding processors, they are treated on each one, but, depending on the flow conditions, this is done correctly only on one processor. Therefore, the missing information must be given by a local communication.

Second - even more severe - it is possible that path lines leave the subdomain area (!! in fig. 3.32). For special algorithms and problem classes, techniques can be developed to solve this efficiently (see HINKELMANN (1997 [108])).

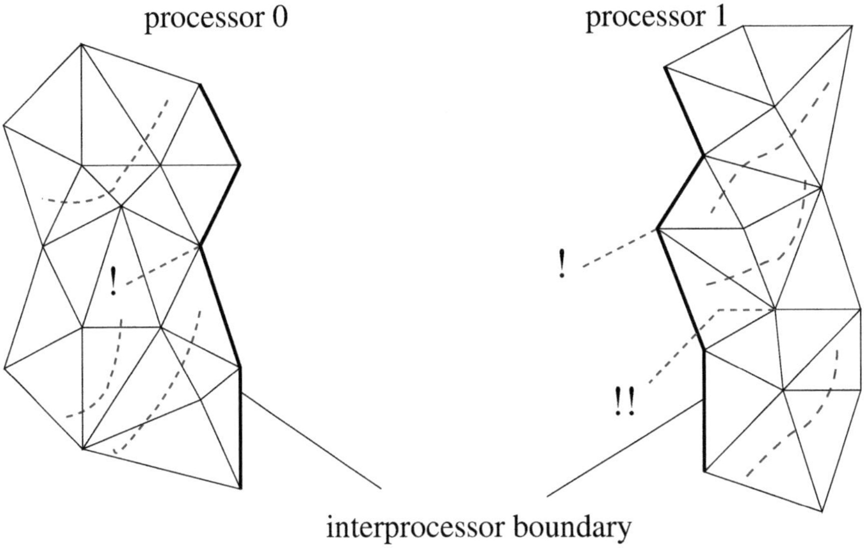

Fig. 3.32. Parallel tracing of path lines on two subdomains / processors, after HINKELMANN (1997 [108])

Generally, an efficient particle-tracking algorithm is only achieved on a distributed-memory parallel computer if each processor has the complete mesh information and velocity field. However, this contradicts the distributed-memory philosophy according to which each processor has only the mesh information of its subdomain, and this in turn may also contradict the distributed-memory-based computation of the velocity field. If the CPU time of a flow and transport simulation is clearly dominated by the particle-tracking algorithm, one solution for parallelization is to compute the flow field on each processor. Other possibilities exist; the choice depends on the different algorithms for the flow and transport simulation.

If particle interactions must be taken into account, such an algorithm possibly cannot be parallelized reasonably. Moreover, Particle Methods cannot be vectorized.

For numerical simulations, huge data sets are often used to steer the model, i.e. to serve as the boundary conditions. There are some advantages in developing these data sets in *Fourier series*, and the computation of such series can easily be parallelized. Each processor is assigned only a part of the coefficients, and eventually a global summation leads to the correct value. Further information is found in HINKELMANN (1997 [108]).

3.2.6 Recommendations

Nowadays, if people want to develop parallel algorithms or extend existing algorithms for the use on parallel computers, I personally recommend the message-passing programming model - not only for the problem classes discussed in this context -, as it enables the portability to nearly all existing kinds of parallel High-Performance Computing architectures and offers the best prerequisites for obtaining scalable algorithms (see sec. 4.2). There are many public- domain tools (see secs. 3.2.2, 3.2.4) which support this work a lot and reduce the amount of individual effort to a minimum; compilers exist for nearly all standard programming languages, e.g *C/C++, F90/F95, Java.* The major part of parallel computers uses a UNIX or LINUX operating system, a few parallel machines run under WINDOWS. Explicit algorithms and those which operate on structured grids can be parallelized more easily and efficiently than implicit algorithms and those which are based on unstructured grids. An algebraic parallelization strategy should be preferred to the parallelization strategy based on Domain-Decomposition methods, especially if hyperbolic parts in the problems have a dominant influence. Numerical simulation tools based on parallel, adaptive Multigrid Methods including dynamic load balancing are state of the art today. Coupled explicit / implicit or *Eulerian / Lagrangian* methods often can be parallelized efficiently. If Multigrid Methods or certain explict methods, e.g. particle tracking, are chosen, one should take into account that they cannot reasonably be vectorized, if at all. Therefore, the use of parallel vector computers cannot be recommended. Finally, one can think about data parallelism, if the problems considered are based on structured grids and if portation to other platforms is definitly not envisaged.

3.2.7 Examples

In this section, two different parallel algorithms, one based on the FDM and the other on the FVM, are considered for the simulation of groundwater flow processes.

Parallel Finite-Difference Method for groundwater flow processes

The problem already investigated in section 3.1.3 is again chosen here (see HINKELMANN (2001 [110])). The parallelization is very easy. First, the one-dimensional computational domain is subdivided in such a way that the same number of connected elements and their data are assigned to each processor. Then, each processor evaluates equation 3.38 with its data. Using MPI, a local communication is required for the nodes on the interprocessor boundaries after each time step to obtain the results.

In figure 3.33 and table 3.2, the *parallel efficiency* (see sec. 4.2) and *parallel speedup* (see sec. 4.2) which have been achieved on the T3E (see sec. 4.2) are shown for different spatial and temporal discretizations, 6400 elements of $\Delta x = 10^{-3}m$ and $\Delta t = 10^{-5}s$, 64000 elements of $\Delta x = 10^{-4}m$ and $\Delta t = 10^{-7}s$ and 640000 elements of $\Delta x = 10^{-5}m$ and $\Delta t = 10^{-9}s$. As this FD algorithm is explicit, the *Neumann*-stability condition (eq. 3.36) must be fulfilled. Here, the *Neumann* number is $Ne = 0.1$ in all three simulations. For a given problem size, the parallel speedup and the parallel efficiency decrease with an increasing number of processors. For a given number of processors, the parallel speedup as well as the parallel efficiency increase.

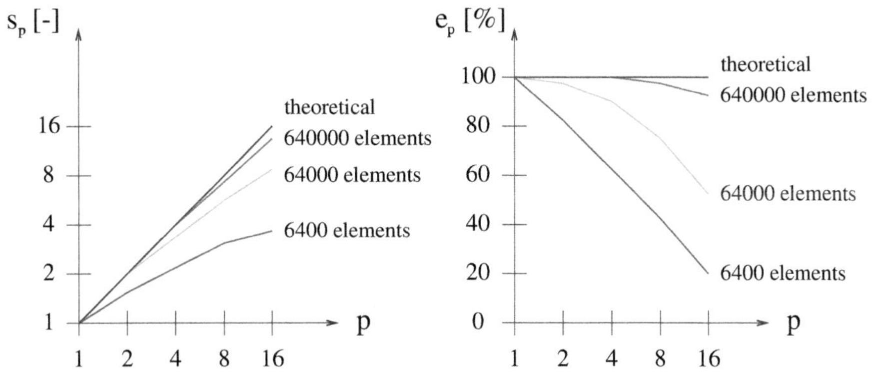

Fig. 3.33. Parallel speedup and parallel efficiency for different problem sizes on the T3E, after HINKELMANN (2000 [109])

	1P	2P	4P	8P	16P
speedup [–]					
6400 elements	1.00	1.62	2.34	3.15	3.80
64000 elements	1.00	1.96	3.62	6.06	8.32
640000 elements	1.00	1.99	3.95	7.69	14.9
efficiency [%]					
6400 elements	100.0	81.0	58.5	39.4	23.8
64000 elements	100.0	98.0	90.5	75.8	52.2
640000 elements	100.0	99.5	98.8	96.1	93.3

Table 3.2. Parallel speedup and parallel effiziency for different problem sizes on the T3E, after HINKELMANN (2000 [109])

Parallel Finite-Volume Method for groundwater flow processes

The FV algorithm and the problem already discussed in section 3.1.5 are investigated here concerning parallel computing (see WITTE (2000 [262]), WITTE et al. (2001 [263])). As this algorithm is implicit, more steps must be carried out for parallelization than when using the explicit FDM. An algebraic parallelization strategy is applied together with a message-passing programming model based on MPI. Due to the chosen structured mesh, its partitioning is easy, and blocks with the same number of elements are assigned to the processors. In the parallel FVM, two major steps must be carried out, here using inconsistent storaging for the system matrix. First, this matrix must be set up by evaluating equation 3.81, 3.82 or 3.83. This can be done fully in parallel without communication. Second, a linear symmetric system of equations is solved iteratively with the parallelized *PCG Method* (see sec. 3.4.2) and diagonal preconditioning (see sec. 3.4.3). These tasks are based on the parallel algebraic tasks introduced in section 3.2.3.

The simulations are carried out on two different parallel hardware architectures, a massively parallel T3E and a cluster of standard PCs (see sec. 4.2). Two different problem sizes are analyzed, $\Delta x = 1m, \Delta z = 2m \rightarrow 251001$ nodes (variants (1) in fig. 3.34) and $\Delta x = 2m, \Delta z = 4m \rightarrow 63001$ nodes (variants (2) in fig. 3.34). In figure 3.34, the results of the parallel efficiency are presented. For more than 4 processors, the results are as expected. For a constant problem size, the parallel efficiency decreases with an increasing number of processors. For a constant number of processors, the parallel efficiency increases with increasing problem size. Due to the faster communication network (see sec. 4.2), the efficiency is higher on the T3E then on the PC cluster. Efficiencies higher than 1.0 which are obtained in variant (1) on the T3E are caused by cache effects (see sec. 4.2.3). The increase of the parallel efficiency on the cluster, variant (1), from 2 to 4 processors is probably also effected by the cache.

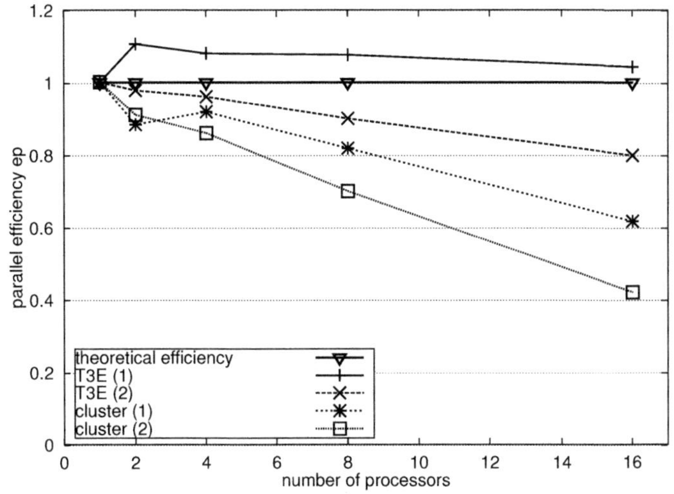

Fig. 3.34. Parallel efficiency for different problem sizes and parallel architectures, after WITTE et al. (2001 [263])

3.3 Adaptive methods

For many problems, for example *moving sharp fronts*, it is neither necessary nor efficient to carry out a uniform mesh refinement in order to achieve a high accuracy of the solution, as significant temporal and spatial changes of the solution functions occur only in small areas of the computational domain. *Adaptive methods* aim at numerical solutions with high accuracy, optimal computational effort and storage requirement by automatically adjusting the mesh or the solution method to the temporally and spatially variable solution function. Therefore, they contribute to the reliability of numerical modeling and computation. Many of the methods and techniques described in this section were first developed in structural mechanics and were then applied to fluid mechanics. There are various methods and techniques for carrying out an adaptation. The adaptation is steered by different *error estimators* or *indicators*, and different rules for *refinement* and *coarsement* must be defined. These topics are introduced in the following.

Attention is drawn to the fact that, for many aspects of adaptive methods, e.g. error estimators and indicators (see sec. 3.3.2), refinement and coarsening (see sec. 3.3.3) or data structures, public-domain toolboxes exist, e.g. Sumaa3D (see JONES, PLASSMANN (1995 [139])), PadFEM (see DIEKMANN et al. (1996 [71])) or UG (see BASTIAN (1997 [19])). Some of these toolboxes are applicable to parallel methods (see sec. 3.2) and are equipped with fast solvers (see sec. 3.4).

3.3.1 Different methods and techniques of adaptation

First, one can distinguish between *a priori* and *a posteriori methods* depending on whether they are applied before or after a computation. In areas where it is generally known that the solution accuracy is poor or where sharp gradients of the solution function can be expected, an *a priori* mesh refinement can be carried out (see BARLAG (1997 [13])). Such areas are found around sinks and sources, corners, discrete fractures and changes of layers with different physical coefficients, e.g. permeability. Figure 3.35 shows an a priori mesh refinement which has been carried out with the mesh generator ART (see FUCHS (1999 [84]), sec. 4.1.3) around a source and a fracture.

The methods discussed in the following are a posteriori. For time-dependent problems, space and time adaptation must be considered. For space-time FE discretizations, for example the Discontinous Galerkin Methods (see sec. 3.1.4), space and time adaptation is carried out by the same means (see JOHNSON (1990 [136], 1992 [137])) while, for the semidiscrete FEM and FVM (see secs. 3.1.4, 3.1.5), spatial and temporal adaptation are treated separately. In

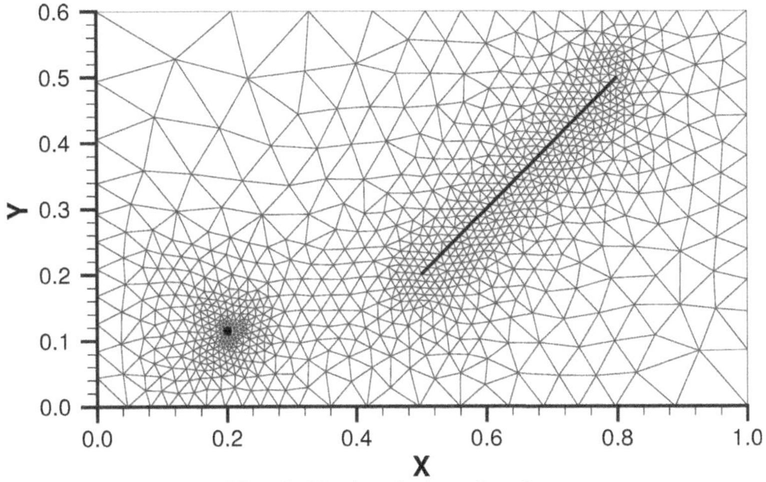

Fig. 3.35. A priori mesh refinement

the following, adaptive methods for semidiscrete formulations are explained.

In recent years, *h-adaptive methods* in which the number of marked elements is increased or decreased have established themselves in many numerical simulation fields, not only in civil and environmental engineering (see JOHNSON (1990 [136]), BORNEMANN et al. (1993 [34]), BASTIAN (1996 [17]), KORNHUBER (1997 [155]), BARLAG (1997 [13]), BARLAG et al. (1998 [14]), THIELE (1999 [247]), KAISER (2001 [140]), LANG (2001 [162]), fig. 3.36, hanging nodes are explained in sec. 3.3.3). H-adaptive methods are the most frequently used adaptive methods.

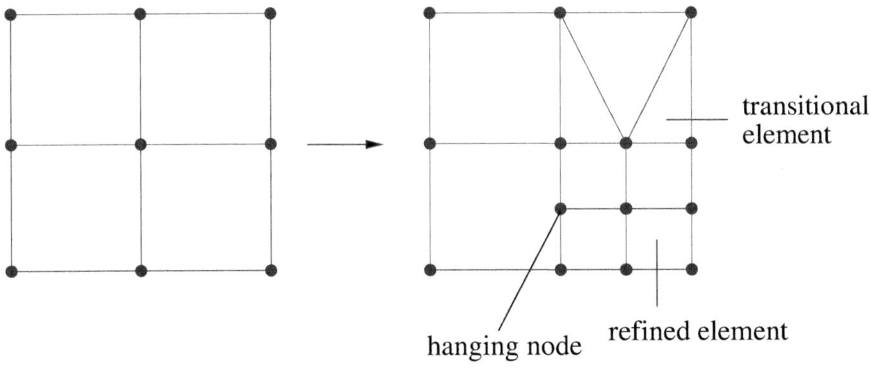

Fig. 3.36. H-adaptive method

When *p-adaptive methods* are applied, the polynomial degree of the shape functions (see sec. 3.1.4, fig. 3.9) is increased; for example, a quadratic shape function is used instead of a linear one (see fig. 3.37). P-adaptive methods are very suitable for problems which are characterized by a high regularity of the solution. However, this is not the case for hyperbolic or mixed hyperbolic / parabolic problems. When simulating a moving sharp front, for example, p-adaptive methods lead to over- and undershooting, so that the large additional effort in CPU time and storage for the higher polynomial degree only causes a small increase in the solution's accuracy. In such cases, h-adaptive methods are more economical. Consequently, p-adaptive methods are hardly used in hydro- and environmental engineering. P-adaptive methods or combinations of them with h-adaptive methods, so-called *hp-adaptive methods*, are frequently found in structural mechanics (see RANK (1987 [212]), OHNIMUS (1996 [198]), AINSWORTH, SENIOR (1997 [3]), RANK, DÜSTER (2001 [213])).

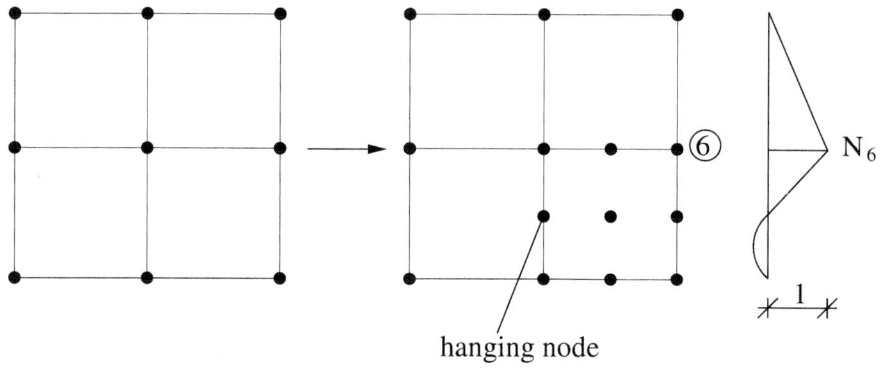

Fig. 3.37. P-adaptive method

In the *r-adaptive method*, also called moving-mesh adaptation, the nodes of the computational domain are shifted on the streamlines (see fig. 3.38). With such a method, the numerical dispersion of a simulation is reduced. As the determination of the streamlines requires comparatively much CPU-time, the r-adaptive method is often only economical for stationary problems (see CIRPKA et al. (1999 [60]), PAPASTAVROU (1998 [201])). In a similar way, nodes can be moved to sharp fronts, but this also has a number of disadvantages (see SCHÖNUNG (1990 [231])). Overall, it must be mentioned that r-adaptive methods have lost importance in recent years.

When a *d-adaptive method* is chosen, the dimensionality of the problem is adapted. Imagine a river which flows into the sea: upstream, the vertical flow profil is constantly parabolic. However, downstream in the interaction area of tides and varying inflow and outflow of saline water, the flow profile shows

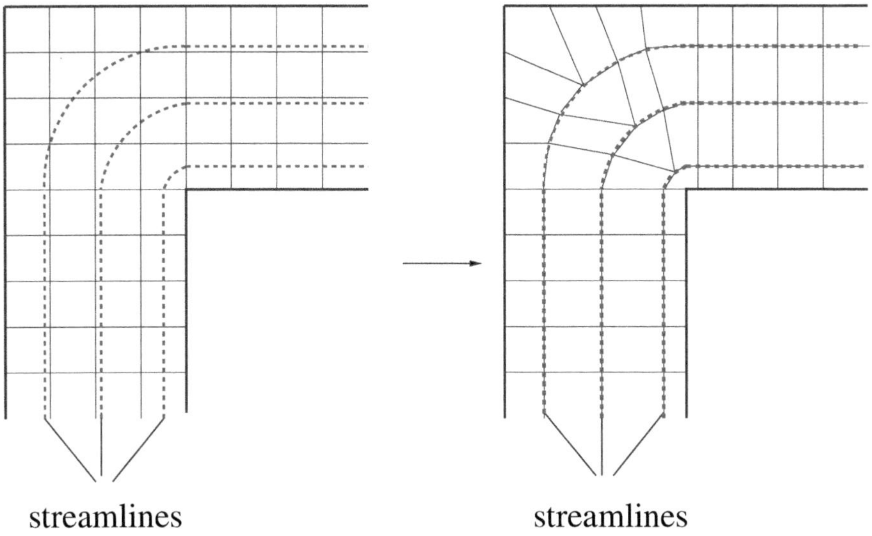

Fig. 3.38. R-adaptive method

strong variations in the vertical direction (see 3.39). As the interaction area changes depending on the tidal conditions, d-adaptive methods can be applied. It is easy to reduce a 3D-model to a 2D-model. However, it is difficult for the lower-dimensional model to detect where its dimensionality must be increased. D-adaptive methods are not widespread in hydro- and environmental engineering. They are found more frequently in structural mechanics (see OHNIMUS (1996 [198])).

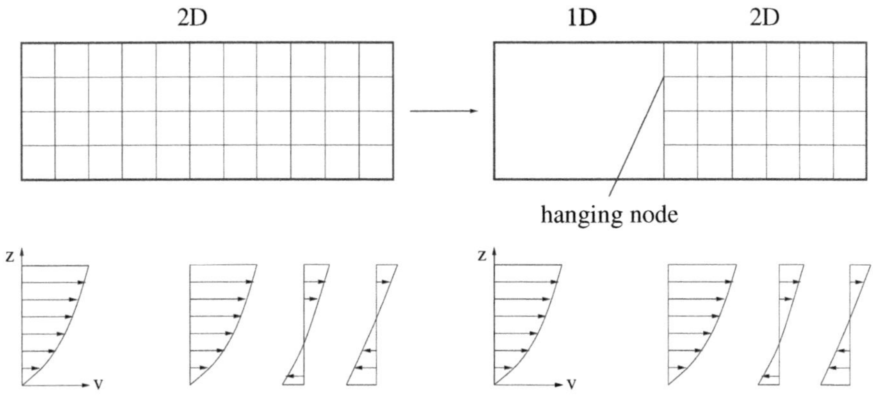

Fig. 3.39. D-adaptive method

The *m-adaptive method* is characterized by the fact that an adaptation of the model concept or the process is carried out. The *process adaptation* can be considered a subordinate part of a model adaptation. If, for example, an advective / dispersive transport problem is considered and advection occurs only in small parts of the domain, the advection-related solution procedures can be ignored there. However, it is difficult to activate advection again if, for example, the processes vary over the time. The idea of m-adaptive methods is that the model concept can be reduced; for example, a two-phase / two-component (water-air) problem is reduced to single-phase / two-component one in areas where only this phase occurs (see fig. 3.40). As already mentioned before, it is very difficult to extend the model concept from the viewpoint of the lower model concept. M-adaptive methods originate in structural mechanics (see OHNIMUS (1996 [198])), and they are also used for modeling hydro- and environmental systems (see THORENZ (2001 [248])). The major drawback of the m-adaptive methods for time-dependent problems is that they lead to free moving boundary problems which cannot be solved without difficulties even today.

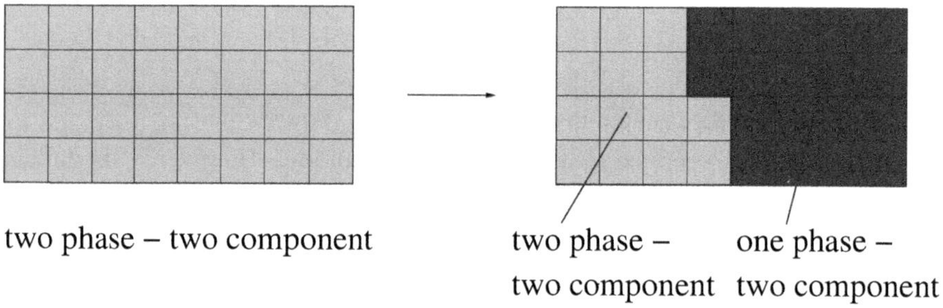

Fig. 3.40. M-adaptive method

CLASS (2000 [62]) has developed a method for non-isothermal multiphase / multicomponent processes in porous media where all processes are treated by the higher model concept and the primary variables are switched depending on the local active processes on each node (see fig. 3.41). Thus, different model concepts are treated in a similar way, and no free moving boundaries must be traced. However, in combination with Multigrid Methods, special attention must be paid to the restriction and prolongation (see CLASS et al. ([63])).

Further adaptive methods are based on the *solutions of local problems*, *hierarchical approaches* or *overlaid grids*. When the local problem technique is chosen, the problem is solved again in small subdomains with higher-order shape functions, for example quadratic instead of linear ones (see VERFÜRTH (1996 [253]), PAPASTAVROU (1998 [201]), ELLSIEPEN (1999 [75])). Hierarchical methods compare the solution with a more accurate solution which

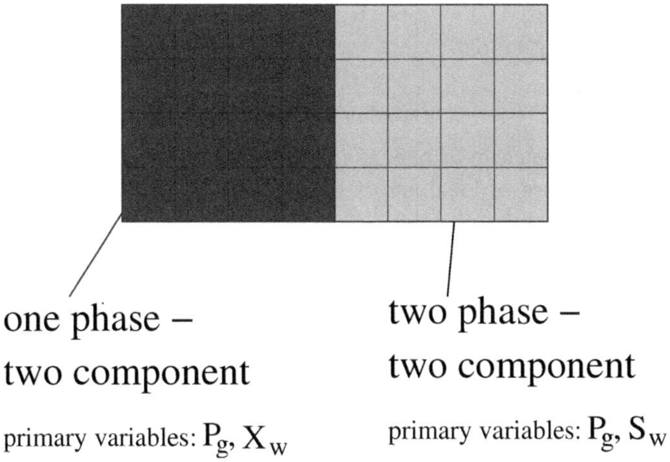

Fig. 3.41. Variables substitution

can be obtained with *hierarchical bases*, i.e. higher-order shape functions, or mesh refinement (see VERFÜRTH (1996 [253]), PAPASTAVROU (1998 [201]), ELLSIEPEN (1999 [75]), see sec. 3.4.4). The solutions of local problems and hierarchical approaches are discussed further in the next subsection in the context of error estimators and indicators. It is also possible to overlay an area of poor accuracy with any other refined grid where no nodes of the original and the refined grids coincide (see BERGER, OLIGER (1994 [29])). Overall, it must be mentioned, that neither methods which are based on the solution of local problems, nor hierarchical approaches, nor overlaid-grid methods are frequently used.

Adaptive methods are also applied today to *multiphysics problems*. Imagine a dike system where an interaction of free-surface and subsurface flow with stresses and strains in the porous medium occurs (see fig. 5.29, sec. 5.2.3). Deformation processes or crack propagations cause moving boundaries and interaction interfaces for the 'single-physics' problems. Therefore, the different computational subdomains must be adapted to the variable conditions, e.g. the geometry.

For semidiscrete methods, a temporal adaptation is generally carried out by a time-step adaptation, i.e. by varying the time step.

3.3.2 Error estimators and indicators

In adaptive methods, *error estimators* and *error indicators* play an important role. Error estimators η give a measure of the magnitude of the error e in a certain norm, generally the *energy norm* or *L^2-norm*, and they are mathematically substantiated. Two positive error constants C_1 and C_2 must exist so that the following holds:

$$C_1 \eta \leq ||e|| \leq C_2 \eta \tag{3.93}$$

C_1 and C_2 should be close to one, as the error tends towards zero. Error indicators give a measure of local errors by *heuristic means*. Therefore, they are mathematically not substantiated.

Spatial discretization error estimators

First, the basic equation for groundwater flow eq. 2.24 and the *Darcy* law eq. 2.17 are recalled:

$$div\,(-\underline{\underline{K}}_f\; grad\, h) = q_w/\rho_w \tag{3.94}$$

$$\underline{v_f} = -\underline{\underline{K}}_f\; grad\, h \tag{3.95}$$

with the piezometric head h, the hydraulic conductivity tensor $\underline{\underline{K}}_f$, the water source term q_w, the density of water ρ_w and the filter velocity $\underline{v_f}$. For an element i, the estimated error in the energy norm e^2 based on BABUSKA, RHEINBOLDT (1978 [10]) reads:

$$e_i^2 = C_3\, l_i \int\limits_{\Gamma_i} \underline{J}(\underline{v}_f)^2 d\Gamma + C_4\, l_i^2 \int\limits_{\Omega_i} div\,(\underline{\underline{K}}_f\; grad\, \tilde{h} - q_w/\rho_w) d\Omega \tag{3.96}$$

In this equation, C_3, C_4 denote error constants, l_i the discretization length of element i, $\underline{J}$ the jump in the *Darcy* velocity $\underline{v}_j$ along the boundary Γ of the domain Ω (see fig. 3.42) and $\tilde{h}$ the approximated piezometric head. The first term in the last equation represents the jumps in the *Darcy* velocities integrated along the edges of the element, while the second term stands for the residual in the element, the inequilibrium of equation 3.94. For linear shape functions and $q_w = 0$, the second term vanishes (see RUST (1991 [227])); otherwise this term can be neglected for smooth solutions (see AINSWORTH et al. (1989 [4])).

For certain special cases, the error constants can be determined. However, they generally depend on the problem, which means that these constants must be computed again for every new problem.

A similar error estimation was developed by ZIENKIEWICZ, ZHU (1987 [270]) and was applied to groundwater flow by RANK, ZIENKIEWICZ (1987

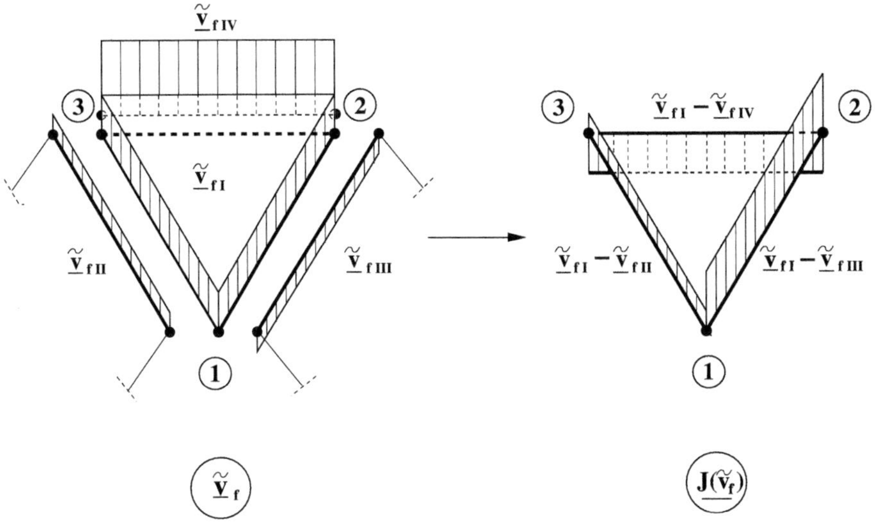

Fig. 3.42. Jumps in the FEM solution used for the error estimator of BABUSKA, RHEINBOLDT (1978 [10])

[214]). The estimated error in the energy norm for an element i is given by RANK, ZIENKIEWICZ (1987 [214]):

$$e_i^2 = \int\limits_{\Omega_i} (\underline{v}_f - \tilde{\underline{v}}_f)^t \underline{\underline{K}}_f^{-1} (\underline{v}_f - \tilde{\underline{v}}_f) d\Omega \tag{3.97}$$

Here, $\underline{v}$ stands for the exact velocity, $\tilde{\underline{v}}$ for the approximated velocity and $\underline{\underline{K}}_f^{-1}$ for the inverse hydraulic conductivity tensor. However, it must be mentioned that e_i^2 in equation 3.97 does not have the unit of energy which is caused by analogously applying the ZIENKIEWICZ, ZHU error estimator determined from elasticity problems to groundwater flow.

The error in the L^2-norm l is obtained from equation 3.96 if the inverse hydraulic conductivity tensor is omitted:

$$l_i^2 = \int\limits_{\Omega_i} (\underline{v}_f - \tilde{\underline{v}}_f)^t (\underline{v}_f - \tilde{\underline{v}}_f) d\Omega \tag{3.98}$$

For the error estimator, the 'exact' velocity $\underline{v}_f$ must be known or approximated. In ZIENKIEWICZ, ZHU (1987 [270]) different methods are explained. The simplest one merely consists of computing an arithmetical average of the values at the considered node (see fig. 3.43) which leads to a smoothed course of the velocity v_f^*.

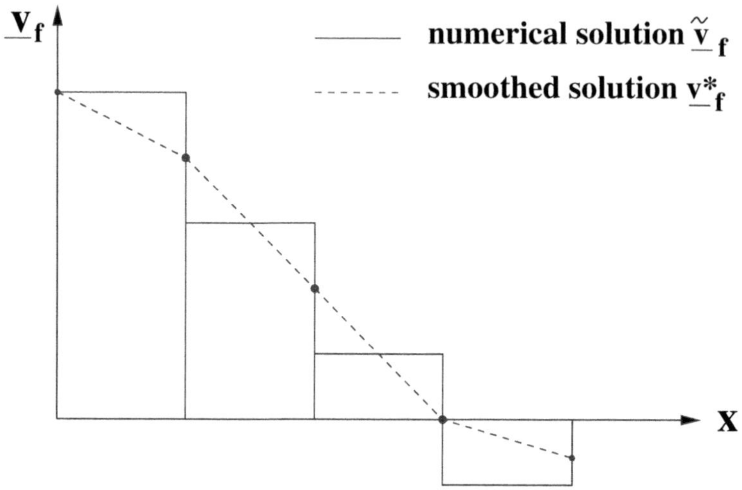

Fig. 3.43. Numerical and smoothed velocity, after RANK, ZIENKIEWICZ (1987 [214])

The error estimator η^2 in the energy norm for element i is determined by:

$$\eta_i^2 = \int_{\Omega_i} (\underline{v}^* - \underline{\tilde{v}})^t \underline{\underline{K}}_f^{-1} (\underline{v}^* - \underline{\tilde{v}}) d\Omega \tag{3.99}$$

The error estimator in the energy norm of the whole domain is computed by summing up over the elements n:

$$\eta^2 = \sum_{i=1}^{n} \eta_i^2 \tag{3.100}$$

The error estimators based on the energy or L^2-norm are widespread for elliptic and parabolic problems because such problems are deduced by a minimising principle or a variational functional (see JOHNSON (1990 [136]), BORNEMANN et al. (1993 [34]), BASTIAN (1996 [17]), VERFÜRTH (1996 [253]), KORNHUBER (1997 [155]), PAPASTAVROU (1998 [201]), ELLSIEPEN (1999 [75])). However, such energy error estimators cannot in general be used for hyperbolic or mixed hyperbolic / parabolic equations.

An exception is given by JOHNSON (1990 [136]), who developed an error estimator in the L^2-norm for advection / diffusion problems which are treated with the FEM. The basic idea for estimating the error consists of adding an artifical diffusion term to such an extend that it dominates advection. However, this error estimator has some disadvantages. On the one hand, the determination of the error constants again depends on the example. On the other hand, the error estimator requires high computational and storage effort compared to heuristic indicators which are introduced in the following (see BARLAG (1997 [13])). Consequently, the *Johnson* estimator is not widely used in hydro- and environmental engineering.

Recently, OHLBERGER (2001 [196], [197]) developed an a posteriori error estimator in the L^1-*norm* for convection / diffusion / reaction equations which are treated with the FVM. This error estimator is based on an *entropy-weak solution*, i.e. a weak solution which fulfills different entropy inequalities. The result and formulation of this error estimator are quite extensive; for this reason, see the abovementioned literature. BÜRKLE, OHLBERGER (2001 [51]) extended this estimator to the equations of two-phase flow in porous media. This estimator will possibly be applied frequently to advection / diffusion / reaction problems in the near future.

If a mesh is refined within a time step, it is generally recommended not to compute the same time step again with the refined mesh, but to proceed with the next time step. The first way causes a lot of additional computing time and storage for generally very small different results. However, a correction scheme is then required for coarsened elements, e.g. a mass correction scheme (see sec. 3.3.3, eq. 3.110).

Spatial discretization error indicators

Heuristic indicators are generally based on differences, gradients or curvatures of the solution function. They are denoted in the same way as the estimators with η. They are comparatively easy to compute and they are widely used in structural engineering (see RUST (1991 [227]), OHNIMUS (1996 [198])) as well as in hydro- and environmental engineering (see ELLSIEPEN (1999 [75]), BARLAG (1997 [13]), BARLAG et al. (1998 [14]), KAISER (2001 [140])). *Difference* and *gradient indicators* are very suitable for locating sharp fronts. In addition, *curvature indicators* detect the top and bottom region of a front. Heuristic indicators are introduced here for the transport equation in groundwater without sink and source terms (eq. 2.30):

$$\frac{\partial c}{\partial t} + \underline{v}_a grad\, c - div\, (\underline{\underline{D}}_{hyd} grad\, c) = 0 \qquad (3.101)$$

Here, c stands for the exact concentration, $\underline{v}_a$ for the pore velocity and $\underline{\underline{D}}_{hyd}$ for the hydrodynamic dispersion tensor.

The difference indicator η is defined for element i as follows:

$$\eta_i = \max_{j \neq k} \mid \tilde{c}_j - \tilde{c}_k \mid \quad (j, k = 1, 2, ..., n_e) \qquad (3.102)$$

In this equation, $\tilde{c}$ denotes the approximated concentration and n_e denotes the number of nodes per element.

The gradient indicator is given by:

$$\eta_i = l \mid grad\, \tilde{c}_i \mid \tag{3.103}$$

Here, l stands for the characteristic length (see sec. 3.1.4, fig. 3.12). The gradient should be computed in the center of gravity.

The curvature or jump indicator is determined by:

$$\eta_i = l^2 \max \mid \frac{\partial \tilde{c}}{\partial \underline{n}_\tau} \mid \quad (\tau = 1, 2, ..., e_e) \tag{3.104}$$

In this equation, $\underline{n}$ denotes the normal vector perpendicular to the edge and e_e the number of edges per element.

BARLAG (1997 [13]) has compared these heuristic indicators with the *Johnson* estimator for a one-dimensional sharp concentration front problem. The heuristic indicators require much less CPU time and storage then the *Johnson* estimator. The curvature indicator achieves almost the same accuracy as the *Johnson* indicator.

When the discrete model concept is applied to fracture-matrix systems (see sec. 2.1.2), elements of different dimensions are coupled. In 3D, the fractures are represented by 2D- or 1D-elements embedded in 3D-matrix elements; in 2D, the fractures are discretized by 1D-elements in the 2D-matrix. JOHN (1994 [135]) has developed simple formulas for the heuristic gradient (eq. 3.105) and curvature indicators (eq. 3.106) to take elements of different dimensions into account:

$$\eta_i = C\, l^{(3-d)/2} \mid grad\, \tilde{c}_i \mid \tag{3.105}$$

$$\eta_i = C\, l^{(2-d)/2} \max \mid \frac{\partial \tilde{c}}{\partial \underline{n}_\tau} \mid \quad (\tau = 1, 2, ..., e_e) \tag{3.106}$$

Here, C stands for a problem dependent constant and d for the dimension of the element.

BARLAG (1997 [13]), BARLAG et al. (1998 [14]) and KAISER (2001 [140]) have successfully applied these indicators to 1D/2D/3D coupled fracture-matrix systems.

Temporal discretization errors

One possibility of estimating the temporal discretization error is the *Richardson extrapolation* (see ELLSIEPEN (1999 [75])). The basic idea consists of comparing the approximated solution $\tilde{c}^{n+2\cdot 0.5}$ obtained with two sub-steps of $\Delta t/2$ with a solution $\tilde{c}^{n+1}$ obtained with one step of Δt. The estimated error is computed as follows:

$$e = \frac{2^p}{2^p - 1}(\tilde{c}^{n+2\cdot 0.5} - \tilde{c}^{n+1}) \tag{3.107}$$

In this equation, p denotes the order of consistency of the discretization scheme. An improved solution $\tilde{c}^{n+1}_{improved}$ is obtained by:

$$\tilde{c}^{n+1}_{improved} = \tilde{c}^{n+2\cdot 0.5} + \frac{\tilde{c}^{n+2\cdot 0.5} - \tilde{c}^{n+1}}{2^p - 1} \tag{3.108}$$

However, it must be mentioned that the *Richardson* extrapolation requires the CPU time and storage of about three time steps to determine the temporal discretization error of one time step. As a result, it is not used very often.

In ELLSIEPEN (1999 [75]), further methods for computing temporal discretization errors are given for *Runge-Kutta* Methods (see sec. 3.1.2).

Often, a time-step adaptation is carried out with *heuristic* means. For explicit methods, the time-step size can be chosen dynamically in such a way that it is close to the stability condition, e.g. the *Courant* (eq. 3.45) or *Neumann* number (eq. 3.36). For some transport models in the subsurface, it is advantageous to restrict the maximal *Peclet* number (eq. 3.64) to 2 or 10 in order to avoid oscillatory solutions (see HINKELMANN, HELMIG (2002 [111])). BARLAG (1997 [13]) adapts the time step for an implicit transport model in such a way that the *Courant* criterion $Cr = 1$ (eq. 3.65) is fulfilled in the area of sharp fronts. However, own simulations have shown that similar accuracies can be obtained with larger time steps. HINKELMANN (1997 [108]) determines 'optimal iteration numbers' of a linear BiCGSTAB solver (see sec. 3.4.1) for an Operator-Splitting Method applied to the shallow-water equations (see sec. 5.4). However, the optimal iteration numbers are problem-dependent. BASTIAN, HELMIG (1999 [21]) restrict the time step of a two-phase flow model in the subsurface by limiting the number of non-linear iterations (see KORNHUBER (1997 [155])).

Spatial and temporal discretization errors

The most common procedure for treating spatial and temporal errors and adaptations is to consider them separately, i.e. the methods described in this section are carried out consecutively.

However, there are a few exceptions. With the error estimator in the L^1-norm of OHLBERGER (2001 [196], [197]), spatial and temporal errors are computed. For the total error, the spatial and temporal parts are multiplied by different constants which must be determined. This method is also applicable to other time-dependent methods. If a space-time FEM is chosen instead of the semi-discrete method (see sec. 3.1.4), a closed error analysis can be formulated, and the spatial and temporal errors can be quantified with the same measure (see JOHNSON (1990 [136], 1992 [137])), i.e. the space and time dimensions and adaptations are treated in the same way.

3.3.3 Refinement and coarsening

Generally, the best way of refining an element is to maintain the shape or the length ratios of the edges. This is shown for a quadrilateral in figure 3.36 and for a hexahedron and tetrahedron in figure 3.44. In figure 3.44, just the refinements on the visible surfaces of the hexahedron and tetrahedron are visualized, not the ones within the three-dimensional bodies.

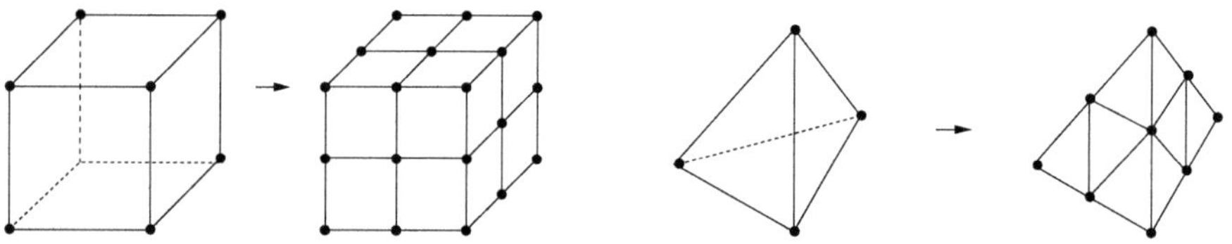

Fig. 3.44. Refined hexahedron and tetrahedron

The transition to the surrounding elements can be carried out with hanging nodes, also called irregular nodes, or transitional elements (see fig. 3.45). Hanging nodes are eliminated or *condensed* from the system matrices (see BARLAG (1997 [13])). RUST (1991 [227]), for example, has applied and developed transitional elements which are based on special shape functions. Finally, it is also possible to construct transitional elements with regular nodes and elements. This may lead to a combination of quadrilaterals and triangles in 2D and to the fact that the length ratios of the edges are not maintained. Additionally more elements than marked by an estimator or indicator are refined, and therefore, considerably more computational effort is required, especially in 3D. Nevertheless, this method is recommended, especially in the context of Multigrid Methods (see sec. 3.4.4). Other refinement and transition possibilities are discussed, for example in RUST (1991 [227]) and VERFÜHRT (1996 [253]), transitions between different dimensions in OHNIMUS (1996 [198]).

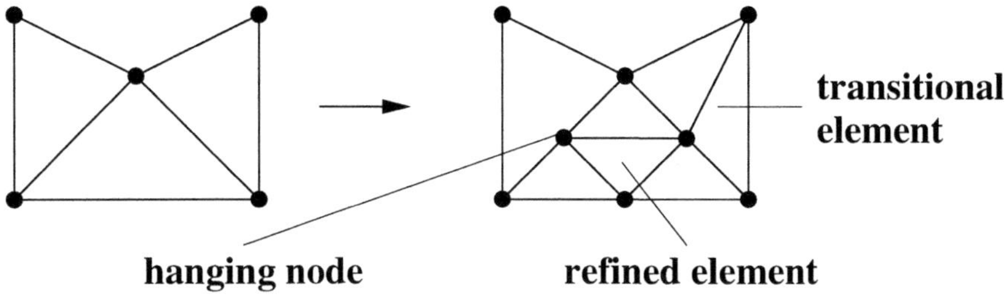

Fig. 3.45. Refined triangle and transitions

If systems are considered in which elements of different dimensions are coupled, e.g. fracture-matrix systems, one should avoid irregular nodes at the coupling interface. If, for example, a one-dimensional element has been marked for refinement, the adjacent two-dimensional element should be refined as well (see fig. 3.46)).

For certain questions, an *anisotropic* refinement is more suitable then an isotropic one, i.e. a refinement in a 2D-problem is only carried out in one direction and in a 3D-problem only in one or two directions (see fig. 3.47). An anisotropic refinement can be required for discrete fractures and fault zones, for example, in the equidimensional modeling approach (see sec. 5.1.6) or for 3D-problems in order to reduce the computational effort (see OHNIMUS (1996 [198]), SIEBERT (1996 [238]), KUNERT (2001 [160])). However, it should be made sure that the shape of anisotropic elements does not degenerate too much.

After the errors from the spatial discretization have been determined in each element with an estimator or indicator, there are different ways of marking elements for refinement or coarsening. Generally, tolerances are given for refinement tol^{refine} and coarsening $tol^{coarsen}$ and are related to a decisive error estimator or indicator η which is explained in the following. This procedure leads to an equal distribution of the error over the elements:

$$\begin{array}{lcl} \eta > tol^{refine} & \longrightarrow & refine \\ \eta < tol^{coarsen} & \longrightarrow & coarsen \end{array} \tag{3.109}$$

The decisive error estimator or indicator can be the averaged error of all elements $\eta^{averaged}$; typical tolerances are approximately $tol^{refine} = 2\eta^{averaged}$

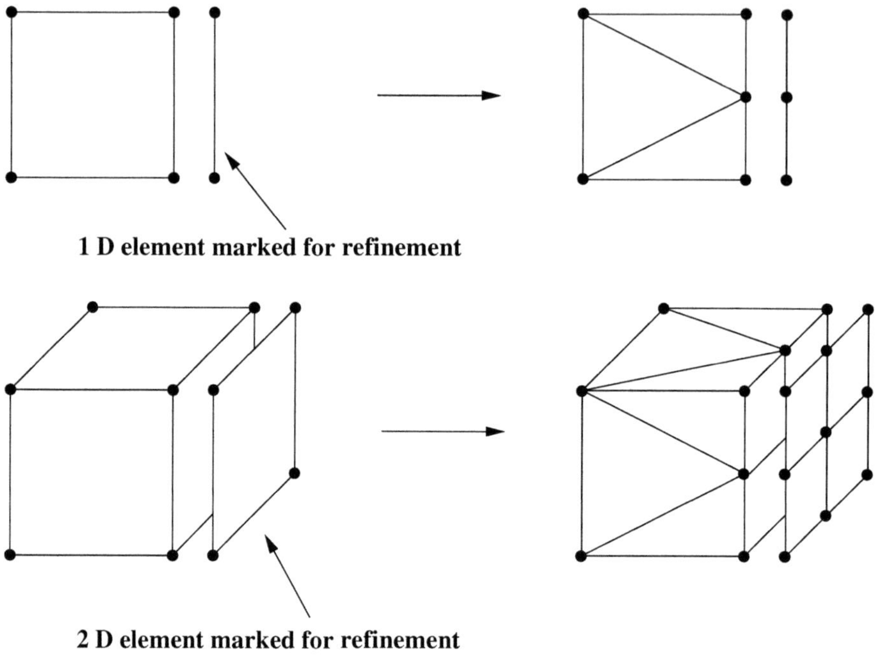

Fig. 3.46. Refinement of coupled elements of different dimensions

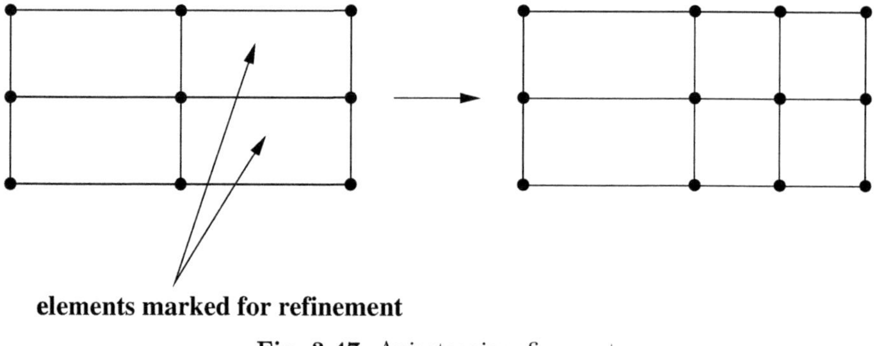

Fig. 3.47. Anisotropic refinement

and $tol^{coarsen} = 0.5\eta^{averaged}$. The decisive error estimator or indicator also can be the maximum error of all elements η^{max}; typical tolerances are approximately $tol^{refine} = 0.9\eta^{max}$ and $tol^{coarsen} = 0.1\eta^{max}$. In both cases, the tolerances are relative values and not absolute ones. Consequently, it is also possible to formulate the decisive error estimator or indicator and the tolerances as absolute values. Further criteria are found in PAPASTAVROU (1998 [201]). Generally, the choices of the decisive error estimator or indicator and the tolerances depend on the error estimator or indicator as well as on the examples. Therefore, one has to investigate the different possibilities for the examples considered. A very strong refinement depth leads to large differences between the sizes of the smallest and largest element and, along with this, to bad condition numbers (see sec. 3.4.3) and bad convergence behavior of the systems of equations to be solved as well as to excessive CPU time. If this should be avoided, the number of refinement steps or levels must be limited.

Coarsening strategies generally cause a more or less small error in mass, momentum or energy. Therefore, a *correction algorithm*, here explained for mass, is required in order to conserve the quantity considered (see fig. 3.48, BARLAG (1997 [13]), BARLAG et al. (1998 [14]), LEYDAG et al. (2001 [169])). If a patch - in 1D, two adjacent elements - is marked for coarsening, the mass in the triangle on the left of figure 3.48 is lost. However, this amount of mass can be redistributed over the coarsened element and its new neighbors, see figure 3.48, right.

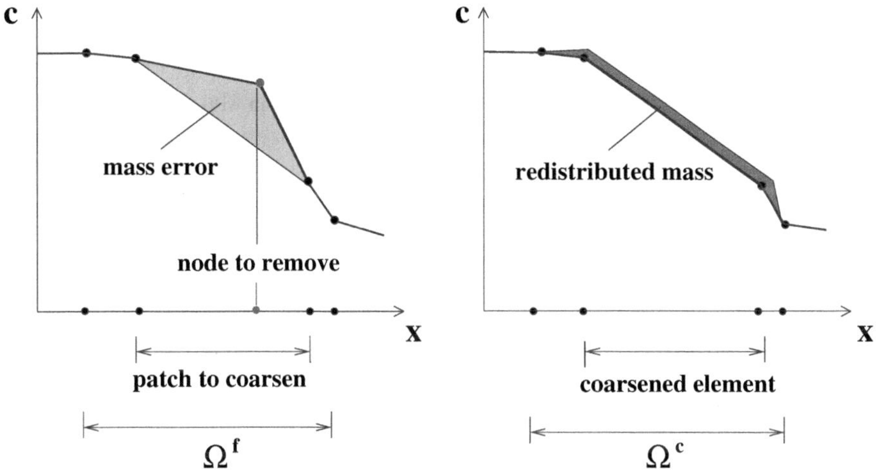

Fig. 3.48. Mass-correction algorithm, after BARLAG et al. (1998 [14])

The condition for the mass redistribution is given by:

$$\int_{\Omega^f} \Delta c^f d\Omega^f = \int_{\Omega^c} \Delta c^c d\Omega^c \tag{3.110}$$

In this equation, $\Delta c^f, \Delta c^c$ denote the concentration of the fine and coarsened grid and Ω^f, Ω^c the subdomains of the fine and coarsened mesh (see fig. 3.48). Generally, the coarsening of elements is limited to the initial grid.

It should be mentioned that it is possible to carry out a grid smoothing after an adaptation step, i.e. the topology of the refined mesh remains the same, but nodes are shifted to 'optimal positions', e.g. the center of a patch. However, this is not recommended in the context of Multigrid Methods, because it leads to problems concerning prolongation and restriction (see sec. 3.4.4). Such a technique is very suitable in the preprocessing stage during a mesh generation (see sec. 4.1.7).

It often occurs that the data set of some parameters, for example the bathymetry, is much finer than the initial grid. Then the initial grid is determined by element-wise averaging (see fig. 3.49, left). If a refinement is carried out in such a case, two possibilities exist. First, the refined elements can get the inherited information of the coarse element (see dashed green line in fig: 3.49, right). However, in this case the finer data set is not used. Second, the refined elements obtain the information of the finer data set which must be computed (see dashed-dotted blue lines in fig: 3.49, right). It is obvious that the second way is a better representation of the reality. However, one has to take into account that the second way influences the convergence behavior, because the *grid convergence* is only proven for constant parameter distributions.

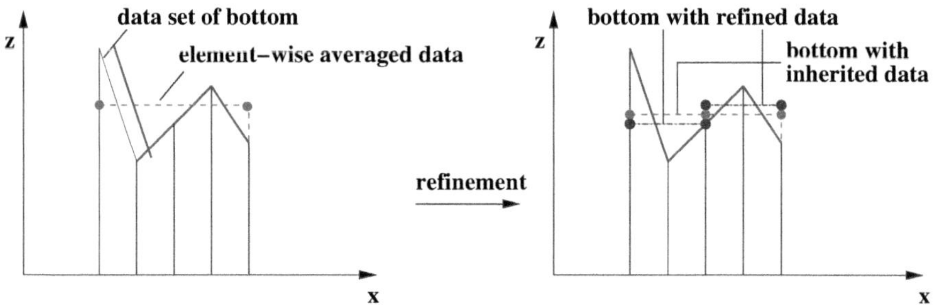

Fig. 3.49. Data set and averaging

Different criteria for time-step adaptations are given in sec. 3.3.2 and sec. 3.3.2.

Finally, it is mentioned that adaptive methods can be combined very well with Multigrid Methods (see sec. 3.4.4), because both require the same data structures. As far as the parallelization is concerned, adaptive methods impose additional difficulties which can be overcome by dynamic load-distribution techniques (see sec. 3.2.4). However, adaptive methods, which are generally implemented with dynamic data structures, are not very suitable for vectorization, because the vector pipelines must be computed again and the data independencies must be checked again after each adaptation step.

3.3.4 Examples

In this section, two different adaptive algorithms based on the FVM are investigated for surface-water and subsurface two-phase flow processes.

Adaptive Finite-Volume Method for flow processes in surface water

Here, a one-dimensional FVM, which can also be called IFDM (see sec. 3.1.5), is considered in combination with adaptive methods for the conservative form (see sec. 2.6.1) of the *Saint-Venant* equations (see sec. 2.6.1, LEYDAG (2001 [168]), LEYDAG et al. (2001 [169])). This algorithm is similar to the one described in section 3.1.3. The space is discretized with the FVM (see sec. 3.1.5); for the temporal derivative, the *Diffusive Lax* Method (see sec. 3.1.3), which is explicit, is chosen. A difference indicator (see sec. 3.3.2) related to an averaged element error (see sec. 3.3.3) is applied to refinement and coarsening in space, and a time-step adaptation for the explicit algorithm is carried out in such a way that the *Courant* condition $Cr \leq 1$ is fulfilled. One should take into account that halving an element causes an approximate halving of the time step for an explicit algorithm. Further, a mass and momentum correction algorithm (see sec. 3.3.3) was implemented for the adaptive FVM.

The same example as discussed in section 3.1.3 is chosen here for the investigation of the performance of the adaptive algorithm. In figure 3.50, the propagating wave is shown on different meshes. On the coarse initial mesh with $\Delta x = 25m, \Delta t = 2s$, the front of the water level is smeared a lot (see fig. 3.50, top). The mesh uniformly refined to level 3 ($\Delta x = 3.125m, \Delta t = 0.25s$) shows a very steep front which is a better approximation of the exact sharp front solution (see fig. 3.50, center). A solution accuracy comparable to that achieved with the uniformly refined mesh is determined with the adaptive refined (maximum level 3) and coarsened mesh (see fig. 3.50 bottom). The nodes are set at the centers of gravity. A high-grid resolution is only seen around the sharp front; the mesh has already been coarsened to the initial stage in the left part of the mesh.

In figure 3.51, the propagating front is shown after $270s$ for various maximum refinement levels. It is obvious that the steepness of the front increases with increasing refinement levels.

In table 3.3, a comparison of run times and numbers of computed state variables, which are an indicator for the required storage, is presented for adaptively and uniformly refined grids, i.e. for comparable solution accuracies, obtained on different grid levels. As the algorithm is explicit, the CPU time and computed state variables related to uniformly refined meshes are nearly

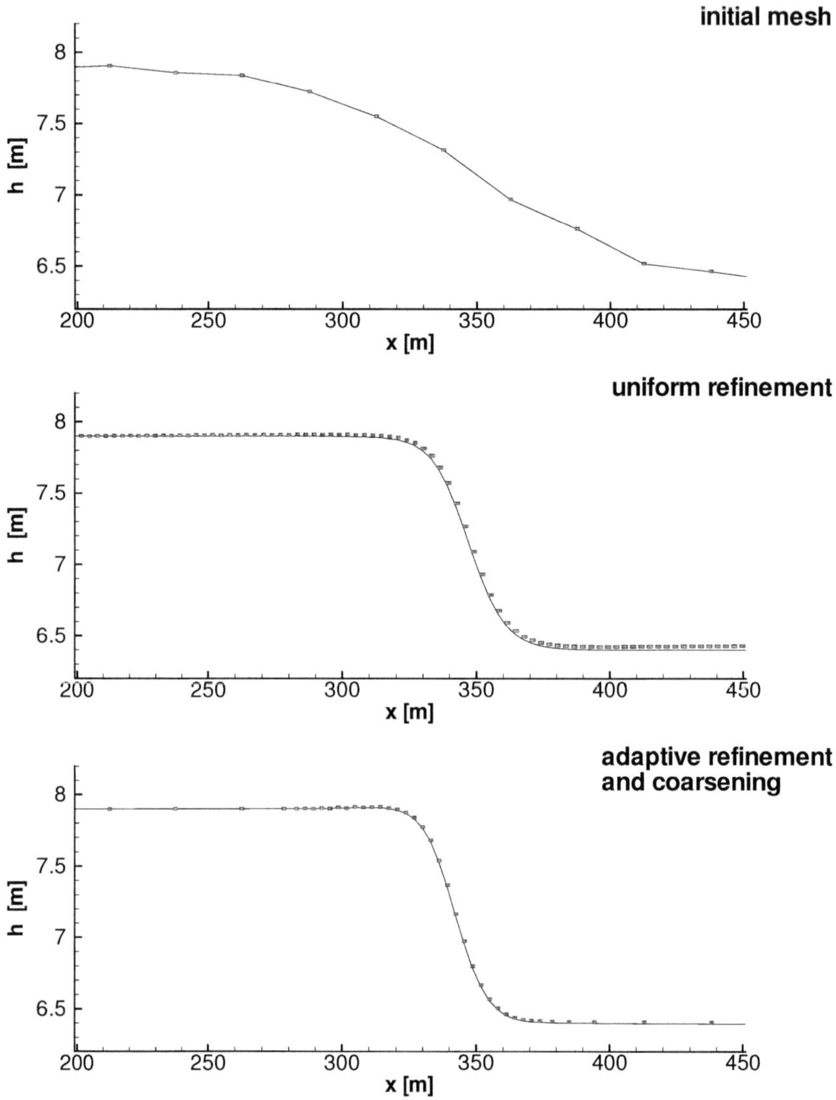

Fig. 3.50. Propagating wave after 40s on different meshes, after LEYDAG et al. (2001 [169])

the same. For refinement level 4, only about 10% of the CPU time and storage compared to a uniformly refined solution are required. The superiority of adaptive methods is clearly demonstrated.

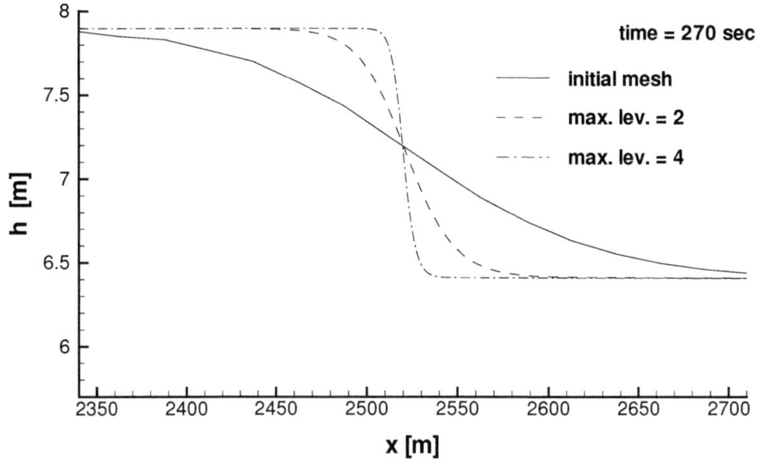

Fig. 3.51. Propagating wave after 270s for different refinement levels, after LEYDAG et al. (2001 [169])

refinement	CPU time [s]	related to uniform [%]	number of computed state variables [-]	related to uniform [%]
initial mesh	0.433	-	18000	-
adaptive max. lev. = 2	1.610	29.5	89183	31.0
uniform lev.= 2	5.463	-	288000	-
adaptive max. lev. = 3	3.620	16.9	197373	17.1
uniform lev.= 3	21.483	-	1152000	-
adaptive max. lev. = 4	8.653	10.4	481892	10.5
uniform lev.= 4	83.035	-	4608000	-

Table 3.3. Comparison of run times and computed state variables, after LEYDAG et al (2001 [169])

Adaptive Finite-Volume Method for two-phase flow processes in the subsurface

The FV algorithm and the *Buckley-Leverett* problem which have already been discussed in section 3.1.5 are investigated here in the context of adaptive methods. A gradient indicator (see sec. 3.3.2) is applied to the water saturation and related to a maximum error (see sec. 3.3.3). It is used for the refinement and coarsening in space. A time-step adaptation (see sec. 3.3.2) is carried out depending on the number of non-linear iterations of the *Newton-Raphson* Method (see sec. 3.4.5).

In figure 3.52, a comparison of the analytical, the uniformly and the adaptively refined solution of the water saturation is given for the same time step.

The initial grid has 256 elements of $\Delta x = 1.172m$. The uniform refinement is carried out up to level 6, i.e. to 16384 elements. The adaptive refinement is restricted up to level 6 as well. Both the uniform and the adaptive refinement and coarsening show the same steepness of the front. However, the adaptively refined solution proceeds a little bit faster; this is caused by the interpolation of the initial condition, i.e. by numerical dispersion, in the first time step. The adaptive solution requires about 10% of the number of elements and about 20% of the CPU time needed to obtain the uniformly refined solution. Again, the excellent performance of adaptive methods is proven.

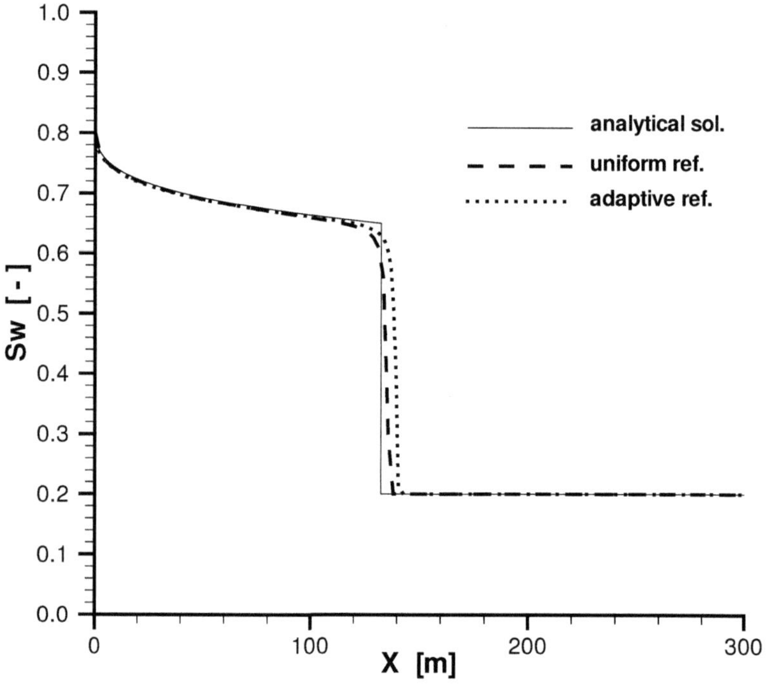

Fig. 3.52. Comparison of the analytical, uniformly and adaptively refined results of the water saturation for the *Buckley-Leverett* problem, after PAUL (2003 [203])

3.4 Fast solvers

3.4.1 Introduction

Most of the discretization methods described in section 3.1 are implicit and lead to the solution of systems of algebraic equations which very often have a large number of unknowns. As the computational effort for the solution is quite high and requires a major part of the whole CPU time of the numerical simulation, fast solvers are of particular importance and have to be considered in the context of parallel and adaptive methods (see secs. 3.2, 3.3, HINKELMANN (1997 [108]), HINKELMANN (2000 [109])).

The system of equations to be solved with the coefficient matrix A, the vector of the unknowns X and the vector of the right-hand side B is given in the form:

$$AX = B \tag{3.111}$$

In index notation, the equation i is determined as follows with the number of unknowns n:

$$\sum_{j=1}^{n} a_{ij} x_j = b_i \tag{3.112}$$

If the coefficient matrix A in equation 3.111 or the coefficients in equation 3.112 are independent of the vector of the unknowns X, the system of equations is *linear* and discussed in sections 3.4.1 - 3.4.4; otherwise it is *non-linear* and treated in section 3.4.5.

For linear systems, *direct* and *iterative* methods are distinguished. Direct methods are generally based on a successive elimination of the unknowns with the *Cholesky algorithm* for *symmetric, positive definite* matrices and with the *Gauss algorithm* for *regular* matrices (see SCHWARZ (1991 [235]), MEISTER (1999 [176])), and thus directly determine the solution. Iterative methods operate with an iteration scheme: after an initial vector has been chosen, the solution is the limit of a series of approximations which is terminated if a stopping criterium which considers the approximation to be sufficient good is fulfilled. Iterative methods are dealt with in sections 3.4.2 and 3.4.4.

The effort for solving a system of equations directly with a *dense* coefficient matrix, i.e. the major part of the entries are non-zero terms, and n unknowns is in the order $O(n^3)$. The discretization methods which are introduced in section 3.1 generally lead to sparse coefficient matrices, i.e. only a small part of the entries are non-zero terms. Therefore, only sparse matrices are treated in the following. There are different methods for storing such matrices efficiently, e.g. by renumbering the mesh and just storing the entries in thin bandwidths (see CUTHILL, MCKEE (1969 [66]), GIBBS et al. (1976 [93])) or by only addressing and storing the non-zero terms (see BARRETT et al. (1994)). In two-dimensional problems, the computational effort is in the order $O(n^2)$, and in three-dimensional problems $O(n^{2.33})$ (see AXELSSON, BARKER (1984 [8])). As the computational effort of direct solvers is comparatively high, they are only suitable for small-scale problems with up to only about one or a few thousand unknowns. Therefore, direct solvers as the only solution procedure are generally not suitable for high performance computing. Additionally, direct solvers cannot be parallelized or vectorized well, because the operations to be carried out are characterized by high data locality which results in large communication times for parallel computers and small vector pipelines for vector computers (see sec. 4.2). However, direct solvers are interesting for composed iterative solvers, e.g. as a coarse-grid solver in a Multigrid Method (see sec. 3.4.4) or as a solver for the interface problem in a Domain-Decomposition Method (see sec. 3.2.2). Special measures must be taken with the direct solvers for parallel computing (see sec.3.4.4).

Iterative methods use a single or multiple grids to determine the solution. Single-grid methods only require the grid of the computational domain for the solution (see sec. 3.4.2), while *Multigrid Methods* additionally need several grids for the iteration process (see sec. 3.4.4) in general. For sparse matrices, the computational effort of single-grid solvers *can* be reduced up to the order $O(n^{1.5})$, sometimes even more (see BASTIAN (1999 [18])) and that of Multigrid solvers to $O(n^1)$ (see secs. 3.4.2 and 3.4.4). Furthermore, iterative solvers have the advantage compared to direct ones that rounding errors do not accumulate.

Domain-Decomposition Methods, which are explained in section 3.2.2 and which have gained importance in the field of parallel computing in recent years, can also be applied as solvers on serial computers. However, there are no obvious advantages when compared to other methods, e.g. Multigrid (see sec. 3.2.2). Domain-Decomposition Methods are discussed again in the context of preconditioners in section 3.4.3.

Finally, it is mentioned that many of the solvers introduced in section 3.4 are available via different public-domain toolboxes, e.g. TEMPLATES (see BARRETT et al. (1994 [16])), ScaLAPACK (see BLACKFORD et al. (1997 [32])) or UG (see BASTIAN et al. (1997 [19])). Some of these toolboxes are also applicable to parallel (see sec. 3.2) and adaptive methods (see sec. 3.3).

3.4.2 Single-grid solvers

As mentioned before, iterative methods are based on an iteration scheme. They start with an initial guess of the solution vector, which is often the zero-vector or the result of the old time level for time-dependent problems. The solution is then computed by a series of iteration steps which terminates if a stopping criterion is fulfilled. A stopping criterium should be much greater than the machine precision. It can be an absolute value, e.g. 10^{-10}, or a relative value, e.g. determined from the *Euclidian norm* of the right-hand-side vector. In the context of parallel computing, it should be mentioned that rounding errors can accumulate in such a way that the numbers of iterations vary for different numbers of subdomains or processors, although this should theoretically not be the case. Generally, such effects are of minor importance. However, if algorithms operate close to the machine precision and a large number of processors is involved, one should keep this in mind. If the machine precision is not sufficient, further means can be applied, such as *long number* or *interval arithmetic* (see HANSEN (1992 [96])). As such methods require a lot of CPU time, their use must be considered very carefully.

Iterative single-grid solvers are faster than direct solvers for systems of equations with more than a medium number of unknowns, i.e. a few thousand. BRUSSINO, SONNAD (1989 [48]) have shown that for sparse non-symmetric systems of equations solved on serial computers and KREIENMEYER (1996 [158]) on parallel computers. An overview is given in SCHWARZ (1991 [235]), HACKBUSCH (1991 [94]) or BARRETT et al. (1994 [16]). The classical methods of *Jacobi* and *Gauss-Seidel* are introduced including their *overrelaxation* variants. Other methods are superior, e.g. the *Conjugate Gradient Methods* described later. However, the *Jacobi* and *Gauss-Seidel* Methods are used nowadays in composed iterative solvers as the Multigrid solvers.

Jacobi Method and Gauss-Seidel Method

The *Jacobi Method* is derived by the following formula with the iteration index k:

$$x_i^{(k)} = (b_i - \sum_{j=1, j\neq i}^{n} a_{ij} x_j^{(k-1)})/a_{ii} \tag{3.113}$$

In matrix and vector notation, the *Jacobi* Method looks as follows with the diagonal matrix D, the strictly lower triangular matrix $-L$, the strictly upper triangular matrix $-U$ and the relation $A = D - L - U$:

$$X_i^{(k)} = D^{-1}[B + (L+U)X^{(k-1)}] \tag{3.114}$$

As the diagonal terms appear in the denominator, they must differ from zero, $a_{ii} \neq 0$, otherwise columns or rows must be exchanged. Generally, the convergence speed is poor. With the following overrelaxation *Jacobian* technique (JOR), the convergence speed is accelerated:

$$x_i^{(k)} = (1-\omega)\, x_i^{(k-1)} + \omega\, (b_i - \sum_{j=1, j\neq i}^{n} a_{ij} x_j^{(k-1)})/a_{ii} \tag{3.115}$$

The factor ω is in the range of $0 < \omega \leq 2$. This factor should be taken from the literature or estimated.

The *Gauss-Seidel* Method is similar to the *Jacobi* Method. It differs in that the results previously computed in the iteration step k are used for the determination of the further unknowns in the same iteration step k:

$$x_i^{(k)} = (b_i - \sum_{j=1}^{i-1} a_{ij} x_j^{(k)} - \sum_{j=i+1}^{n} a_{ij} x_j^{(k-1)})/a_{ii} \tag{3.116}$$

$$X^{(k)} = D^{-1}(B + LX^{(k)} + RX^{(k-1)}) \tag{3.117}$$

Therefore, the convergence behavior of the *Gauss-Seidel* Method is better than that of the *Jacobi* Method. Again, the convergence speed can be improved, here with a successive overrelaxation technique (SOR):

$$x_i^{(k)} = (1-\omega)\, x_i^{(k-1)} + \omega\, (b_i - \sum_{j=1}^{i-1} a_{ij} x_j^{(k)} - \sum_{j=i+1}^{n} a_{ij} x_j^{(k-1)})/a_{ii} \tag{3.118}$$

If the matrix A is symmetric, a symmetric successive overrelaxation technique (SSOR) is carried out (see BARRETT et al. (1994 [16])).

The methods just described are also called the *Point-Jacobi* and *Point-Gauss-Seidel* Methods. Additionally, the *Block-Jacobi* and the *Block-Gauss-Seidel* Methods exist. If the matrix A is subdivided into a number of blocks, the iteration schemes in equations 3.114 and 3.117 are also valid for the block variants. For the inversion of a block, a small system of equations, in which the number of unknowns equals the number of equations in the block, must be solved (see HACKBUSCH (1991 [94])).

The convergence behavior of the *Gauss-Seidel* Method depends on the ordering of the equations and blocks as well as on the numbering of the computational domain. In the field of parallel computing, this means that the convergence depends on the number of subdomains or processors and that the convergence speed decreases with an increasing number of subdomains or processors. Because of the successive integration of results within an iteration step, the *Gauss-Seidel* Method cannot reasonably be vectorized.

Conjugate Gradient Methods

Conjugate Gradient Methods are very fast solvers for medium-scale problems with a number of unknowns of about a few thousand up to tens of thousands. The effort for solving sparse matrix systems with n unknowns can be kept in the order of $O(n^{1.5})$ or even better with optimal preconditioning (see BASTIAN (1999 [18]), sec. 3.4.3)). The convergence is generally only proven for symmetric positive definite matrices; in other cases, e.g. non-symmetric matrices, the solver is more expensive. For medium-scale problems, even Multigrid Methods (see sec. 3.4.4) are generally not superior.

The *CG Method (Conjugate Gradient Method)*, which is called *PCG Method (Preconditioned Conjugate Gradient Method)* in a preconditioned form (see sec. 3.4.3), is a very efficient solver for *symmetric positive definite* matrices. In recent years, it has established itself as the standard solver for this kind of matrix. Originally, it was developed by HESTENES, STIEFEL (1952 [106]). The problem of solving $AX = B$ is transformed into the equivalent problem of searching for the minimum of the quadratic functional $F(X) = \frac{1}{2}X^T AX - B^T X$ (see fig. 3.53). The system $AX = B \leftrightarrow AX - B = 0$ is a necessary condition for minimising $F(X)$. The positive definiteness of A leads to a positive definite *Hesse matrix* of $F(X)$ and thus to a sufficient condition for a minimum:

$$F(X) = \frac{1}{2}X^T AX - B^T X$$

$$gradF(X) = AX - B = 0 \longrightarrow extremum \tag{3.119}$$

$$grad[gradF(X)] = A \rightarrow positive\ definite \longrightarrow minimum$$

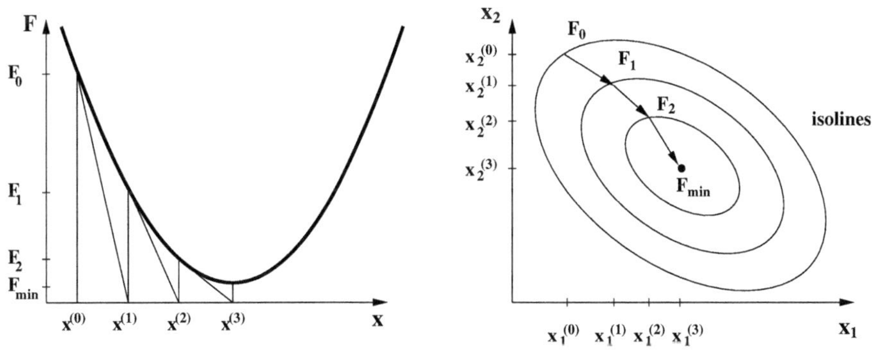

Fig. 3.53. Idea of the CG Method

After an initial guess of the solution, optimal search directions are determined in such a way that the minimum is reached with a minimum number of steps. A sequence of vectors which are orthogonal or conjugate to the matrix A is generated. These vectors form a *Krylov subspace*; for this reason, the name *Krylov* Methods is also used in the context of Conjugate Gradient Methods. The convergence is mathematically proven, i.e. the minimum is reached after n steps at the latest (see REID (1971 [215])). Therefore, the CG Method also belongs to the group of direct solvers. Generally, convergence, i.e. a sufficiently accurate approximation, is achieved with much fewer than n steps.

A pseudocode of the parallel PCG Method is given in figure 3.54. (see LEPEINTRE (1992 [166]), BARRETT et al. (1994 [16]), MEISTER (1999 [176])). From the computational point of view, it is interesting that only recursions, which need little storage, occur within the iteration loop of the solver. For the storage, the matrix A, the vectors X, B and additional 4 vectors and a preconditioner matrix C (see sec. 3.4.3) are required. If the solver is parallelized or / and vectorized, the measures introduced in section 3.2.2 must be taken. For the parallelization, communication is required for 1 matrix-vector product (mvp), 3 scalar products (sp) and possibly for 1 preconditioner matrix (preco).

If the system matrix is symmetric, but possibly indefinite, the Conjugate Gradient Methods *MINRES* or *SYMMLQ* should be applied (see BARRETT et al. (1994 [16])).

Among a number of different *Krylov* Methods for non-symmetric matrices, the *BiConjugate Gradient Stabilized* or *BiCGSTAB Method* and the *Generalized Minimal Residual* or *GMRES Method* have become accepted as the standard solvers. For regular non-symmetric matrices, the problem that the vectors of the residuals cannot be orthogonalized by short recursions occurs.

The BiCGSTAB Method is a further development of some other methods. The sequence of vectors which are orthogonal or conjugate to the matrix A is generated by a product of two polynomials ($\rightarrow$ Bi in BiCGSTAB), of which one locally minimizes the vector of the residuals and points in the direction of the steepest descent. The acronym STAB indicates that it is a stabilized variant which improves the convergence behavior. Theoretically, there are two possibilities for the BiCGSTAB Method to fail. However, this happens very rarely in modeling hydro- and environmental systems.

A pseudocode of the parallel BiCGSTAB Method is shown in figure 3.55 (see LEPEINTRE (1992 [166]), BARRETT et al. (1994 [16]), MEISTER (1999 [176])). In the iteration loop, again only recursions occur. Storage for 2 matrices and 10 vectors must be provided. For parallelization and vectorization, the same measures as for the PCG Method must be taken (see sec. 3.2.2). For the parallelization, communication is needed for 2 matrix-vector products

pseudocode PCG Method

initial phase

$r^{(0)} = A\,X^{(0)} - B$	for an initial guess $X^{(0)}$	**mvp**
solve: $C\,g^{(0)} = r^{(0)}$		**preco**
$d^{(0)} = g^{(0)}$		
$t^{(0)} = A\,d^{(0)}$		**mvp**
$\lambda_0 = \langle r^{(0)} g^{(0)} \rangle$		**sp**
$\varrho_0 = \lambda_0 / \langle t^{(0)} d^{(0)} \rangle$		**sp**
$X^{(1)} = X^{(0)} - \varrho_0\, d^{(0)}$		

iteration loop: k = 1, 2, ...

$t^{(k)} = A\,d^{(k-1)}$		**mvp**
$r^{(k)} = r^{(k-1)} - \varrho_{k-1}\, t^{(k)}$		
$\epsilon = \langle r^{(k)} r^{(k)} \rangle;$	check convergence, continue if necessary	**sp**
solve: $C\,g^{(k)} = r^{(k)}$		**preco**
$\lambda_k = \langle r^{(k)} g^{(k)} \rangle$		**sp**
$d^{(k)} = g^{(k)} + \lambda_k / \lambda_{k-1}\, d^{(k-1)}$		
$\varrho_k = \lambda_k / \langle d^{(k)} t^{(k)} \rangle$		**sp**
$X^{(k+1)} = X^{(k)} - \varrho_k\, d^{(k)}$		

storage: 2 matrices: A, C
6 vectors: X, B, r, g, d, t

communication in iteration loop: 1 mvp, 3 sp, if necessary 1 preco

Fig. 3.54. Pseudocode of the parallel PCG Method, after LEPEINTRE (1992 [166])

(mvp), 5 scalar products (sp) as well as possibly for 2 preconditioning steps (preco).

The GMRES Method is an alternative to the BiCGSTAB Method with the advantage that (possibly) indefinite matrices can be treated as well. Unfortunately, the generation of the sequence of vectors which are orthogonal to the matrix A cannot be carried out in a recursion, i.e. in each iteration step, all vectors previously computed have to be taken into account. Intermediate values are stored in a *Hessenberg matrix*. The convergence is mathematically proven, i.e. the minimum is reached after n steps at the latest (see SAAD, SCHULZ (1985 [228])). Therefore, the GMRES Method also belongs to the

pseudocode BiCGSTAB Method

initial phase

$$\begin{array}{lll} r^{(0)} = B - A\,X^{(0)} & \text{for an initial guess } X^{(0)} & \textbf{mvp} \\ \text{chose } \bar{r} \; (e.g.\ \bar{r} = r^{(0)}) & & \end{array}$$

iteration loop: k = 1, 2, ...

$$\begin{array}{rll} \lambda_{k-1} &= \langle \bar{r}\, r^{(k-1)} \rangle & \textbf{sp} \\ \text{if } \lambda_i &= 0: \text{ stop} & \\ \text{if } k &= 1: \; p^{(k)} = r^{(k-1)} & \\ \text{if } k &> 1: \; \beta_{k-1} = (\lambda_{k-1}/\lambda_{k-2})(\delta_{k-1}/\Omega_{k-1}) & \\ p^{(k)} &= r^{(k-1)} + \beta_{k-1}(p^{(k-1)} - \Omega_{k-1}\, v^{(k-1)}) & \\ \text{solve: } C\,\bar{p} &= p^{(i)} & \textbf{preco} \\ v^{(k)} &= A\,\bar{p} & \textbf{mvp} \\ \delta_k &= \lambda_{k-1}/\langle \bar{r}\, v^{(i)} \rangle & \textbf{sp} \\ &= r^{(i-1)} - \delta_k\, v^{(k)} & \\ \text{solve: } C\,\bar{s} &= s & \textbf{preco} \\ t &= A\,\bar{s} & \textbf{mvp} \\ \Omega_k &= \langle t\, s \rangle / \langle t\, t \rangle & \textbf{2sp} \\ X^{(k)} &= X^{(k-1)} + \delta_k \bar{p} + \Omega_k\, \bar{s} & \\ r^{(k)} &= s - \Omega_k\, t & \\ \epsilon &= \langle r^{(k)} r^{(k)} \rangle; \quad \text{if necessary continue} & \textbf{sp} \\ & \quad \text{necessay for continuation: } \Omega_k \neq 0 & \end{array}$$

storage: 2 matrices: A, C
10 vectors: $X, B, r, \bar{r}, p, \bar{p}, s, \bar{s}, v, t$

communication in iteration loop: 2 mvp, 5 sp, if necessary 2 preco

Fig. 3.55. Pseudocode of the BiCGSTAB Method, after LEPEINTRE (1992 [166])

group of direct solvers. Generally, convergence, i.e. a sufficiently accurate approximation is achieved for fewer than n steps.

A pseudocode of the parallel GMRES Method is found in BARRETT et al. (1994 [16]) or MEISTER (1999 [176]). From the computational point of view, it is advantageous that only 1 matrix-vector product must be computed in the iteration loop. The major disadvantage lies in the fact that the computational and storage efforts which are caused by taking the vectors determined previously into account increases linearly with the number of iterations.

Further improvement is given by restarted GMRES variants. If a solution is not obtained after m iteration steps, the storage is set free and the last solution is taken as the initial solution for a new computation. This is also called the GMRES(m) Method. Theoretically, convergence is not guaranteed. The determination of a suitable m is difficult and depends to a great extent on the problem. One- or two-digit numbers are found in the literature. For the parallelization and vectorization, the same measures must be taken as for the PCG and BiCGSTAB Methods (see sec. 3.2.2); additionally, a further problem results from the *Hessenberg* matrix, which is a dense upper triangular matrix with a few unknowns. KREIENMEYER (1996 [158]), for example, stores the *Hessenberg* matrix and solves it directly on every processor.

Finally, the author recommends using GMRES instead of BiCGSTAB in the field of modeling hydro- and environmental systems only if the system matrix is (possibly) indefinite or if the advantages concerning CPU time and storage are obvious.

3.4.3 Preconditioners

The convergence behavior of iterative solvers greatly depends on the *spectral* properties or the *condition* of the matrix. The *condition number* κ is defined as the ratio of the largest to the smallest eigenvalue of matrix A, $\kappa(A) = \lambda_{max}/\lambda_{min}$; the ideal condition number is $\kappa = 1$. During the preconditioning, the system is pretreated to improve the convergence behavior. The overall aim consists of reducing the number of iterations and the CPU time which is required for the solution including the CPU time spent during preconditioning. Therefore, a preconditioner must be 'cheap' with regard to the CPU time or - if this is not the case - the number of iterations must be reduced considerably.

The preconditioning can be carried out in two different ways:

$$AX = B \quad \longrightarrow \quad C^{-1}AX = C^{-1}B \tag{3.120}$$

$$AX = B \quad \longrightarrow \quad C_1^{-1}AC_2^{-1}(C_2X) = C_1^{-1}B \tag{3.121}$$

The second possibility (eq. 3.121) should be preferred because here, certain properties of the system matrix are maintained, e.g. the symmetry. Furthermore, the transferred system should have the same structure as the initial system, i.e. a sparse matrix structure should be obtained. The inverse of A is the ideal preconditioner; however, the computation of the inverse requires at least the same amount of CPU time as the solver. Consequently, a preconditioner approximates the inverse of A with simpler means.

Generally, the convergence behavior of iterative methods cannot be predicted exactly. Sometimes, limits can be given, e.g. for the CG Method. The difference between the solution in the kth iteration step $X^{(k)}$ and the exact solution $\overline{X}$ can be estimated from the difference between the initial solution $X^{(0)}$ and the exact solution in the A norm (eq. 3.123) as (see MEISTER (1999 [176])):

$$\|X^{(k)} - \overline{X}\|_A \leq 2 \left[\frac{\kappa - 1}{\kappa + 1}\right]^{(k)} \|X^{(0)} - \overline{X}\|_A \tag{3.122}$$

$$\|Y\|_A = (Y^t A Y)^{1/2} \tag{3.123}$$

Equation 3.122 indicates an asymptotic convergence behavior as well as the dependence on the condition number and the initial solution.

In the following, important preconditioners are introduced. In parallel computations, different values may occur on the interprocessor boundaries (see 3.2.2); arithmetically averaging these values often improves the convergence. Further preconditioners are given e.g. in BARRETT et al. (1994 [16]) or MEISTER (1999 [176]).

Jacobi preconditioners

The simplest preconditioner according to equation 3.121 consists of just the diagonal of the matrix:

$$c_{1ij} = c_{2ij} = \begin{cases} \sqrt{a_{ii}} & if \quad i = j \\ 0 & otherwise \end{cases} \tag{3.124}$$

This method is also called *diagonal* or *Point-Jacobi preconditioner* (see sec. 3.4.2) or *scaling*. Attention is drawn to the different meaning of scaling in the context of parallel computing ('scalable algorithm', see sec. 4.2.3). For a positive definite matrix, it is proven that the condition number at least equals the quotient of the largest and smallest diagonal term (see SCHWARZ (1991 [235])). With a diagonal preconditioner, this lower limit is set to 1. Instead of the diagonal matrix of A, a *lumped* diagonal matrix which consists of row- or column-wise sums of entries can be chosen. The improvement of the convergence achieved by the two methods is equivalent. They are computationally easy, but do not reduce the number of iterations in the ranges of orders of magnitude as other more sophisticated preconditioners. For parallelization with the inconsistent storage technique (see sec. 3.2.3), a local communication to obtain the full values should be carried out for values of the interprocessor-boundary nodes, because this leads to a convergence behavior which is independent of the number of processors. The use of at least a diagonal preconditioning is recommended. More sophisticated preconditioners often start with a scaling.

A *Block-Jacobi preconditioner* (see sec. 3.4.2) is derived if the number of variables is subdivided into non-overlapping subsets and an estimation or computation of an inverse is done for each subset (see $Cg = r$ in fig. 3.54). For parallel computations, the subsets can coincide with the subdomains and the computations of the inverses are done fully in parallel with the local available data of the subdomain or processor A_p (see fig. 3.54, here $C_p = A_p$). If the subsets are small, a direct solver can be chosen to compute the inverse. If the subsets are large, iterative solvers should be used. As the problems in the subsets are independent of each other, they require different numbers of iterations, and thus a load imbalance occurs (see sec. 3.2.4) during preconditioning. A further consequence of the independence of the problems in the subsets is the fact, that the convergence of the solver depends on the number of subdomains or processors (see sec. 3.2.2). The subsets of variables can also be smaller than the subdomains, e.g. groups of elements or just one element called *Element-By-Element (EBE) preconditioner* (see HUGHES et al. (1987 [124])). If problems with multiple variables per node are treated, subsets can be formed by grouping the equations. *Block-Jacobi* preconditioners are generally easy to parallelize; vectorization depends on the solution technique for the inverse. A moderate reduction of the number of iterations and CPU time can be expected.

Domain-Decomposition preconditioners

Domain-Decomposition preconditioners resemble *Block-Jacobi* preconditioners. They are based on overlapping or non-overlapping subdomains and are further treated with *Schwarz* or *Schur Complement Methods.*

The *Additive Schwarz preconditioner* is given in the form of 3.120 as follows:

$$C_{AS}^{-1} = R_c^t A_c^{-1} R_c + \sum_{i}^{p} R_i^t A_i^{-1} R_i \tag{3.125}$$

In this equation, A_c denotes a coarse mesh with the nodes of the interprocessor boundaries (called 'interface problem' in sec. 3.2.2) and A_i a fine mesh with the inner nodes of a subdomain plus nodes from an overlap (called 'local problem' in sec. 3.2.2). R_c is a restriction operator which maps a function on the coarse mesh onto the fine mesh by interpolation, and R_i is a restriction operator which maps a function on the fine mesh to the coarse mesh by extension. p is the number of processors. The second term on the right-hand side can be computed fully in parallel with the three steps restriction, inversion and interpolation. Due to the independence of the local problem, an iterative computation will cause a load imbalance (see sec. 3.2.4). Special measures are required for the first term on the right-hand side; this is treated in a similar way to the coarse-grid solver in the multigrid method (see sec. 3.4.4). The simplest one is just to skip it out, further are discussed in BARRETT et al. (1994 [16])). The *Additive Schwarz* preconditioner can considerably reduce the number of iterations and the CPU time depending on the local solver and the abovementioned special considerations.

For non-overlapping subdomains, the Additive *Schwarz* Method coincides with the Block-*Jacobi* Method and the Multiplicative *Schwarz* Method with the Block *Gauss-Seidel* Method (see sec. 3.4.2), HACKBUSCH (1991 [94])).

Schur-Complement Methods are based on a decomposition of the mesh into a coarse and a fine mesh as mentioned before, without, however, taking overlapping into account. The *Schur Complement* S is determined by a blockwise *Gauss* elemination:

$$AX = B \Leftrightarrow \begin{bmatrix} A_i & A_{ic} \\ A_{ic}^t & A_c \end{bmatrix} \begin{bmatrix} X_i \\ X_c \end{bmatrix} = \begin{bmatrix} B_i \\ B_c \end{bmatrix} \tag{3.126}$$

$$\Leftrightarrow \begin{bmatrix} A_i & A_{ic} \\ 0 & S \end{bmatrix} \begin{bmatrix} X_i \\ X_c \end{bmatrix} = \begin{bmatrix} B_i \\ B_c - A_{ic}^t A_i^{-1} B_i \end{bmatrix} \tag{3.127}$$

$$S = A_c - A_{ic}^t A_i^{-1} A_{ic} \tag{3.128}$$

For preconditioning, the *Schur Complement* is not computed exactly, but with some approximations. In this context, the BPS preconditioner is mentioned (see BRAMBLE et al. (1989 [38]), BARRETT et al. (1994 [16])). *Schur* Complement preconditioners improve the convergence in a similar measure to *Schwarz* preconditioners.

Incomplete factorization preconditioners

The system matrix A can be factorized by a lower matrix L and upper matrix U:

$$A = LU \tag{3.129}$$

Such a factorization occurs in the context of direct solvers (see sec. 3.4.1, SCHWARZ (1991 [235])). An exact factorizations leads to the direct solution of the equation and requires much CPU time (see sec. 3.4.1) and storage because, even for a sparse matrix, L and U are dense. Incomplete factorizations obtain the structure of the matrix, i.e. for a sparse matrix, only entries of the original structure are taken into account for the preconditioner. The *ILU (Incomplete LU factorization)* and the *MILU (Modified ILU)* preconditioners belong to this group (see BRUSSINO, SONNAD (1989 [48]), BARRETT et al. (1994 [16])). Generally, a significant improvement of the convergence and CPU time is achieved. However, they are not very suitable for parallelization and vectorization, because many non-locally available data are required for the factorization. Further, it is possible to carry out block factorizations, i.e. not the whole matrix is factorized. For parallel computations, the blocks can coincide with the subdomains. This improves the parallelization, but at the price of worse convergence which then additionally depends on the number of processors.

Multigrid preconditioners

Multigrid preconditioners improve the convergence behavior very efficiently. They can be combined very well with CG Methods (see BRAESS (1986 [36]), RUST (1991 [227]), KREIENMEYER (1996 [158]), KLAAS (1996 [144])). The *Additive Multigrid Method* of the *hierarchical bases* (see YSERENTANT (1986 [266])) and *BPX* (see BRAMBLE et al. (1991 [39])), which are explained in the next section 3.4.4, are often used as preconditioners in a CG Method (see sec. 3.4.2). The *Multiplicative Multigrid Methods hierarchical bases Multigrid* (see BANK et al. (1988 [11])) and *Local Multigrid* (see BASTIAN (1996 [17])), which are also explained in the next section 3.4.4, are used as preconditioners as well. Often, there is only a slight advantage for the Multigrid solver compared to a solver with Multigrid preconditioner. Further aspects are discussed in the next section 3.4.4.

3.4.4 Multigrid solvers

Multigrid Methods (MG) are very fast solvers for large-scale problems with more than about ten thousands of unknowns. The computational effort for the solution of sparse matrix systems can be reduced up to the order $O(n^1)$, with the number of unknowns n. This is proven for symmetric elliptic problems (see HACKBUSCH (1991 [94])); in other cases, e.g. non-symmetric matrices, the MG solver is more expensive. The error between the iterative and the exact solution can be decomposed into a *high-frequency* and *low-frequency* part. It can be shown that the low-frequency error can be diminished much better on a coarser mesh than the initial one. In figure 3.56, it0, it1, it2 denote the error after the 0th, 1st and 2nd iteration step on the fine grid and it2^{mg} the error after two iteration steps on the fine grid and one switch to the coarse grid.

Therefore, two grids are required for a two-grid solver and multiple grids for a Multigrid solver. The grids should be determined by uniform refinement (see figs. 3.57, 3.44)), and the subspaces $\Omega_m, 0 \leq m \leq l$ of the shape functions should be nested into each other:

$$\Omega_0 \subset \Omega_1 \subset \Omega_2 ... \; \Omega_{l-1} \subset \Omega_l = \Omega \tag{3.130}$$

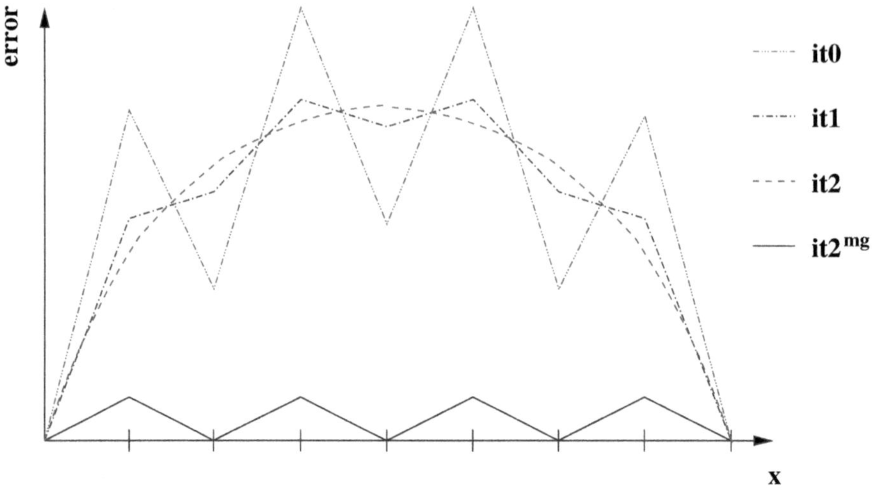

Fig. 3.56. Principle error reduction for single- and two-grid solver, after MEISTER (1999 [176])

Transfer operators which are called *restriction* R_m^{m-1} for the transfer between fine and coarse grids and *prolongation* P_{m-1}^m for the transfer between the coarse and fine grids are needed:

$$X_{m-1} = R_m^{m-1} X_m \tag{3.131}$$

$$X_m = P_{m-1}^m X_{m-1} \tag{3.132}$$

Generally, the grids are derived by uniform refinement. Therefore, it is possible to compute the coarse-grid matrices with the existing coarse-grid information. If the coarse-grid information is not available, the coarse-grid matrix can be computed as shown in equation 3.133. This is further explained in the context of *algebraic Multigrid Methods*:

$$A_{m-1} = P_{m-1}^m \, A_m \, P_{m-1}^m \tag{3.133}$$

A linear restriction is shown for a one-dimensional example in figure 3.58 and a linear prolongation for the same example in figure 3.59 (see RUST (1991 [227]), MEISTER (1999 [176])).
Finally, a solver on the fine grid(s) ($1 \leq m \leq l$ in eq. 3.131), which is called *smoother (smo)*, and a solver on the coarse grid ($m = 0$ in eq. 3.134), which is called *coarse-grid solver (cgs)*, are required:

$$A_m X_m = B_m \tag{3.134}$$

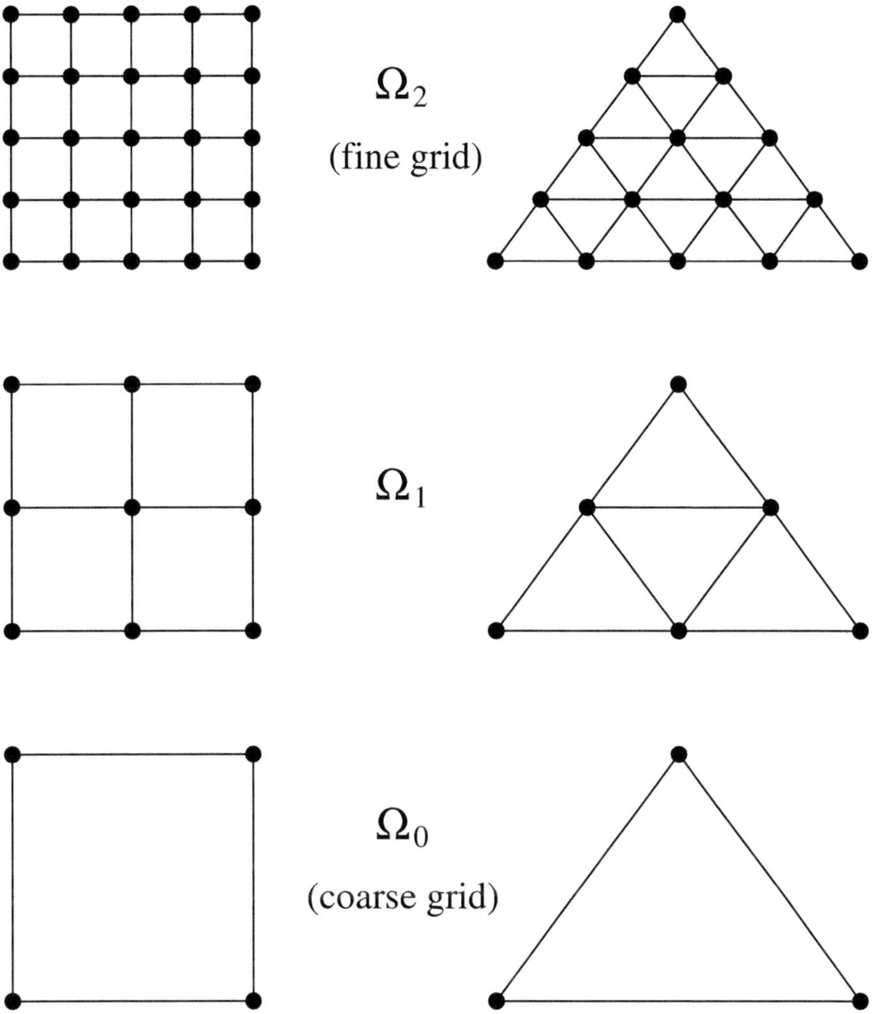

Fig. 3.57. Multigrids for three levels

The methods introduced in section 3.4.2 or other methods, especially *Jacobi*, *Gauss-Seidel* and ILU Methods, are frequently chosen as smoothers. An appropriate smoother should damp out the high-frequency errors. Classically, the coarse-grid solver is a direct solver, i.e. a *Cholesky* or *Gauss* algorithm (see sec. 3.4.1). However, an iterative solver or an *algebraic* Multigrid Method, which is explained later, can also be chosen. Before restriction and after prolongation, so-called *pre-* and *postsmoothing steps* can be carried out, i.e. iteration steps with the smoother are computed. To obtain a good convergence, the number

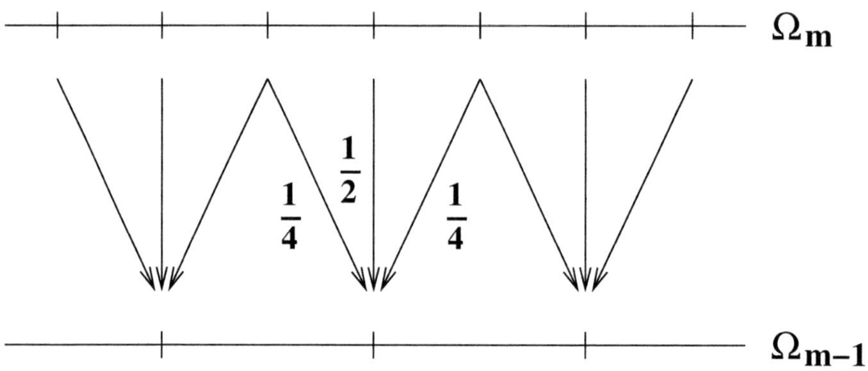

Fig. 3.58. Linear restriction, after MEISTER (1999 [176])

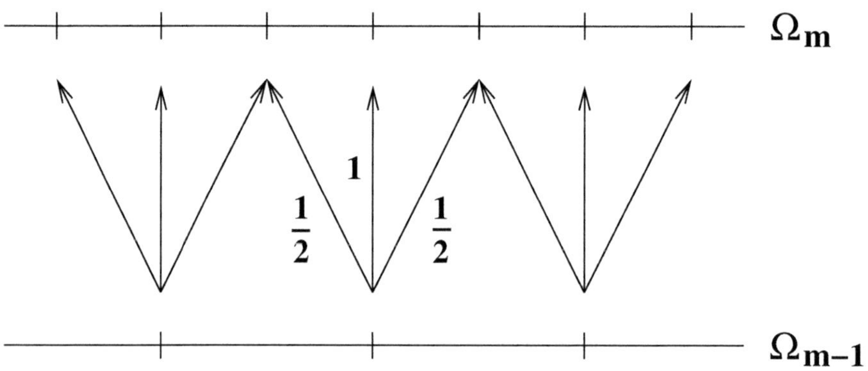

Fig. 3.59. Linear prolongation, after MEISTER (1999 [176])

of pre- and postsmoothing steps should be larger than one each (see BASTIAN et al. (1998 [20])). A pseudocode and a course for a two-grid method are given in figures 3.60 and 3.61.

For large systems of equations, the coarse-grid solver also has many unknowns and requires much CPU time. In such cases, it can be shown that the error can be reduced in a more efficient way if the equation is not solved directly on the coarse grid, but iteratively with another two-grid cycle (see HACKBUSCH (1991 [94]), KORNHUBER (1997 [155]), MEISTER (1999 [176]), BRIGGS et al. (1999 [44])). Depending on the size of the problem, this procedure can then be repeated recursively. The sequence of passing the grid levels can be carried out in a $V-$ or $W-$ *cycle*, as shown in figure 3.62.

solver:	$\mathbf{2GS}^{(\alpha,\beta)}(\mathbf{X_1}, \mathbf{B_1})$
system:	$A_1\, X_1 = B_1$
initial solution:	$X_1^{(0)}$
iteration loop:	$k = 1, 2, 3, ...$
α presmoothing steps:	$X_1^{(k)} = smo_1(X_1^{(k-1)}, B_1)$
restriction:	$r_0 = R_1^0\,(A_1 X_1^{(k)} - B_1)$
coarse-grid solver:	$e_0 = A_0^{-1}\, r_0$
prolongation:	$X_1^{(k)} = X_1^{(k)} - P_0^1\, e_0$
β postsmoothing steps:	$X_1^{(k)} = smo_1(X_1^{(k)}, B_1)$

Fig. 3.60. Pseudocode for a two-grid solver $2GS^{(\alpha,\beta)}(X_1, B_1)$, after MEISTER (1999 [176])

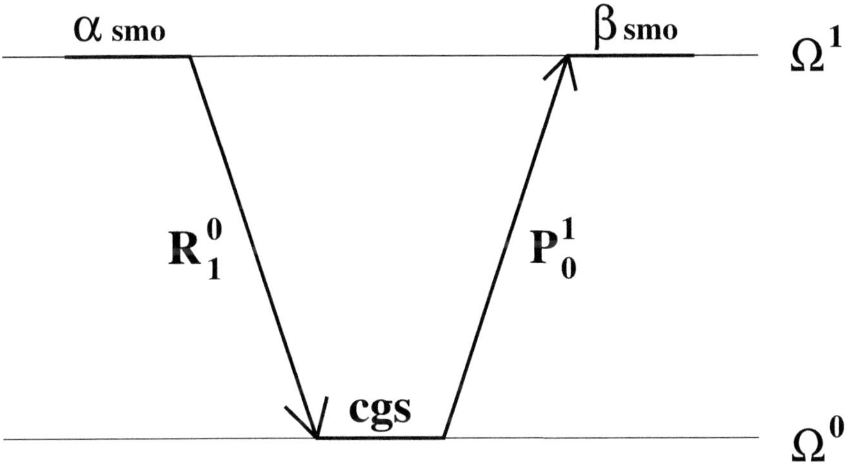

Fig. 3.61. Course for a two-grid solver $2GS^{(\alpha,\beta)}(X_1, B_1)$, after MEISTER (1999 [176])

A pseudocode for a Multigrid solver is given in figure 3.63. The pre- and postsmoothing can be carried out on each level except the coarse one.

The *Full Multigrid Method (FMG)*, which is also called *nested iteration*, improves the initial solution on the fine grid by starting on the coarse mesh with a direct solution and then using prolongation to obtain the results on the fine mesh. From here, a $V-$ or $W-$cycle proceeds (see MEISTER (1999 [176]), BRIGGS et al. (1999 [44])). The FMG is shown with a cycle and three levels in figure 3.64.

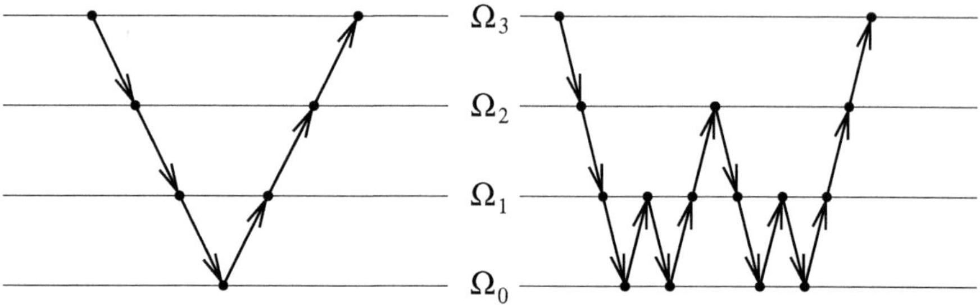

Fig. 3.62. $V-$ and $W-$Multigrid cylce, after MEISTER (1999 [176])

solver: $\mathbf{MGS}_l^{(\alpha,\beta,\gamma)}(\mathbf{X}_l, \mathbf{B}_l)$

system: $A_l\, X_l = B_l$

initial solution: $X_l^{(0)}$

outer iteration loop: $k = 1, 2, 3, ...$
α presmoothing steps: $X_l^{(k)} := smo_l(X_l^{(k-1)}, B_l)$
restriction: $r_{l-1} := R_l^{l-1}\,(A_l X_l^{(k)} - B_l)$
$e_{l-1}^0 = 0$

inner iteration loop: $i = 1, 2, 3, ..., \gamma$
$e_{l-1}^i := MGS_{l-1}^{(\alpha,\beta)}(e_{l-1}^{i-1}, r_{l-1})$

prolongation: $X_l^{(k)} := X_l^{(k)} - P_{l-1}^l\, e_{l-1}^\gamma$
β postsmoothing steps: $X_l^{(k)} := smo_l(X_l^{(k)}, B_l)$

Fig. 3.63. Pseudocode for a Multigrid solver $MGS_l^{(\alpha,\beta,\gamma)}(X_l, B_l)$, after MEISTER (1999 [176])

A linear iteration scheme is formulated in an abstract form with the approximated inverse of A, i.e. $C \approx A^{-1}$:

$$X^{(k+1)} = X^{(k)} + C(B - AX^{(k)}) \tag{3.135}$$

The iteration procedure can be carried out in an *additive* and *multiplicative* way (see BASTIAN (1996 [19])). In the additive variant, the approximated inverse C is determined as the sum of C_m starting from the coarse grid $m = 0$ up to the maximal grid level l; C_m is computed in a similar way as A_{m-1} shown in equation 3.133:

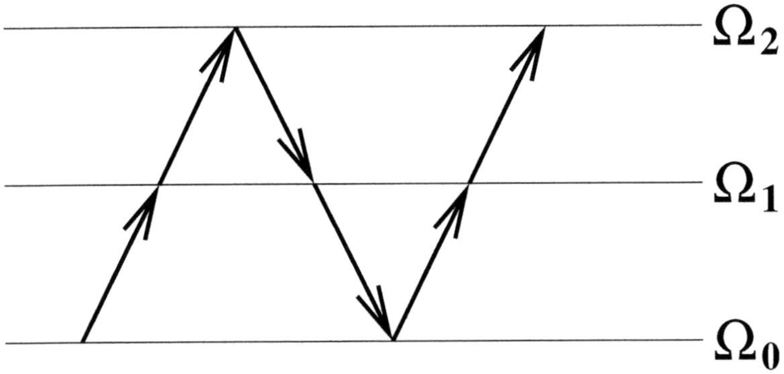

Fig. 3.64. Full Multigrid Method with a $V-$cycle, after MEISTER (1999 [176])

$$C = \sum_{m=0}^{l} C_m \tag{3.136}$$

In the multiplicative variant, the following recursion holds (see BASTIAN (1996 [19])):

$$X^{(k+1+\frac{m+1}{l+1})} = X^{(k+\frac{m}{l+1})} + C_{l-m}(B - AX^{(k+\frac{m}{l+1})}) \tag{3.137}$$

The Multigrid Methods explained in figures 3.60 and 3.63 are multiplicative.

If Multigrid Methods are applied together with adaptive methods, this is called Multigrid Methods with locally refined grids, and the smoothing process is only carried out in the refined regions. On such adaptively refined grid levels, only a small number of elements or nodes occurs (see fig. 3.65). The minimum number of points to be smoothed consists just of all the new nodes on a certain grid level. This leads to the method of *hierarchical bases* which was analyzed for an Additive Multigrid Method by YSERENTANT (1986 [266]) and for a Multiplicative Multigrid Method by BANK et al. (1988 [11]). The last is called *hierarchical bases Multigrid Method* . If the number of points to be smoothed is slightly increased by some neighboring nodes, an optimal complexity is achieved for an Additive Multigrid Method after BRAMBLE et al. (1990 [40]), which is called *BPX method*, and for a multiplicative method by BRAMBLE et al. (1991 [39]), see also BASTIAN (1996 [17]), which is called *Local Multigrid Method.* For a uniform refinement, the Local Multigrid Method is identical to the Multiplicative Multigrid Methods explained in the figures 3.60 and 3.63. Mostly, the convergence behavior of Multiplicative Multigrid Methods is much better than that of Additive Multigrid Methods (see BASTIAN et al. (1998 [20])).

The Multigrid Methods explained up to now are called *geometric* because they use geometric information for refinement and coarsening. *Algebraic Multigrid*

Methods (AMG) are used for a mesh coarsening. They do not need any geometric information, but determine the coarse mesh and the transfer operations with only the algebraic information of the system matrix and the connectivity of the unknowns. AMG is applied if a fine mesh has not been generated by the uniform refinement of coarser meshes or if the boundaries and the inner structures of the coarse mesh are so complicated that they cannot be reasonably reproduced with only a few nodes (see RUGE, STÜBEN (1986 [226]), BRAESS (1995 [36]), BANK, XU (1996 [12])).

In principle, Multigrid Methods are very suitable for structured grids and elliptic problems. Problems occur in the context of unstructured meshes and complex geometries, advection (see RENTZ-REICHERT (1996 [217])), jumps in coefficients (see BASTIAN (1996 [17])) and moving boundaries (see HOPPE (1993 [121])).

In many cases, Multigrid Methods can be parallelized well. However, for convection-dominated or anisotropic problems, robust smoothers which are not so suitable for parallel computations, e.g. *Gauss-Seidel* (see sec. 3.4.2), must be chosen. If the coarse-grid solver is a direct one, the coarse-grid data should be given to each processor and the direct solution should be carried out on each processor. Multigrid Methods, especially in combination with adaptive methods (see sec. 3.3), require a dynamic load balancing (see sec. 3.2.4). A mesh partitioning is not only carried out in the horizonal, but also in the vertical, e.g. by the cluster technique described in section 3.2.4. Although Additive Multigrid Methods are more suitable for parallelization, multiplicative ones should be preferred because of the better convergence behavior (see BASTIAN et al. (1998 [20])). For the combination with adaptive methods, Local Multiplicative Multigrid Methods are recommended. As already mentioned in section 3.2.3, Multigrid Methods are not suitable for vectorization.

Finally, it is mentioned that Domain-Decomposition Methods (see secs. 3.4.3, 3.2.2) can also be interpreted as Multigrid Methods because they consist of a hierarchy of grids. The interface problem corresponds to the coarse-grid solver and the local problem to the fine grid solver.

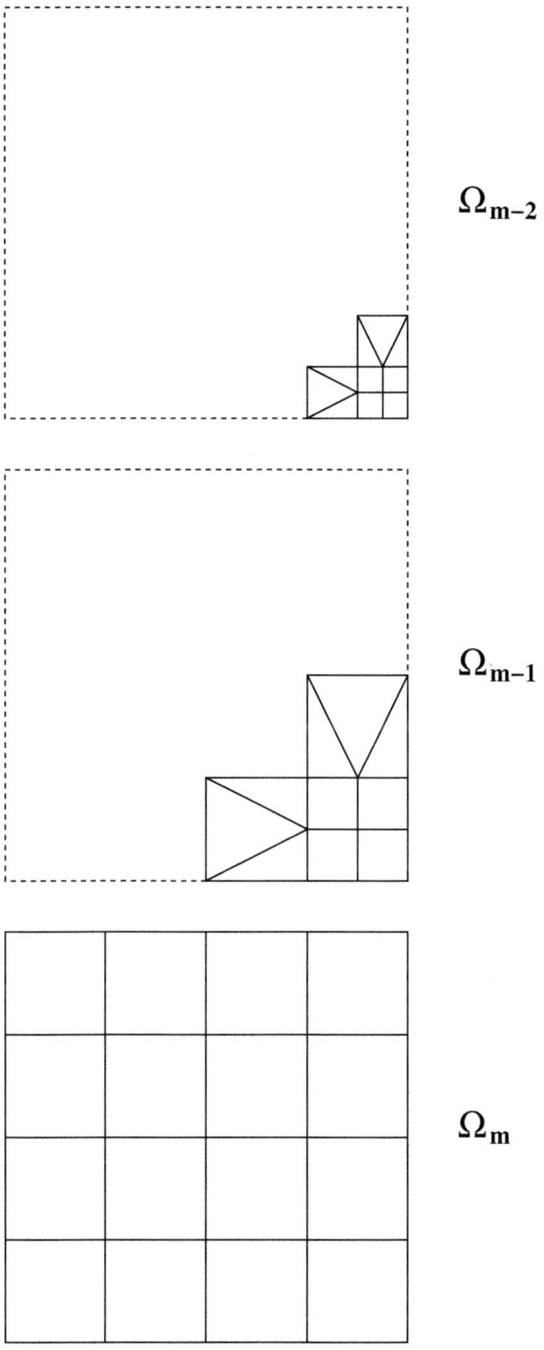

Fig. 3.65. Local Multigrid Method

3.4.5 Non-linear solvers

The coupled balance equations explained in sections 2.3 - 2.6 are characterized by a number of possibly strong non-linearities which exemplarily result from the coefficients in the equations, such as the constitutive relationships (eqs. 2.39, 2.43) or the equations of state (eqs. 2.12, 2.46). Non-linearities also occur, for example, in advection terms in the shallow water equations (eq. 2.62) and in terms with products of unknowns in the continuity equation of the multiphase / multicomponent model (eq. 2.53). These circumstances lead to the following system to be solved where the system matrix $A(X)$ depends on the unknowns X:

$$A(X)X = B \tag{3.138}$$

It is necessary to distinguish between *weak* and *strong* non-linearities. Non-linearities are often weak, e.g. in the shallow water equations (see HINKELMANN (1997 [108])) and in pressure- or temperature-dependent equations of state in subsurface multiphase / multicomponent modeling (see HELMIG (1997 [98])). In such cases, a linearization can be carried out at the discretization stage (see sec. 3.1), e.g. by assuming a linear behavior of a function e between the current and the new time level:

$$e^{n+0.5} \approx \frac{1}{2}(e^{n+1} + e^{n}) \tag{3.139}$$

Strong non-linearities are caused, for example, by the constitutive relationships in subsurface multiphase / multicomponent modeling, i.e. the capillary pressure-saturation and the relative permeability-saturation relationships (see sec. 2.4.3). These functions vary strongly depending on the saturations and geological structures.

For the linearization of the non-linear algebraic equations which result from the discretization methods of section 3.1, there are two commonly used methods, the *Picard* and the *Newton-Raphson* linearization (see PRESS (1992 [208]), HELMIG (1997 [98])). In the *Picard* Method, the coefficients of the matrix A are determined by the values from the previous iteration step (k) to obtain the new iteration step $(k+1)$:

$$A(X^{(k)})\,X^{(k+1)} = B^{(k)} \tag{3.140}$$

The *Picard* linearization has a linear convergence behavior. As the *Newton-Raphson* Method has a superior, quadratic convergence behavior, it is often preferred. The system of equations 3.138 is written in a functional form:

$$F(X) = A(X)X - B = 0 \tag{3.141}$$

The *Newton-Raphson* linearization is based on a *Taylor*-series expansion (eq. 3.12). The new iterate is determined by the following scheme:

$$X^{(k+1)} = X^{(k)} - J^{-1}(X^{(k)})\,F(X^{(k)}) \tag{3.142}$$

In this equation, J denotes the *Jacobian matrix* or *tangent-stiffness matrix*. It should be mentioned that the *Jacobian* matrix is used in a similar context dealing with the transformation relations between natural and *Cartesian* coordinates (see sec. 3.1.4, eq. 3.58). The *Jacobian* matrix here is computed as follows:

$$J = \frac{\partial F}{\partial X} \quad \leftrightarrow \quad j_{il} = \frac{\partial f_i}{\partial x_l} \tag{3.143}$$

If the matrix $A(X)$ is sparse, the *Jacobian* matrix is sparse, too. As a consistent linearization often causes very much effort, the differentiation for the *Jacobian* matrix can also be carried out numerically (see BASTIAN (1999 [21])). Alternatively, the tangent matrix can be replaced by a secant matrix, which can be computed much faster, or the tangent matrix is not updated in each iteration step (see RUST (1991 [227])). If robustness problems occur, methods with *incremental steering* may be advantageous (see RUST (1991 [227])).

For time-dependent problems, the time step must be chosen according to the linearization method, i.e. it must be reduced if the number of non-linear iterations increases too much (see BASTIAN (1999 [21])). In multiphase / multicomponent modeling, the *Newton-Raphson* Method can lead to robustness problems, which can be overcome by the *Picard* linearization (see THORENZ (2001 [248])).

If a Multigrid Method is to be applied to non-linear problems, two different variants exist, the *Newton-Multigrid* and the *non-linear Multigrid Method* (see RUST (1991 [227]), BRIGGS et al. (1999 [44])). In the *Newton*-Multigrid Method, the outer iteration loop is determined by the *Newton-Raphson* Method which leads to a linear system in each outer iteration step. These linear systems can be solved with the Multigrid Methods described in section 3.4.4 in an inner iteration loop. In the non-linear Multigrid Method, a non-linear system is treated on each grid level, i.e. a non-linear smoother and a non-linear coarse-grid solver are required. MOLENAAR (1995 [180]) and BASTIAN (1999 [18]) show that the *Newton*-Multigrid Method is superior, at least for subsurface multiphase flow modeling.

3.4.6 Examples

In this section, the convergence and parallel run-time behavior of two different fast solvers, the PCG and the Multigrid Method, are analyzed.

PCG Method: Parallel efficiency and different preconditioners

The parallel efficiency e_p (see sec. 4.2.3) of the PCG Method (see sec. 3.4.2) with a diagonal preconditioner (see sec. 3.4.3), which is applied to the diffusion step of the 2D-transport equation (see sec. 5.4.1), is shown in figure 3.66 (see HINKELMANN (1997 [108])). For the coarse mesh with about 2200 nodes and one degree of freedom, 100 time steps of $\Delta t = 15s$ are investigated and, for the fine mesh with about 8900 nodes, 100 time steps of $\Delta t = 6s$. The mesh was partioned with the Recursive-Spectral Bisection in combination with *Kernighan-Lin* heuristic (see sec. 3.2.4). The simulations were undertaken on a *nCUBE2S* parallel computer. The absolute hardware data are no longer up-to-date. However, the ratio of the data transfer rate to the processor performance, which is the most important criterion for a *fine granular* algorithm (see sec. 4.2.2), is representative for modern parallel computers with $1.8MB/s/Mflops$. Further information is given in section 5.4.1.

For a given problem size, the parallel efficiency e_p decreases with an increasing number of processors; for a constant number of processors, the parallel efficiency increases with increasing problem size (see fig. 3.66). Further, the *scaling* of the parallel algorithm is obvious: if the problem size is increased by the factor 4 and the number of processors by the factor 2, the parallel efficiency remains approximately the same; this is illustrated by the dashed horizontal lines in figure 3.66.

In table 3.4, numbers of iterations of the PCG Method are given for different preconditioners on 4 and 16 processors. A simple diagonal preconditioner significantly reduces the numbers of iterations here which is caused by the diagonal dominance of the system matrices. The Element-By-Element (EBE) and the *Block-Jacobi* (BJ) preconditioners (see sec. 3.4.3) further reduce the number of iterations. Both depend on the number of processors, but EBE only weakly. For small numbers of processors, the BJ preconditioner requires fewer iterations than the EBE Method, for large numbers of processors vice versa. For increasing problem size, the dependence of the number of iterations on the number of processors becomes weaker for the BJ preconditioner. Finally, it must be mentioned that the CPU time for the BJ preconditioner could hardly be reduced, and sometimes even increased compared to the CPU time for the diagonal preconditioner.

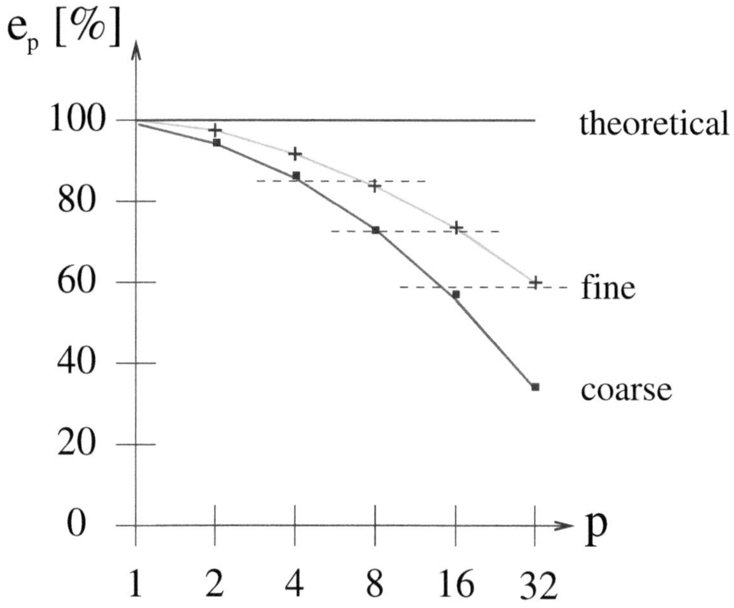

Fig. 3.66. Parallel efficiency of the PCG Method, after HINKELMANN (1997 [108])

	without	diagonal	EBE (4p/16p)	BJ (4p/16p)
coarse grid	3306	669	410/417	328/612
fine grid	6637	580	361/371	306/453

Table 3.4. Numbers of iterations of the PCG Method for different preconditioners on 4 and 16 processors, after HINKELMANN (1997 [108])

Multigrid Method: Different Multigrid levels and parallel speedup

The Multigrid Method (see sec. 3.4.4) is investigated for the two-phase flow equations (see sec. 5.2.4, ÖLMANN et al. (2002 [194])). It is a preconditioner (see sec. 3.4.3) of the BiCGSTAB Method (see sec. 3.4.2) which is the inner solver of an outer *Newton-Raphson* iteration (see sec. 3.4.5). A $V-$Multigrid cycle (see sec. 3.4.4) is chosen with a direct solver (see sec. 3.4.5) for the coarse grid and an ILU smoother with 2 pre- and postsmoothing steps. The simulation period is $1h$ with an initial time step of $60s$. A time step adaptation depending on the number of non-linear iterations (see sec. 3.3.2) is carried out. A static load distribution is determined with the Recursive-Inertial Bisection (see sec. 3.2.4, see fig. 5.35). The coarse-grid solution is computed on each processor. The simulations are carried out on the *Hitachi* parallel vector computer (see sec. 4.2.1). However the Multigrid Methods hardly use the vectorization (see sec. 3.4.4).

In table 3.5, the CPU times computed on 32 processors of the Multigrid Method explained above are listed for different grid levels. In this case, the minimal CPU time is obtained with a four-grid solver, not with the five-grid solver. This circumstance is caused by the parallel computations, because the 'sequential' Multigrid theory proposes decreasing effort with increasing numbers of grid levels. For a direct coarse-grid solver, the number of grid levels should be chosen in such a way that the number of unknowns on the coarse grid is in the range of a few hundred. If the number of unknowns on the coarse grid is much larger, the direct solver requires too much CPU time.

Multigrid levels	1	2	3	4	5
number of unknowns on the coarse grid	263682	66306	16770	4290	1122
CPU time [s]	1749	1251	657	**365**	574

Table 3.5. CPU time for a Multigrid Method with different grid levels, after ÖLMANN et al. (2002 [194])

In figure 3.67, the parallel speedup s_p (see sec. 4.2.3) is investigated for the four-grid solver. It has an overlinear course for 2 and 4 processors probably caused by cache effects (see sec. 4.2.3). For more than 8 processors, the inclination of the speedup decreases with increasing number of processors. It must be taken into account that the convergence of the smoother depends on the number of processors, i.e. the number of iterations increases with an increasing number of processors and, the speedup is thus worsened.

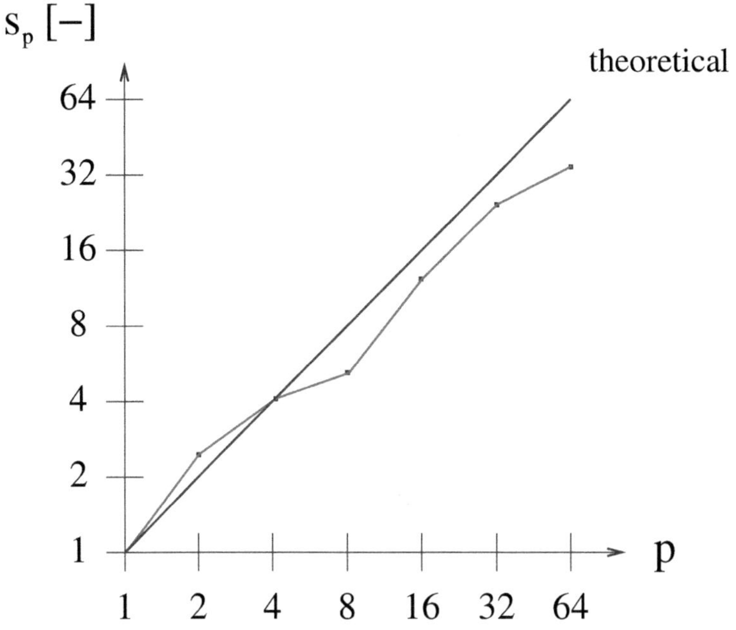

Fig. 3.67. Parallel speedup of the Multigrid Method, after ÖlMANN et al. (2002 [194])

4

Efficient information-processing techniques

This chapter deals with efficient information-processing techniques which are applied in the context of modeling hydro- and environmental systems and which belong to modeling systems. Numerous techniques are explained, most of them, however, only very briefly. Several preprocessing tools which are used for the model set-up, such as CAD, databases and GIS, are explained. The processing part concentrates on High-Performance Computers and the modeling system MUFTE-UG. Aspects of postprocessing and visualization are introduced. Finally, methods and techniques of WWW-based Collaborative Engineering are discussed.

4.1 Preprocessing

4.1.1 Introduction

Preprocessing is concerned with all the tasks which must be carried out before a numerical simulation, i.e. all available data must be processed to determine the *model set-up*. The necessary steps consist of the geometry approximation, the assignment of physical parameters to geometric units, the mesh generation and the determination of the initial and boundary conditions. For problems with a simple structure with regard to the geometry, parameter distributions and boundary conditions, the model set-up is determined at the input stage of the numerical simulator. However, most problems in hydro- and environmental engineering are characterized by complex structures. Therefore, special preprocessing tools as well as the corresponding interfaces are required; these are explained in the following (see fig. 4.1).

The data are space- and / or time-dependent. In many cases, the geological structures of an aquifer or the bathymetry of a river are only space-dependent. However, if effects like land subsidence or morphological changes are taken into

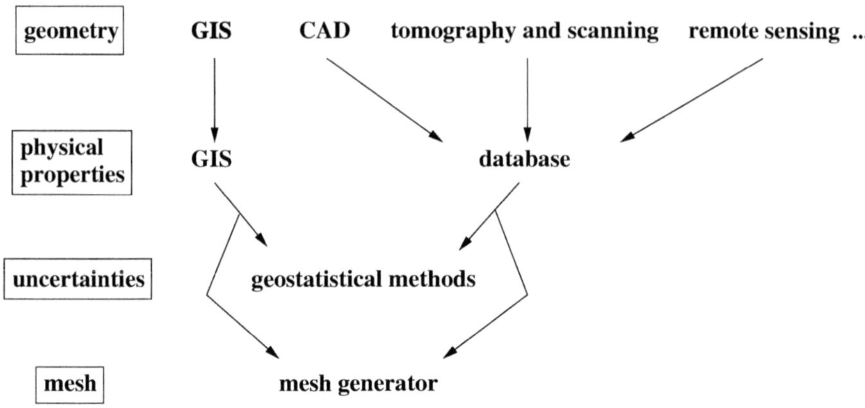

Fig. 4.1. Steps during preprocessing

account, they are also time-dependent. Classical examples of time-dependent data are contamination distributions in the subsurface or measured gauge levels in a surface-water system. The data can be based on a *raster* or *vector model* (see fig. 4.2). In a raster model, the basic element is a pixel or a grid cell. Raster models are rather simple. Generally, they require more storage than vector models, and logical connections between the basic elements can hardly be established. In a vector model, the basic elements are points, lines and areas together with their topological connections which enable logical connections such as a neighborhood relationship. In both, the raster and vector models, the information under consideration is mapped onto the basic elements, and the data must be checked for plausibility and measurement or regionalization errors before they are treated further.

For the geometry approximation, the possibly very complex geological structures of a subsurface system, e.g. spatially intersecting discrete fracture planes or fault zones, or the possibly complex structures of water bodies in a surface-water system, e.g. taking into account waterways and shallow areas as well as flooding areas, must be determined. At one time, the geometry was determined by hand or, for example, by digitizing maps. Nowadays, there are a number of tools which support this work, such as *CAD systems* (see sec. 4.1.2), *Roentgen tomography* (see sec. 4.1.4) and *GIS systems* (see sec. 4.1.5), *remote sensing* (see BAUMGARTNER et al. (1997 [24]), SCHULTZ, ENGMAN (2000 [234])) and *geophysical methods* (see KNÖDEL et al. (1997 [145])). Remote sensing in this context is mainly applied to detecting structures on the surface of the earth, e.g. shore lines, river banks, precipitation or vegetation distributions. However, the bathymetry of a surface-water system, for example, cannot be recognized. This is carried out cross-section-wise or continuously with digital

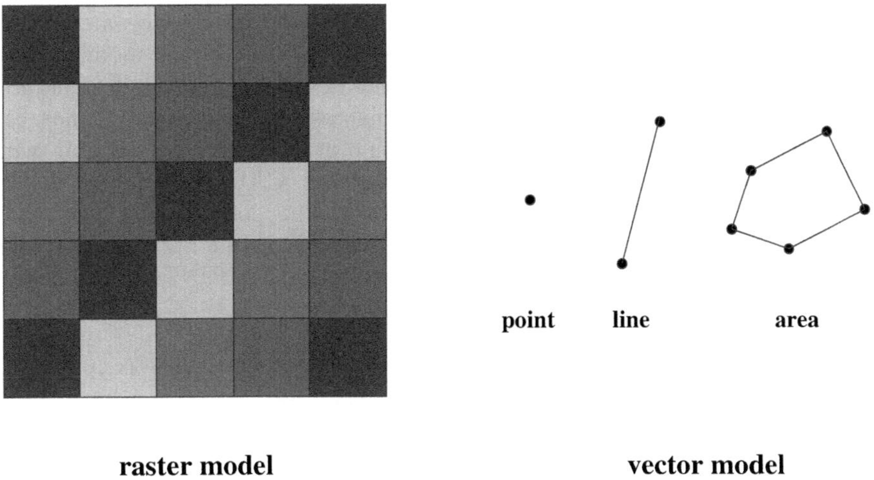

Fig. 4.2. Raster and vector models

depth measurements from a special boat. With geophysical methods, the water table in the subsurface, for example, or the transitions between layers can be measured. However, these methods cannot be used for the direct determination of, for example, permeabilities.

The assignment of physical data, such as the distribution of the bottom friction in a surface-water system or the distribution of the permeabilty in the subsurface, to the geometric units can be handled with the aid of a database (see sec. 4.1.3) which is linked to a CAD system or within a GIS (see sec. 4.1.5). Often, only point information which must be inter- or extrapolated onto the domain is available, e.g. borehole data or local field measurements. The influence of small-scale heterogeneities can be estimated with geostatistical methods (see sec. 4.1.6).

Finally, a mesh must be generated for the computational domain (see sec. 4.1.7), and the initial and boundary conditions must be prescribed. It should be mentioned that the temporal effort for pre- and postprocessing is continuously increasing compared to the processing or simulation time and will take up the major part of the whole modeling procedure within the next years. Further reading can be found in ZIELKE et al. (1999 [267]), HELMIG et al. (1999 [102]) and ROTHER (2001 [225]).

4.1.2 CAD systems

Computer Aided Design or *CAD systems* enable an interactive geometric construction in two and three dimensions based on the vector model. There are many ways of developing, changing and visualizing geometric structures using layering techniques, self-definable macros etc. Additionally, they can be connected over interfaces with different other tools, for example with databases, statics or cost-calculation programs. CAD systems are widely used in civil and building engineering. In building engineering, the *Auto-CAD system* (see AUTOCAD (2001 [2])) is quasi-standard; in civil engineering, the PRO/ENGINEER system (see VOGEL, BUNTE (2001 [255]), PRO/ENGINEER ([209])) is in widespread use.

For modeling hydro- and environmental systems, the market leaders mentioned above are only used in a few cases. Very often, special CAD tools for different systems in environment water were developed in the context of the development of the corresponding modeling systems. For surface-water modeling, TICAD (HOLZ et al. (1990 [120])), MATISSE as the preprocessor of the TELEMAC system (see GALLAND et al. (1991 [86]), TELEMAC ([246]), sec. 5.4) and JANET (see SELLERHOFF (2002 [236]), JANET ([130])) can be mentioned and, for subsurface modeling, the preprocessors of the SPRING system (see KÖNIG (1991 [154]), SPRING (2000 [242])) and the MODFLOW system (see MCDONALD, HARBAUGII (1984 [175]), MODFLOW ([179])). There are a number of useful features and automatisms; one very important function consists of adding, deleting and moving nodes due to the engineer's modeling experience. Such actions are often required in critical zones, e.g. where there are steep bottom gradients in a surface-water system or narrow flow passages in multilayered aquifers. A large number of modeling systems for water-related and environmental problems has been developed in recent years, and many of them have their own CAD or preprocessing tools. Unfortunately, only a few of these CAD systems use standard interface, such as DXF, and are compatible to other systems.

Figure 4.3 presents an AutoCAD plan of a vertical section through a subsurface system located in the Ruhr area in the western part of Germany, showing different layers as well as the course of the surface of the earth (see sec. 5.2.2).

In figure 4.4, the bathymetry of the Hüttenbühl reservoir which is located about 80 km north-east of Stuttgart, Germany, is visualized. This CAD plan was made with MATISSE, the preprocessor of the TELEMAC system. The different colours indicate the different mean water levels. The inflow region in the front, the lateral boundaries and the reservoir wall at the right end are shown.

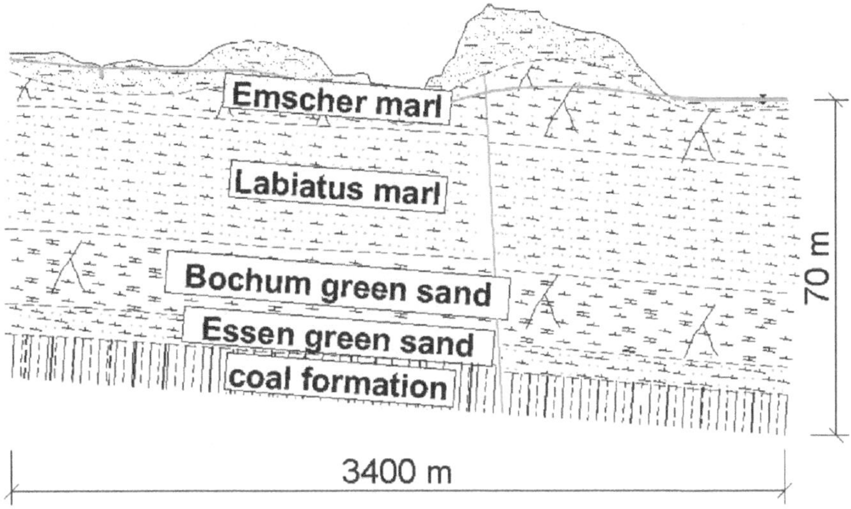

Fig. 4.3. AutoCAD plan of a vertical section through a subsurface system

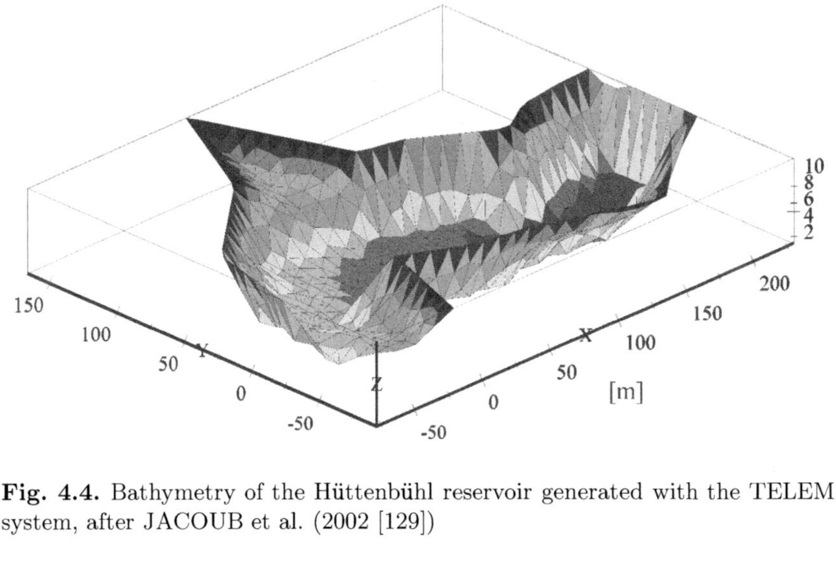

Fig. 4.4. Bathymetry of the Hüttenbühl reservoir generated with the TELEMAC system, after JACOUB et al. (2002 [129])

4.1.3 Database-management systems

Database-management systems (DBMS) are a means of processing information which has continuously gained importance in recent years. They consist of one or more *databases* and a *management system*. The *database* contains structured and interlinked data which are often stored in tables, while the *management system* offers the tools for adding to, changing, linking and controling the data. DBMS enable the storage and administration as well as the handling and analysis of data embedded in a user-friendly and windows-steered software environment. Very often, the data are organized with the *Entity-Relationship (ER) Model*. Entities are objects or abstractions, and they are linked by relations which are represented by tables; this is called a *relational DBMS (RDBMS)*.

The data must be stored consistently and independently of the application programs. For multi-user purposes, an access control, i.e. the permission and the sequence, must be specified. As far as the system architecture is concerned, a *3-level architecture* has become standard for DBMS. The *internal level* contains the physical data organization, i.e. the location of the stored data and the manner of data access. On the *conceptual level*, the data model is described, i.e. the user data and data-access regularization. The *external level* is concerned with the specific requirements of the users. All architectural levels are independent of each other and describe a different point of view on the DBMS. Working with the data, such as selecting, grouping, deleting, updating etc., is handled by a database language, generally the standard *Structured Query Language (SQL)*, in combination with a product-specific development environment. Overall, a database implementation should be efficient that queries and changes should be carried out fast. There are a number of RDBMS, e.g. ORACLE ([200]) or the open-source MySQL ([184]) database.

In recent years, *distributed DBMS* have been developed. The data are stored on different computers connected by a network. For the user, this is not visible; however, the coordination tasks are much more complicated. For simply-structured data, RDBMS are very suitable. For non-standard applications with complex data structures, such as CAD (see sec. 4.1.2), *CASE (Computer-Aided Software Engineering)* or *Multimedia* (see sec. 4.4), *object-orientated DBMS (OODBMS)* are superior, because they enable the usage of *object-orientated* methods. OODBMS should also be preferred for *multidimensional databases*, i.e. if data are investigated from several points of view. When dealing with huge data volumes, they are called *data warehouses* in the context of storing data and *data mining* in the context of analyzing data. Further information is given, for example, in ABTS, MÜLDER (2000 [1]).

In figure 4.5, the database-management system MySQL which has been connected with the modeling system MUFTE-UG (see sec. 4.2.4) is shown. The soil-dependent parameters as well as the parameters for the constitutive relationships are given for different layers (see sec. 5.2.2).

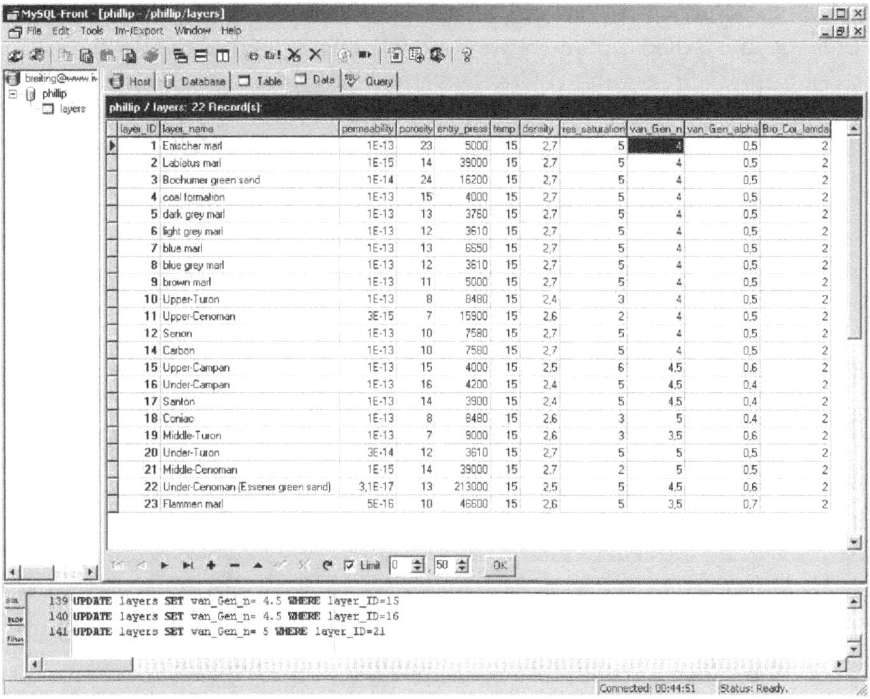

layer_ID	layer_name	permeability	porosity	entry_press	temp	density	res_saturation	van_Gen_n	van_Gen_alpha	Bro_Cor_lamda
1	Emscher marl	1E-13	23	5000	15	2,7	5	4	0,5	2
2	Labiatus marl	1E-15	14	39000	15	2,7	5	4	0,5	2
3	Bochumer green sand	1E-14	24	16200	15	2,7	5	4	0,5	2
4	coal formation	1E-13	15	4000	15	2,7	5	4	0,5	2
5	dark grey marl	1E-13	13	3760	15	2,7	5	4	0,5	2
6	light grey marl	1E-13	12	3610	15	2,7	5	4	0,5	2
7	blue marl	1E-13	13	6650	15	2,7	5	4	0,5	2
8	blue grey marl	1E-13	12	3610	15	2,7	5	4	0,5	2
9	brown marl	1E-13	11	5000	15	2,7	5	4	0,5	2
10	Upper-Turon	1E-13	8	8480	15	2,4	3	4	0,5	2
11	Upper-Cenoman	3E-15	7	15900	15	2,6	2	4	0,5	2
12	Senon	1E-13	10	7580	15	2,7	5	4	0,5	2
14	Carbon	1E-13	10	7580	15	2,7	5	4	0,5	2
15	Upper-Campan	1E-13	15	4000	15	2,5	6	4,5	0,6	2
16	Under-Campan	1E-13	16	4200	15	2,4	5	4,5	0,4	2
17	Santon	1E-13	14	3900	15	2,4	5	4,5	0,4	2
18	Coniac	1E-13	8	8480	15	2,6	3	5	0,4	2
19	Middle-Turon	1E-13	7	9000	15	2,6	3	3,5	0,6	2
20	Under-Turon	3E-14	12	3610	15	2,7	5	5	0,5	2
21	Middle-Cenoman	1E-15	14	39000	15	2,7	2	5	0,5	2
22	Under-Cenoman (Essener green sand)	3,1E-17	13	213000	15	2,5	5	4,5	0,6	2
23	Flammen marl	5E-16	10	46600	15	2,6	5	3,5	0,7	2

Fig. 4.5. Database-management system MySQL connected with the MUFTE-UG modeling system

4.1.4 Tomography and scanning

In recent years, techniques based on information-processing have been newly developed or applied to problems in hydro- and environmental engineering to detect geometric structures of the computational domains in a high resolution. In this context, *tomography* and *scanning* are explained, another technique is remote sensing.

With a tomograph, the pore structure of a porous medium can be reproduced at a high resolution. This is done by measuring the density distribution of the porous medium which can be visualized in form of different gray scales. Such a procedure is similar to the usage of tomographs in medicine. For a sample of $1cm^3$, 10^3 pixels are needed for a resolution of $1mm$ and 10^9 pixels for a resolution of $10\mu m$. In figure 4.6, a cross-section through a cylindrical soil sample with different shades of gray is shown (see VOGEL (1998 [254])). Such geometric information can be chosen to model flow and transport processes on the pore scale based on the solution of the *Reynolds* and transport equations (see sec. 2.6), for example, with *Lattice-Boltzmann Methods* (see KRAFCZYK et al. (2000 [157])). An analysis of the pore-scale processes with e.g. Lattice-Boltzmann Methods, pore-network models or inverse parameter estimation leads to spatially variable hydraulic parameter fields on the continuum scale (see fig. 2.1, VOGEL, ROTH (1998 [256])), e.g. the capillary pressure- and the relative permeability-saturation relationships (see sec. 2.4.3). On this scale, gas-water flow processes can be simulated with the *Darcy* law. Bringing together high-resolution geometry with pore- and continuum-scale simulations is intended to lead to the development of new *upscaling methods* (see MILLER et al. (1998 [178]), MANTHEY et al. (2003 [172])).

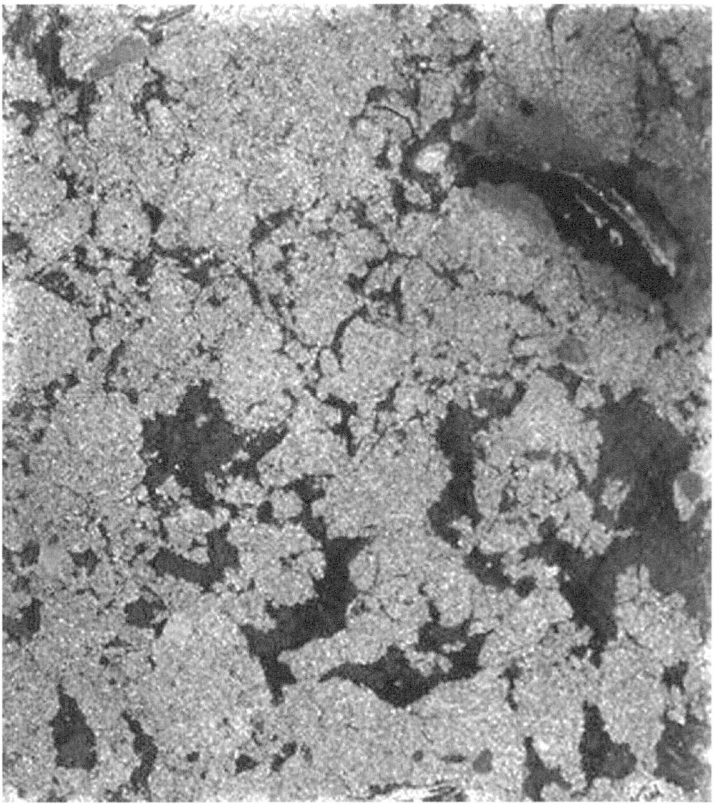

Fig. 4.6. Section through a soil sample of 15x15mm, after VOGEL (1998 [254])

In figure 4.7, a scanned surface of a fracture is visualized (see SILBERHORN-HEMMINGER (2002 [237])). The roughness of the surface is clearly recognizable. The information obtained from this scan is further used to determine the aperture distribution of the fracture.

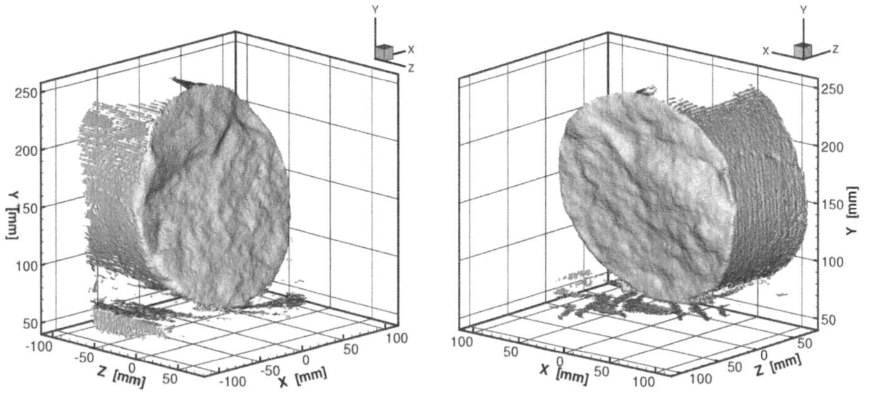

Fig. 4.7. Section through a soil sample, after SILBERHORN-HEMMINGER (2002 [237])

4.1.5 Geographical Information Systems

Geographical Information Systems or *GIS* are software systems which are capable of assembling, storing, manipulating, displaying and analyzing geographically referenced information, i.e. data identified according to their locations. The data can contain geometric information or attributes which are stored, for example, in different layers or themes (see fig. 4.8). Generally, a GIS projects the information onto a surface map, and is thus restricted to two dimensions as well as to time-independent data. However, there are some extensions to three dimensions and time-dependent data. The quasi-standard GIS tools are *ArcVIEW* as an introduction and *ArcGIS* for advanced use (see ArcVIEW, ArcGIS ([5])). GIS has become widespread in recent years, and its use is strongly recommended, especially for multidisciplinary work, because the information exchange is relatively simple. In the following, some basics about GIS are explained. Further information is given in, for example DAVIS (2001 [69]).

In GIS, different kinds of data are processed, such as ASCII, digitized, vector, raster, graphic, multimedia data etc. Interfaces and functions are available for overlaying and linking these data, including, for example, merging and intersection maps and data, extracting features, analyzing proximity relations etc.

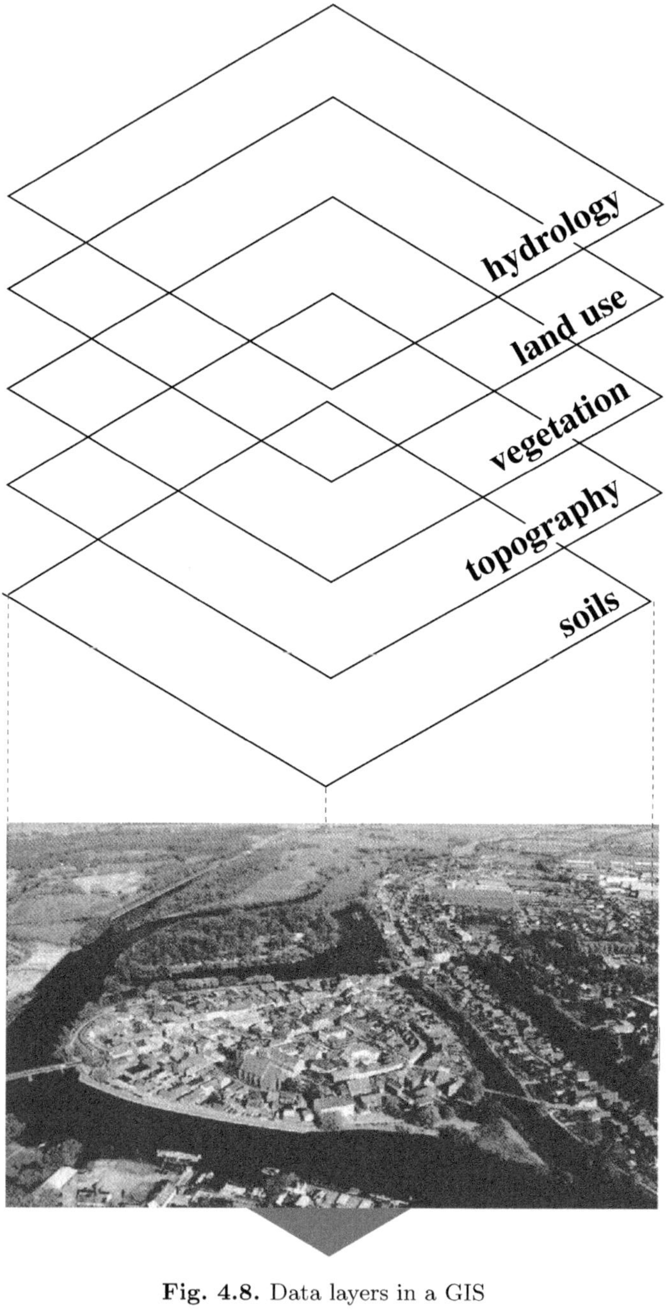

Fig. 4.8. Data layers in a GIS

If a *digital terrain model (DTM)* which contains the vertical coordinates of the surface of the earth is set up, this is based on raster data (see sec. 4.1.1). For complex structures, a *triangular irregular network (TIN)*, which is determined with a vector model (see sec. 4.1.1), is often more suitable. Before the data are mapped onto the DTM or TIN, they must be checked for plausibility and should be thinned out and generalized, i.e. simplified data sets which represent the essential properties must be generated. A further important feature within GIS is *geo-referencing*, i.e. the linking of models which are based on different spatial reference systems.

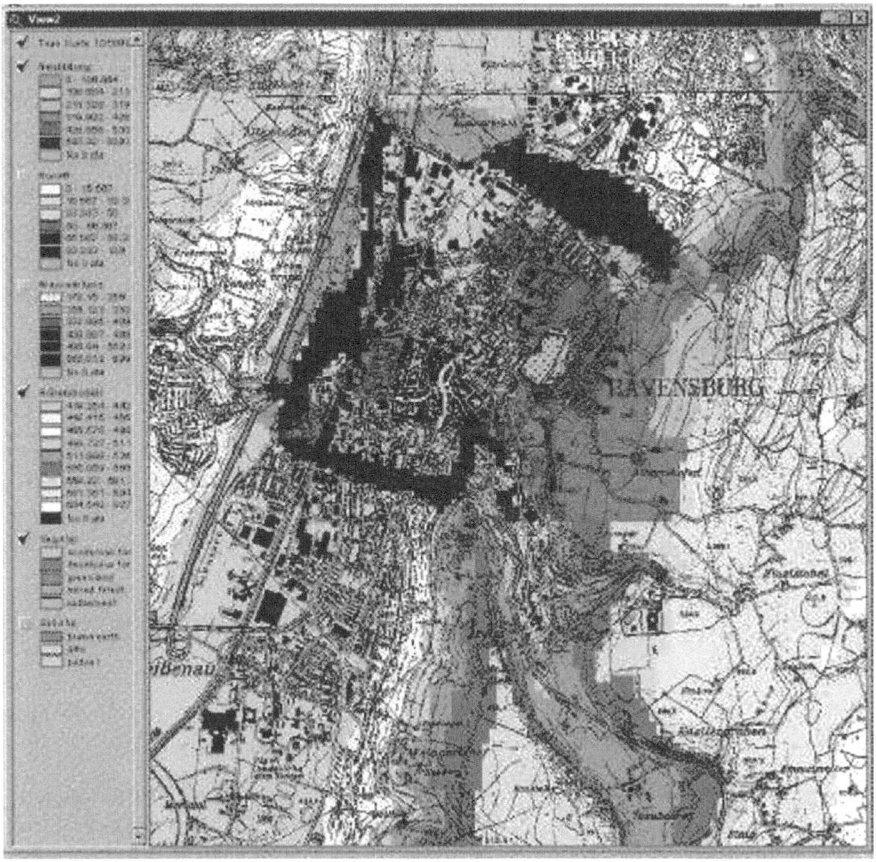

Fig. 4.9. GIS application for recharge determination, after RIEGGER (2001 [218])

For surface-water modeling, GIS can be applied to determine the boundaries of the computational domain at mean water level or to predict flood areas at extreme water levels. For subsurface modeling, GIS can be used, for example to describe the course of the system boundary with the atmosphere and to evaluate the surface-runoff and the recharge which depend on a number of parameters.

In figure 4.9, an application of ArcVIEW is shown for a recharge determination which is computed from the parameters open area, relief, surface runoff, soil, vegetation, precipitation and evapotranspiration (see RIEGGER (2001 [218])).

4.1.6 Geostatistical methods

Numerical simulation in the subsurface is always associated with uncertainties (see sec. 3.1.6). Small-scale heterogeneities whose spatial distribution is generally unknown can influence the flow and transport processes significantly, especially in two- and multiphase systems (see secs. 5.2.2, 5.3.2). With *geostatistical methods*, such influences can be estimated. Geostatistical methods provide information about the spatial variability of a considered property for given *mean value*, *variance* and *correlation length*. For subsurface simulations, the most important property in this context is the permeability field. Various methods exist for determining geostatistical permeability distributions (see BARDOSSY (1992 [15]), WACKERNAGEL (1998 [257]), SILBERHORN-HEMMINGER (2002 [237])). Here, *exponential variogram models* which are a measure of differences between spatial data are briefly introduced.

Variogram functions $var(\underline{x})$ are characterized by the *variance* v, the *correlation length* c and the *nugget effect* ne (see fig. 4.10). The variance for a permeability field considered at i locations is computed with the arithmetic average K_m as follows:

$$v = \frac{1}{n}\sum_{i=1}^{n}(K_i - K_m)^2 \tag{4.1}$$

$$K_m = \frac{1}{n}\sum_{i=1}^{n} K_i \tag{4.2}$$

The correlation length is a measure for the area of influence. If the distance between two points is larger than the correlation length, they are not correlated. To take layering effects into account, the correlation length in the horizontal direction can be chosen to be greater than that in the vertical direction. Often, the horizontal correlation is 5 or 10 times larger than the vertical one. The nugget effect is often neglected in hydro- and environmental engineering.

The variogram function in figure 4.10 is determined by the following formula:

$$var(\underline{x}) = v(1 - e^{\frac{-|\underline{x}|}{a}}) + ne \tag{4.3}$$

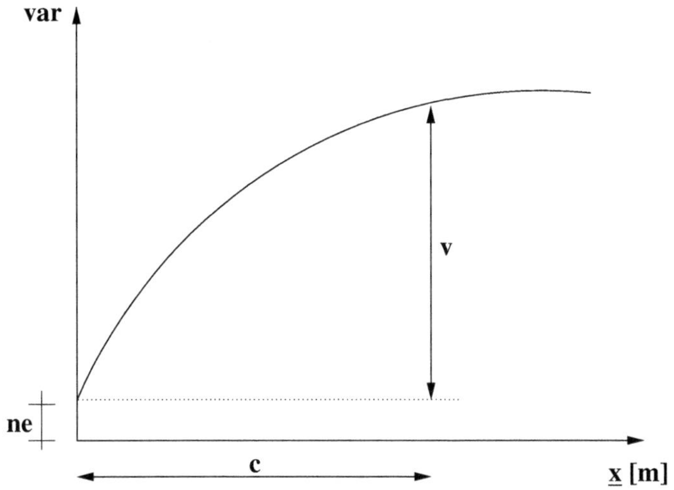

Fig. 4.10. Exponential variogram

If the correlation length is assumed to be the distance where 95% of v is reached, $a = -c\,/ln\,0.05 = c/3.00$.

For two- and multiphase systems, it is mentioned that the entry pressure distribution p_d (eq. 2.39) in a geostatistically varied field can be determined according to the *Leverett function* $J = p_d\sqrt{K/\phi}$ which is given here in a simplified form (see HELMIG (1997 [98]), SHETA (1999 [230])). If the permeability K_m, the porosity ϕ_m (see sec. 2.4) and the entry pressure p_d for the averaged or homogeneous system are known, the entry pressure at location i in the geostatistically varied field can be computed as:

$$p_{di} = p_d\sqrt{K_m\phi_i/(K_i\phi_m)} \tag{4.4}$$

In figure 4.11, geostatistical permeability fields with different variances ($v = 0.25$, left and $v = 1.0$, right) are shown (see KOBAYASHI et al. (2002 [146])). They were generated with the SIMSET model (see BARDOSSY (1992 [15])). The nugget effect is set to zero $ne = 0$. The correlation length in the horizontal direction is $c_h = 20m$, and is five times larger than in the vertical direction, $c_v = 4.0m$. Therefore, layering effects are observed in both graphics. A comparison of the left and right permeability field clearly shows the larger variance on the right, caused by the larger inhomogeneities.

Geostatistical methods can also be applied to generated discrete fracture systems (see *Geostatistical Software Library (GSLIB)* of DEUTSCH, JOURNEL (1992 [70]), SILBERHORN-HEMMINGER (2002 [237])). Further information about geostatistical methods can be found, for example, in BARDOSSY (1992

[15]), WACKERNAGEL (1998 [257]), or SILBERHORN-HEMMINGER (2002 [237]).

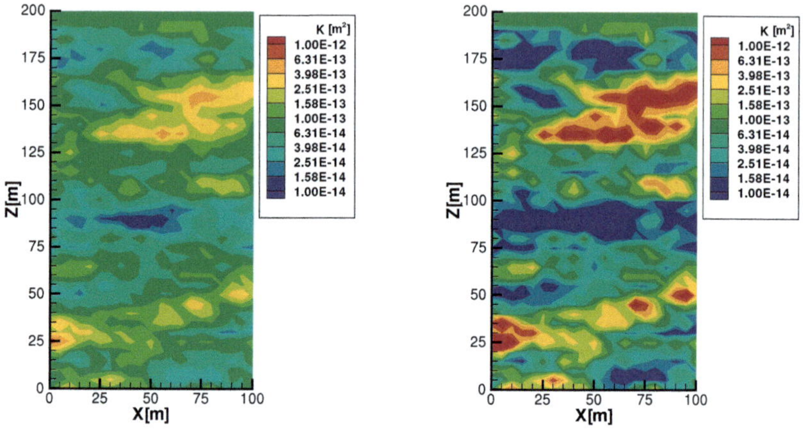

Fig. 4.11. Geostatistical permeability fields with different variances; left: $v = 0.25$, right: $v = 1.0$

4.1.7 Mesh generation

After the geometric and physical information has been prepared, a mesh of the computational domain must be generated. If the domain and the boundaries as well as the spatial distribution of the physical parameters are 'simple', a *structured mesh* can be chosen, i.e. *rectangular* elements in 2D and *bricks (rectangular parallelepipeds)* in 3D. Here, structured means that, in 2D, each inner node has a connection to 4 neighboring nodes and to 4 rectangular elements and, in 3D, each inner node has a connection to 6 neighboring nodes and to 8 brick elements. Such a mesh generation is rather simple, and in combination with, for example, GIS (see sec. 4.1.5), can be easily determined from a raster model (see sec. 4.1.1). Structured meshes are specially designed for Finite-Difference Methods and are also applicable for Finite-Element and Finite-Volume Methods (see fig. 5.23).

For complex domains and boundaries as well as spatial parameter distributions, *unstructured meshes* are recommended, i.e. triangles or quadrilaterals in 2D, and tetrahedra and hexahedra in 3D. For rather complex cases, triangles and tetrahedra should be preferred. 'Unstructured' means that any degree of connectivity to neighboring nodes or elements is possible. Unstructured meshes are related to the vector model (see sec. 4.1.1), CAD systems (see sec. 4.1.2) and TIN in GIS (see sec. 4.1.5). For 3D-simulations, the mesh is often unstructured on the horizontal plane and, mapping it several times over the vertical, it is structured on the vertical plane. Such a mesh generation is called 2.5D (see fig. 5.55). The generation of unstructured meshes requires special methods and is much more time-consuming than that of structured grids. The most popular mesh-generation methods are based on the *Delaunay triangulation*, the *Advancing Front Method* or *block-structured methods* (see KASPER (1999 [143]), FUCHS (1999 [84])). Unstructured meshes are especially suitable for the Finite-Element and Finite-Volume Method (see figs. 5.53, 5.15).

In many practical water-related and environmental problems, geometrically complex systems must be treated and, therefore, unstructured meshes are used. In subsurface systems, the geological structures may be very complex, if spatially intersecting discrete fractures and fault zones (see fig. 4.20), different layers or sink / source areas are dealt with. In surface-water systems, the transitional zones between shipping lanes and shallow areas, regions with steep bottom gradients as well as flooding or drying areas require special attention (see fig. 5.47). In such critical regions, a higher mesh resolution should be chosen depending, for example, on the jump in the permeability (see fig. 4.12) or the steepness of the bottom gradient. This is called an a priori adaptive method and has already been discussed in sec. 3.1.1 (see fig. 3.35). Furthermore, the preprocessing tool should enable adding to, moving or deleting nodes 'by hand', e.g. with a CAD system (see sec. 4.1.2).

A good or high quality of the mesh or the generated elements is desirable, as the quality of numerical results may strongly depend on it. Therefore, elements are to be equilateral, and the angles close to 60° for triangles and 90° for quadrilaterals. In this context, the mesh generator *ART (Almost Regular Triangulation*, see FUCHS (1999 [84])) is mentioned. ART is based on the *Delaunay triangulation* which is shape-optimized with the *Laplace Method.* It generates high-quality triangles or tetrahedra with a regular mesh structure and enables area-wise higher mesh resolutions according to user-defined density functions. ART has been extended to discrete fractured systems; for fractures as elements of lower dimensions (e.g. 1D-fractures in a 2D-matrix), see NEUNHÄUSERER et al. (1998 [187]), for fractures as elements of equal dimension (e.g. 2D-fractures in a 2D-matrix), see NEUNHÄUSERER et al. (2001 [188]) and section 5.1 for further information. Moreover, TANIGUCHI et al. (1996 [245]) have developed a mesh generator for fracture-matrix systems. In figure 4.12, a triangular mesh for a fracture-matrix system generated with ART is shown. The density functions are chosen higher along the fracture-matrix interaction space (see NEUNHÄUSERER et al. (2000 [190])).

The ratio between the smallest and largest element edge should not differ by too many orders of magnitude. However, the systems to be modeled often do not fulfill this requirement. In surface-water as well as in subsurface systems, the length scales on the horizontal plane (kilometers) are much larger than

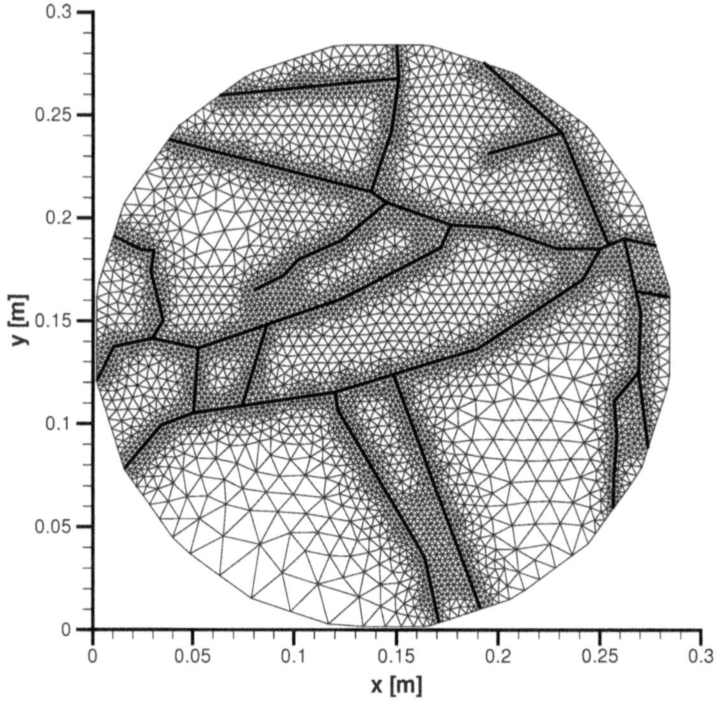

Fig. 4.12. Fracture-matrix system generated with ART, after NEUNHÄUSERER et al. (2000 [190])

in the vertical direction (meters). Therefore, certain compromises have to be made. It is further mentioned that certain numerical schemes require special geometric conditions which must be guaranteed by the mesh generator, e.g. the *Voronoi property* for the node-centered FVM (see FUHRMANN, LANGMACH (1998 [85])).

In the context of parallel computations, the mesh generation can be carried out in two different ways. On the one hand, the mesh is generated sequentially and divided into a number of subdomains afterwards (see secs. 3.2.4, 5.4, HINKELMANN (1997 [108])). On the other hand, the mesh can be subdivided into macro-elements; then a mesh is generated in each macro-element in parallel (see LANG (2001 [162])). The mesh generation in the macro-elements can be carried out with certain constraints at the interprocessor boundaries, e.g. no hanging nodes (see fig. 3.36). Such a mesh was been generated with tetrahedra for an abandoned coal mine (see fig. 4.13, sec. 5.2.2). The different layers are shown with different colors. In the upper graph, a view into the domain is given and some macro-elements have therefore been removed. For a better visualization, the layers with the generated tetrahedra are presented separately in the lower graph.

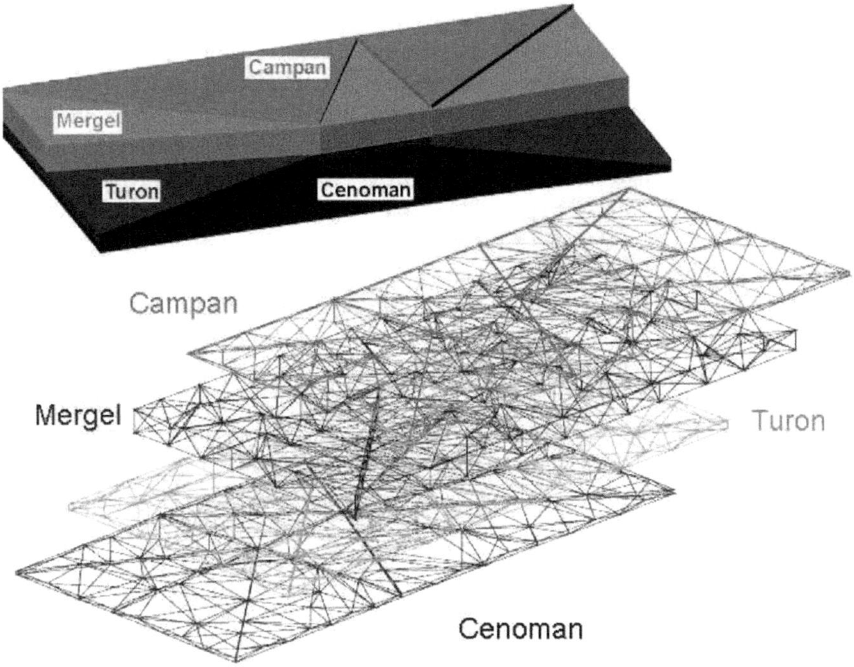

Fig. 4.13. 3D-mesh of a coal mine

Further information is given in, for example, KASPER (1999 [143]), FUCHS (1999 [84]), SELLERHOFF (2002 [236]) and ROTHER (2001 [225]).

4.2 Processing

4.2.1 High-Performance Computational Architectures

Many large-scale computational problems cannot be solved in a reasonable time or at all on a single PC or workstation. High-Performance Computers, such as *parallel* and *vector computers*, are required; they are introduced in this section. Different computer architectures can be distinguished according to the classification of *Flynn*: the *von Neumann* computer is based on the *Single Instruction Single Data principle (SISD)*, i.e. one processor is connected to one memory (see fig. 4.14), and the commands are carried out sequentially with one datum. This is the way a classical PC or workstation operates. On parallel computers, several processors work at the same time in parallel for a task. If the *Single Instruction Multiple Data principle (SIMD)* is chosen, every processor carries out the same command, each with different data, however; this is called *data parallelism*, and the data-parallel programming model can be applied (see secs. 3.2.2, 3.2.3). If the *Multiple Instruction Multiple Data principle (MIMD)* is used, different 'autonomous' processors work on different commands with different data at the same time. The processors can operate fully independently of each other; this is called *functional parallelism*, and the message-passing programming model (see secs. 3.2.2, 3.2.3) must be chosen.

If parallel processors all share the same memory, this is called a *shared-memory (SM) system*. Each processor has direct access to the whole memory (see fig. 4.14). However, the independence of the data and the sequence of the accesses must be ensured. Usually, shared-memory systems operate on the data-parallel programming model (see sec. 3.2.2), sometimes also on the message-passing programming model (see sec. 3.2.2). In a *distributed-memory (DM) system*, each processor has its own memory (see fig. 4.14), and these units are connected by a *communication network* (see sec. 4.2.2). If data from another processor are required, they must be sent over the communication network (see sec. 3.2.3). Distributed-memory systems work on the message-passing programming model, the data-parallel programming model is not possible. Moreover, it is possible that small units of a parallel system operate with shared memory, and several such units are connected and work with distributed memory; this is a *hybrid-memory system* (see fig. 4.14).

Vector computers consist of a *scalar* and a *vector unit*. The scalar unit is made of one or a few usual processors. The performance of vector computers results from one or a few highly developed *vector processors* within the vector unit. Vector processors carry out the same operations for special data structures in an extremely fast way based on the SIMD principle. This is called *pipelining*, and the operations are done in so-called *clock cycles* which are much faster than operations in scalar processors (see fig. 4.15). The performance of vector computers cannot be increased very much further for physical reasons, and

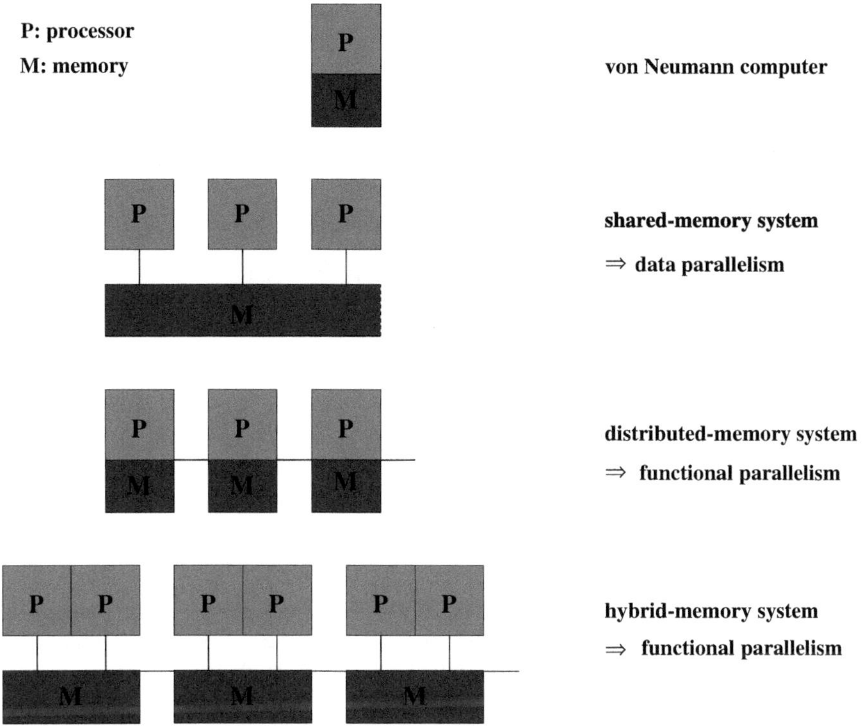

Fig. 4.14. Computer architectures, after HINKELMANN (2000 [109])

the fast data access for the *vector pipelines* becomes a bottleneck and is very expensive. The use of vector computers is interesting for those algorithms (see sec. 3.2.3) where the major part of the computations can be carried out in the vector unit. Moreover, it is advantageous to have large vector lengths.

In the following, the development of high-performance computational architectures in recent years is shown and the current situation discussed (see fig. 4.16). This is based on the *TOP500 Supercomputing Sites* which is a list of sites with the 500 most powerful computer systems installed worldwide and is updated twice a year (see MEUER et al. (2001 [177])). *Massively Parallel-Processing (MPP) systems* consisting of several hundred single processors clearly dominate the TOP500 list with about 50% of the entries. *Constellations*, which are clusters with large SMP (explained later) as building blocks, i.e. more than 16 processors per SMP, make up about 25% of the entries. The rest of the TOP500 list is taken up by *Symmetric Multi-Processing (SMP) systems*, which are computer nodes with shared memory that are part of a larger system, and *Clusters*, which consist of SMP nodes with up to 4 processors. About

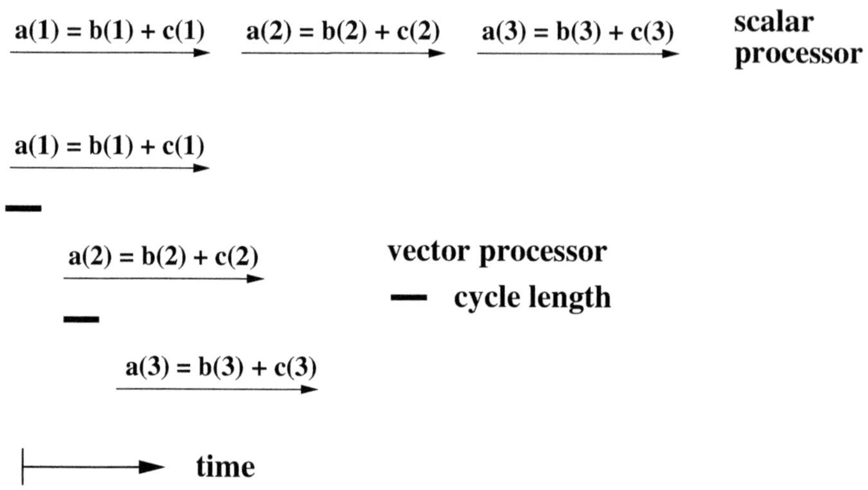

Fig. 4.15. Sequence of operations in a scalar and a vector processor, after HINKELMANN (2000 [109])

50% of the entries in the TOP500 list operate with hybrid memory. Since the TOP500 list came into existence, the performance of the fastest systems has nearly doubled each year (see MEUER et al. (2001 [177])). The fastest parallel system achieves several Teraflops (10^{12} floating point operations per second) in the TOP500 list of November 2001 (see MEUER et al. (2001 [177])).

Most of the HPC systems use UNIX or LINUX as the operating system, a few use WINDOWS NT. In most cases, the pre- and postprocessing (see secs. 4.1, 4.3) is not carried out on the HPC systems. Exceptions are, for example, parallel mesh generation (see sec. 4.1.7) or High-Performance Visualization (see sec. 4.3).

To gain an idea of current computer-performance numbers, some are given for the parallel vector computer *Hitachi SR8000* and the MPP system CRAY T3E, where numerical simulations have been carried out. The SR8000 at the High-Performance Computing Center of the University of Stuttgart, Germany, consists of 16 SMP units, also called nodes, with 8 processors and 8 GigaByte RAM each; overall, 128 processors with 128 GigaByte RAM achieve 128 Gigaflops. The T3E at the High-Performance Computing Center of the University of Stuttgart, Germany, consists of 512 nodes with 64 GigaByte RAM and achieves 461 Gigaflops. Again, it must be mentioned that the given performances are peak performances which cannot be reached for many applications. Number 1 in the TOP500 list of November 2001 aims at 7.2 Teraflops, number 500 at 95 Gigaflops. The LINUX PC-Cluster at the Institute of Hy-

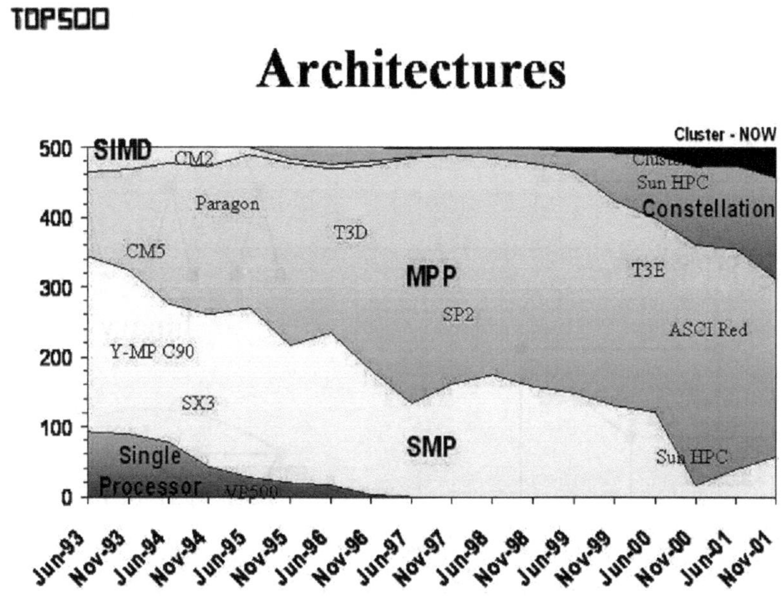

Fig. 4.16. Development of computer architectures over time, after MEUER et al. (2001 [177])

draulic Engineering, University of Stuttgart, Germany, has 56 processors with 1 GigaByte RAM each, and achieved about 56 Gigaflops.

4.2.2 Networks and communication in HPC

In distributed- and hybrid-memory systems, the parallel processors or units must be connected by a communication network. If the number of processors is small, each processor is connected to every other processor; this is called *crossbar* (see fig. 4.17). For large numbers of processors, different *network topologies* are used, and information must be routed over other processors for non-neighboring processors. The topologies are designed in such a way that, on the one hand, the maximal distance between processors in the network is small and, on the other hand, the maximal number of connections per node is small, too. In figure 4.17 different often-used network topologies are shown. Generally, users and developers of parallel models do not have to take the different network topologies into account.

Nowadays, *Fast Ethernet* or *Myrinet* ([183]) are generally chosen as the communication hardware depending on the *granularity* of the parallel model. If

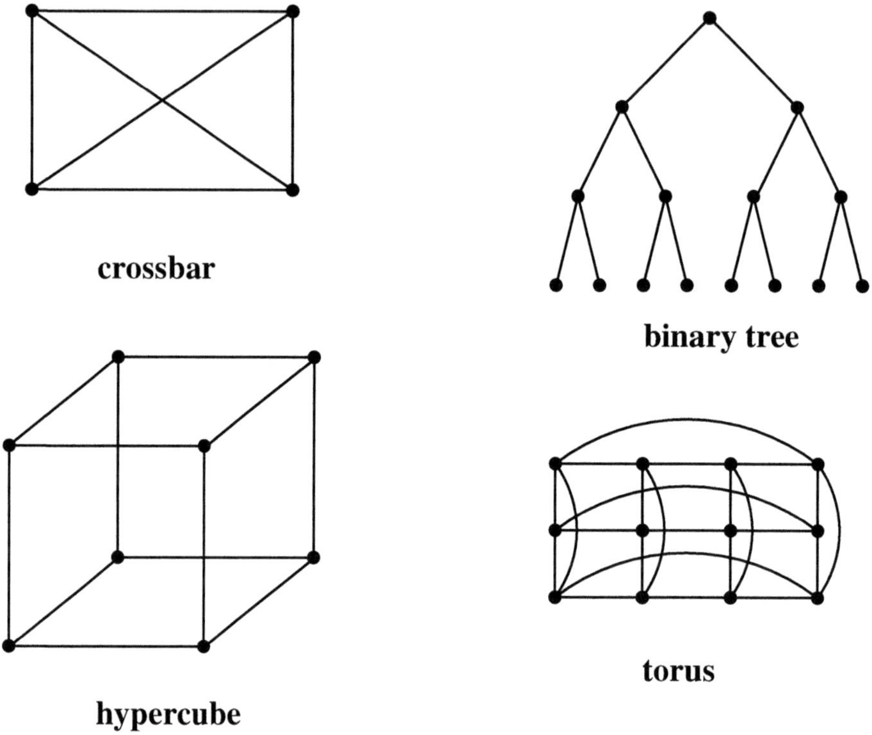

Fig. 4.17. Network topologies, after HINKELMANN (2000 [109])

the communication effort is high / low with respect to the computational effort, this is called *fine / coarse granularity.* The communication speeds are up to 2 Gbit/s for Myrinet and 100 Mbit/s for Fast Ethernet. However, these are *peak performances* and achievable performances are, in most cases, much lower. Myrinet is much faster than Fast Ethernet, but also much more expensive. Fast Ethernet is sufficient for coarse granular problems. For fine granular problems, one should investigate very carefully whether Myrinet should be preferred to Fast Ethernet. Both enable the usage of the standard communication interfaces MPI and PVM (see sec. 3.2.2).

To assess which communication hardware should be chosen, equation 3.92 is recalled, including the units of the variables with the communication time t_{com} [s], the *start-up time* t_{st} [s], the amount of data to be sent m [Mbit], and the data-transmission rate r_{tr} [s/Mbit]:

$$t_{com} = t_{st} + m\, r_{tr} \tag{4.5}$$

In typical communications, such as arise in the majority of parallel applications in modeling hydro- and environmental systems, many 'long' vectors are sent and, consequently, the communication time is dominated by the data-transmission rate, while the start-up time can be neglected. For fine granular problems, such as parallel fast solvers (see sec. 3.4), the product of the processor performance L [Mflops] and the data-transmission rate r_{tr} [s/Mbit] should be comparatively small. The author recommends this ratio to be smaller than 0.1 [Mflops s/Mbit] at least. This can be chosen as an estimation for single-processor parallel systems. For hybrid systems, one has to think further, and it is difficult to make recommendations. Generally, the best way is to check the parallel run-time behavior of the applications on the different processors and communication hardwares.

4.2.3 Performance numbers

There are a few definitions for the performance quality of parallel and vector algorithms which are introduced in the following. The *parallel speedup* s_p is defined as the ratio of the run time on 1 processor to the run time on p processors:

$$\textbf{parallel speedup} \qquad s_p = \frac{\text{CPU time on 1 processor}}{\text{CPU time on } p \text{ processors}} \tag{4.6}$$

The principle course of the parallel speedup is shown in figure 4.18, left, in double-logarithmic scale (see figs. 3.33, 3.67). Generally, the parallel speedup increases with an increasing number of processors, its course is on the right side of the theoretical limes, and the value of the parallel speedup is smaller than the number of processors, i.e. $s_p(p) < p$. Due to the parallel overhead (see sec. 3.2.3), the inclination of the parallel speedup becomes flatter with an increasing numbers of processors. In a few exceptions, $s_p > p$ is possible; this is caused by a different intensive usage of a *cache* on different numbers of processors. A cache is a special unit in a processor with little memory where a limited number of operations is carried out much faster than in the other processor units.

The *parallel efficiency* is the ratio of the run time on 1 processor to the product of the run time on p processors and p:

$$\textbf{parallel efficiency} \qquad e_p = \frac{\text{run time on 1 processor}}{p \text{ x run time on } p \text{ processors}} = \frac{s_p}{p} \tag{4.7}$$

The course of the parallel efficiency is shown in figure 4.18, middle, in a double-logarithmic scale (see figs. 3.33, 3.34, 3.66). The parallel efficiency decreases with an increasing number of processors. It can be considered a scaling of the parallel speedup. Therefore, generally $e_p < 1$ or 100%, and the explanations of the parallel speedup apply as above.

A parallel algorithm is called *scalable* if a simultaneous increase of the problem size and the number of processors in a certain ratio causes the same parallel efficiency. Scalability is a prerequisite for parallelization.

A *vectorized* algorithm consists of a serial (non-vectorizable) part ser and a vectorizable part $(1 - ser)$. The vectorization rate vr is a measure of the acceleration of the algorithm in the vector unit. The *vector speedup* s_v is defined as follows:

$$\textbf{vector speedup} \qquad s_v = \frac{1}{ser + \frac{1-ser}{vr}} \qquad \longrightarrow \qquad \lim_{vr\to\infty} s_v = \frac{1}{ser} \qquad (4.8)$$

In figure 4.18, right, the course of the vector speedup is given for two different serial parts. The vector speedup increases with an increasing vectorization rate. However, the vector speedup is limited to a value of $1/ser$, independent of the vectorization rate. If $ser = 0.1$, the maximum vector speedup is 10. This is a very severe restriction and clearly indicates the advantage of parallelization over vectorization. In a parallel code, non-parallelizable parts can generally be neglected. Of course, there are parts in a parallel algorithm which cannot be parallelized very well; however, they do not lead to such a bottleneck.

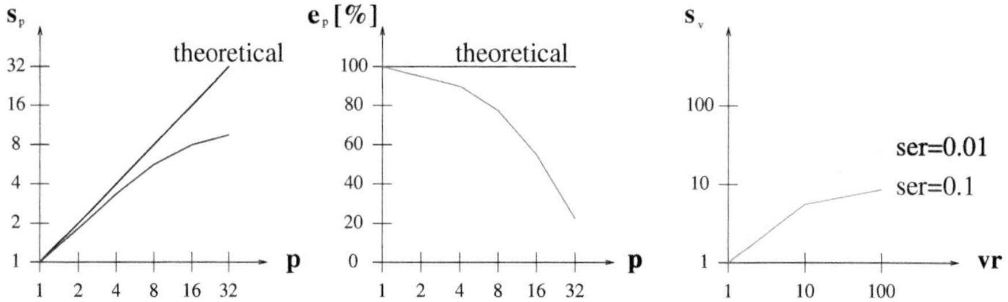

Fig. 4.18. Parallel speedup, left, parallel efficiency, middle, and vector speedup, right; after HINKELMANN (2000 [109])

Further reading is given in, for example, KREIENMEYER (1996 [158]) or HINKELMANN (2000 [109]).

4.2.4 MUFTE-UG

In this section, the modeling system MUFTE-UG, especially the processing part, is introduced as an example of a numerical simulator in hydro- and environmental engineering. MUFTE-UG is a combination of MUFTE and UG. MUFTE stands for **MU**ltiphase **F**low, **T**ransport and **E**nergy model, and this software toolbox mainly contains the physical model concepts and discretization methods for isothermal and non-isothermal multiphase / multicomponent flow and transport processes in porous and fractured-porous media (see secs. 2.3 - 2.5, 3.1; see HELMIG (1997 [98]), HELMIG et al. (1998 [100]), BREITING et al. (2000 [42])). UG is the abbreviation for **U**nstructured **G**rid, and this toolbox provides the data structures and fast solvers for the discretization of partial differential equations based on parallel, adaptive Multigrid Methods (see secs. 3.2 - 3.4; see UG ([250]), BASTIAN (1996 [17]), BASTIAN et al. (1997 [19]), LANG (2001 [162])). MUFTE uses UG as its fast solver. Figures 4.20 and 4.19 present an overview of the modeling system with its pre- and postprocessors as well as interfaces (see sec. 5.2.2). Further information about the pre- and postprocessing tools can be found in sections 4.1 and 4.3.

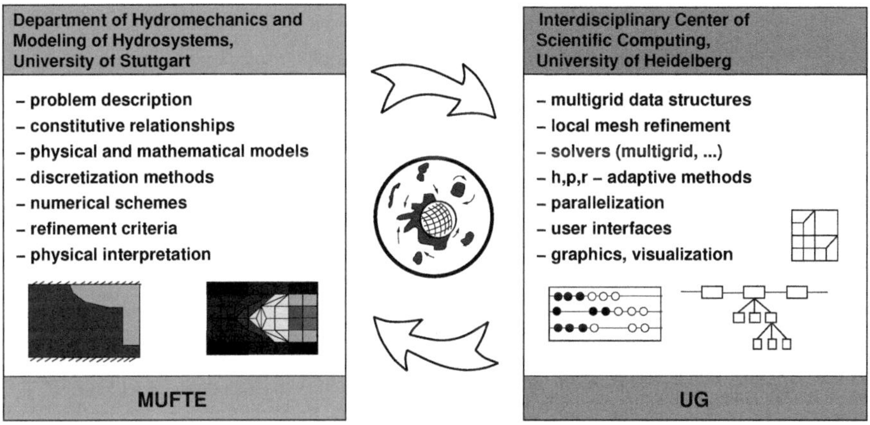

Fig. 4.19. Numerical simulator MUFTE-UG

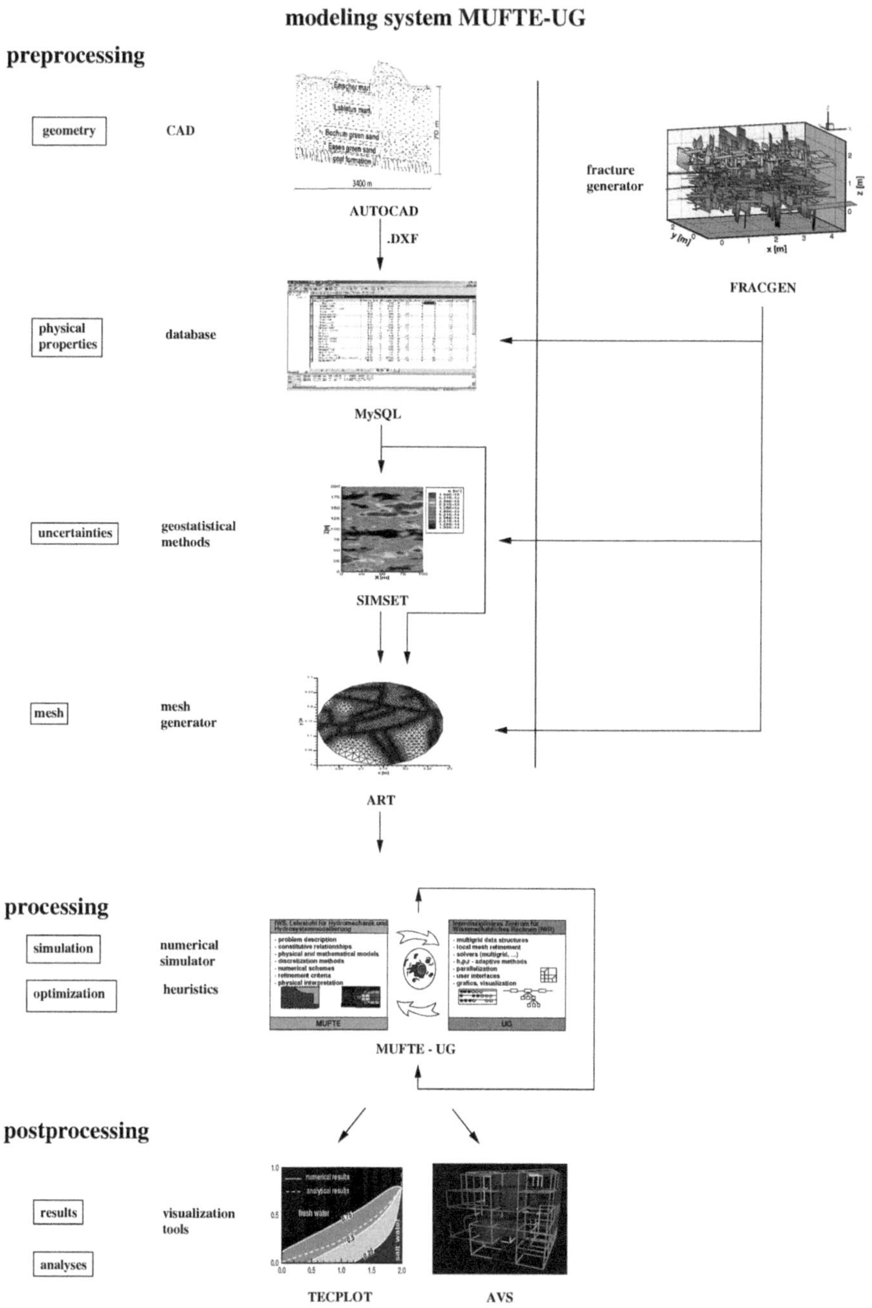

Fig. 4.20. Modeling system MUFTE-UG with its pre- and postprocessors

Software toolbox MUFTE

MUFTE provides several modules for the numerical simulation of isothermal and non-isothermal multiphase / multicomponent flow and transport processes in porous and fractured-porous media:

- single-phase systems
  - single-phase flow: liquids, e.g. water, NAPLs, ...; gases , e.g. air, methane,...;
    incompressible, compressible
  - single- and multicomponent transport: e.g. contaminants, salt, dissolved gases, ...
  - fractures
  - 2D, 3D
- two-phase systems
  - two-phase flow: liquid / liquid, e.g. water / NAPL, water / oil, ...; liquid / gas, e.g. water / air, water / methane, ...;
    incompressible, compressible
  - two- and multicomponent transport: e.g. salt, dissolved gases, contaminants, ...
  - isothermal and non-isothermal (including phase transitions)
  - fractures
  - 2D, 3D
- three-phase systems
  - three-phase flow: liquid / liquid / gas, e.g. water / NAPL / air, ...;
    incompressible, compressible
  - three- and multicomponent transport: e.g. contaminants, wator vapor, dissolved gases, ...
  - isothermal and non-isothermal (including phase transitions)
  - 2D

The model concepts include the extended *Darcy* law for the movement of multiple phases in isotropic or anisotropic, homogeneous or heterogeneous porous media. For the flow through fault zones and fractures, different flow laws and modeling techniques are available. Moreover, a large set of constitutive relationships for the relative permeability and the capillary pressure as well as state equations for densities, viscosities etc. are provided.

The two most common discretization techniques are the Fully Upwind Box Method, which is a Finite-Volume formulation with piecewise linear shape functions including fully upwinding of the upstream mobilities, and a Control-Volume Finite-Element Method, which is a mass-conservative formulation on a discrete patch including a first-order upwinding scheme. The time integration employs the Finite Difference method, and the temporal discretization is carried out fully implicitly. The development of *optimization methods* is underway (see KOBAYASHI et al. (2002 [147])) to extend MUFTE-UG to a *decision-support system.*

A software engineering tool is strongly recommended for the development of large, complex software systems where different groups with a large number of scientists are involved. The *Concurrent Version System (CVS)* is used for the development of MUFTE-UG.

Software toolbox UG

UG is a toolbox for the solution of partial differential equations. Its special advantages are the data structures for unstructured grids, i.e. the ability to deal with complex geometries and boundaries in two and three dimensions, functional parallelization, i.e. especially suitable for MIMD parallel computers, adaptive local-grid refinement in order to minimize the degrees of freedom for a desired accuracy and robust Multigrid solvers for linear and non-linear problems. On the one hand, a large collection of state-of-the-art numerical methods which can be applied to users' purposes are available, i.e. by adding a few lines of code parallel computations can be carried out. On the other hand, UG offers the opportunity for the user's own developments. Moreover, UG has several pre- and postprocessing tools which are also suitable for parallel computers, e.g. online graphics. UG is currently used by a number of different groups in engineering research and application.

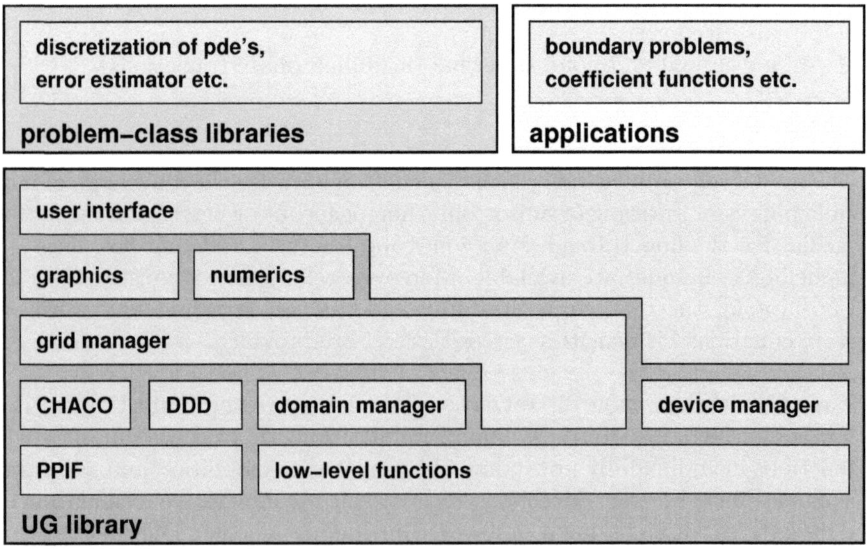

Fig. 4.21. Structure of the UG toolbox, after BASTIAN (1996 [17])

As shown in figure 4.21, UG is divided into three major parts: the *UG library*, *problem-class libraries* and *applications*. The UG library is independent of the partial differential equation, and it contains geometric and algebraic data structures, refinement and coarsening techniques, numerical algorithms, visualization techniques as well as a user interface. The problem class libraries provide discretization methods as well as error estimators and indicators. Finally, the applications describe the system set-up, i.e. the geometry, physical parameters and their functional relationships as well as the initial and boundary conditions. A simulation run is steered by a script file. More detailed information can be found in the literature mentioned above and in several manuals which can be downloaded from the the UG homepage ([250]).

4.3 Postprocessing

After the numerical simulation or processing, the computed data are analyzed, interpreted and presented in the *postprocessing.* Generally, scalars and vectors are visualized in the one, two or three space dimensions and time in form of *time series, isolines, isoareas, vector fields, path lines, tables* or *diagrams.* The geometric model of the *visualization tool* must not coincide with the computational grid, i.e. an unstructured grid computation can be visualized with a tool which works with a square grid. Such circumstances must be taken into account when the results are analyzed.

In 1D, the course of the computed results is plotted along the coordinate axis, for time-dependent problems at a certain time, at different time steps (see figs. 3.5, 3.7) or in an animation or video. The temporal development of a variable at a certain location is shown in a time series (see figs. 5.36, 5.48). In 2D, the results of scalars, such as the water level, concentrations or saturations, are presented with isoline or isoarea plots (see figs. 5.5, 5.18), the results of vectors, such as the flow vector, in vector plots (see figs. 5.40, right, 5.50). For time-dependent problems, time series, *animations* and *videos* are chosen to analyze the results (see fig. 5.19). For particle models, the path lines of certain particles give insight into the numerical results. Sometimes, it is helpful to combine maps of the computational area (see figs. 4.9, 5.47). For forecasting purposes, it is often advantageous to plot the differences between results, e.g. differences between an original situation and the one after an interaction with the system.

The visualization of time-dependent 3D-simulations can become very complex. A first impression can be obtained by looking at the results in certain cuts or cross sections, and all the techniques described in the 2D-context are applicable. For a detailed view of the 3D-results, *advanced visualization tools* are required, e.g. AVS ([7]), the IBM Data Explorer ([127]) or COVISE ([65]). These tools need special interfaces and they enable the creation of user-defined visualization by setting up a visualization network where the user defines the data flow and the methods and techniques which are applied to the data. One has to define, for example, the point of view, the light intensity, the shading, transparent or hidden lines or surfaces etc. of the results and the domain.

In figure 4.22, an infiltration process of a dense non-aqueous phase (DNAPL) into a fully water-saturated porous medium with two low-permeability lenses is shown in the form of DNAPL isoareas of the DNAPL saturation $S_n = 0.1$ after $750s$ and $S_n = 0.38$ after $1000s$ (see BASTIAN, HELMIG (1999 [21]), HELMIG et al. (1999 [102])). The visualization was carried out with AVS, and the simulation with MUFTE-UG (see sec. 4.2.4) on a parallel computer. Due to the higher density, the DNAPL displaces the water. However, the DNAPL does not penetrate into the lenses which are a capillary barrier. Of the porous medium, only the lenses, i.e. the domain with a low permeability, are visualized. The edges of the computational grid are transparent in the parts of the domain and the lenses. However, they are hidden in the areas of the DNAPL saturation isoarea. Due to the different saturation isoareas ($S_n = 0.1$ left; $S_n = 0.38$ right) associated with different grey shades, the lenses are shown in different grey shades as well to distinguish them from the DNAPL. Moreover, the point of view between the left and right figure has changed.

Fig. 4.22. Infiltration of a DNAPL into a system with low permeability lenses after 750s, left, and 1000s, right, visualized with AVS ([7]), after BASTIAN, HELMIG (1999 [21])

The visualization can be carried out *online* or *offline*. The online graphics is carried out during a simulation after a defined number of time steps, and it tends to be simpler. It is useful to observe the simulation and interrupt it, for example, in case of oscillatory results. In the offline variant, the results are analyzed after a numerical simulation; this is recommended if advanced visualization tools are used.

There are two ways of visualizing parallel computations. On the one hand, the results can be presented on the distributed meshes. The grids in the subdomains should be shrunk (see fig. 5.35) or the interface lines should be omitted (see fig. 5.53) if the different subdomains are to be recognized. On the other hand, the results can be collected on one processor and visualized in the usual way. However, one has to take into account that the gathering of the data cannot be parallelized well.

The amounts of data which are to be visualized can be huge, especially when complex, 3D, large-scale and time-dependent processes are investigated. To get a better insight into the results, *Virtual Reality (VR)* based methods and techniques have been developed in recent years. With Virtual Reality, the user can immerge into the visual representation of the results and can interact with the simulation model. For such visualization purposes, special visualization rooms and special technical equipment are required, e.g. an all-round projection in a *CAVE (Cave Automatic Virtual Environment)*. Moreover, High-Performance Computers are urgently needed to visualize the data in a reasonable time.

4.4 WWW-based Collaborative Engineering

4.4.1 Introduction

The engineering work of today and tomorrow is characterized more and more by aspects of *interdisciplinarity* and *globalization*. For example, the growing complexity of single research and application fields shows an increased need for *collaboration* with national and international partners in order to solve complex global problems. Such collaboration has become much more practicable as a result of recent advances in *information* and *communication technologies* using the *World Wide Web (WWW)*. The teamwork can be carried out at the same location or at different locations; this is called *central* or *distributed collaboration*. It can be carried out at the same time or at different times; this is called *synchronous* or *asynchronous collaboration*. Working in such a way requires a well-organized structuring of the subtasks, a clear definition of the contents, milestones and the time schedule, especially for distributed and asynchronous collaboration. The advantages are, on the one hand, that collaboration is made possible or simplified - for example, less travelling is involved - and / or, on the other hand, that individual flexibility is improved. Further information is given in MOLKENTHIN (2000 [181]).

As an example, the collaboration of different expert groups for the modeling and analysis of a complex hydrosystem is explained in figure 4.23. Experiments for the simulation of the free surface in a porous dike system were carried out at ETH Zürich, Switzerland. A mesh was generated with a special generator of UG (see sec. 4.2.4, UG ([250])) at the University of Heidelberg, Germany. The numerical simulation was undertaken with the two-phase flow module of MUFTE-UG (see sec. 4.2.4) at the University of Stuttgart, Germany. Finally, the numerical results were evaluated in a further education course of the *IAHR-EGW (International Association of Hydraulic Engineering and Research - Engineering Graduate School of Environment Water*, [126]) at the University of Calabria, Italy, and compared with the experiments. Further information is given in PAUL et al. (2000 [205]) and section 5.2.2.

The methods and techniques of *Collaborative Engineering* can also be considered in the context of *New Media* in education and further education in the forms of *Teleteaching, Telelearning* and *Open Distant Learning*. With the use of New Media, several improvements can be achieved, e.g. a better learning of network thinking and acting, a better understanding of complex coupled processes and more self-defined and flexible learning.

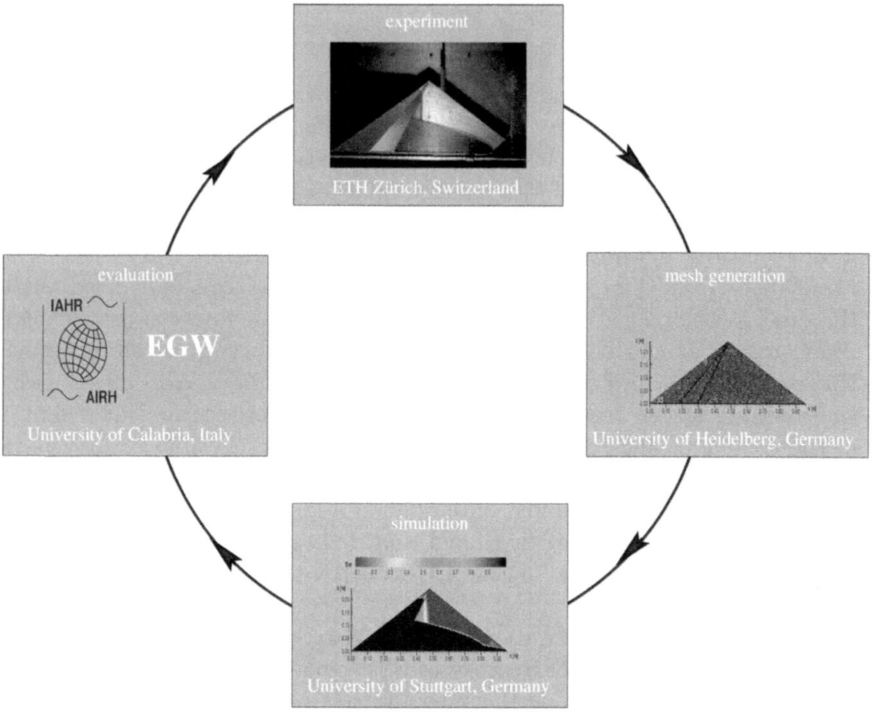

Fig. 4.23. Collaborative Engineering for modeling and analysis of a complex hydrosystem

4.4.2 Software tools and hardware requirements

WWW-based Collaborative Engineering is generally understood as a cooperation with more tools than just email and ftp (file transfer protocol). Special tools for coordination, documentation and communication are required. Using one central server and a groupware tool, e.g. the *Basic Support for Cooperative Work (BSCW)* ([49]), is recommended. BSCW offers a shared workplace for document, group and event management. Moreover, all users should work with the same tools in order to simplify the information exchange.

If the users collaborate to solve an engineering task using a modeling system, two different possibilities exist. On the one hand, the modeling system is installed on the central server, and a multi-user license may be required. The whole modeling procedure - preprocessing, processing and postprocessing - is then carried out on the server; the user's PC is just a terminal and the user or client is in contact with the server via a platform-independent *applet.* Such a procedure can cause several problems, e.g. working over the Internet may be very time-consuming, and the turn-around times can be high if many users are working at the same time. If the modeling system is available on each user's PC, this has the advantages of direct access, small turn-around times and the fact that no applets are required. However, the system must be installed, maintained and possibly paid for by each user.

Over the Internet, direct communications are possible via *chats*, i.e. written text, *audio connections* which are similar to a telephone call, and *videoconferencing systems*, which generally includes audio. For a videoconference between two PCs, only a webcam, a headset, a microphone, a sound card, as well as Internet access or a telephone (with a modem or ISDN) are needed; the costs are low and in the range of 10% of the PC. For a videoconference between two or several groups, special videoconferencing systems and rooms are required. The video conference system consists of a *video projector (beamer)*, a *smart board*, and an *audio system with loudspeakers*. Additionally, such a room can be equipped with a number of PCs and possibly a steering and control system for the PCs, especially when it is used for educational purposes. The media technology and the other equipment is expensive, up to a few hundred thousand Euros. Collaboration via videoconferencing is supported by tools like Microsoft *NetMeeting*, which is a freeware product used in the WINDOWS world and quasi-standard. NetMeeting offers the opportunity of making *chats*, using the *white* or *smart board* and *sharing applications*. NetMeeting enables that people at different locations in the world can communicate and work on the same document and application at the same time. Moreover, an application, e.g. a presentation or camera movement, can be steered from another location. If videoconferences are transmitted over large distances, ISDN telephone connections should be preferred to Internet connections. They enable a constant transmission quality of audio and video and are independent of the Internet-communication trafic.

In figure 4.24, the *Multimedia Lab* at the Department of Hydromechanics and Modeling of Hydrosystems, Institute of Hydraulic Engineering, University of Stuttgart, Germany, is shown during the IAHR-EGW short course *Multiphase Flow, Transport and Bioremediation in the Subsurface* (see HELMIG, CUNNINGHAM (2002 [101])). The Multimedia Lab is equipped with a videoconferencing system and 13 PCs including a teacher-student steering unit. Education and continuing education in modeling of hydrosystems is carried out here in the form of a combination of lectures, computer exercises and videoconferences which enable an integration of expert lectures from other locations and post-course meetings over the Internet.

Fig. 4.24. Continuing education in the Multimdia Lab

4.4.3 HydroWeb

The *HydroWeb* serves as an educational platform for *WWW-based Collaborative Engineering in Hydroscience* (see MOLKENTHIN (2000 [181])). It is given as a course within the IAHR-EGW and is also involved in other international education and continuing education activities. The overall aim of the HydroWeb consists of teaching methods and techniques for collaborating over the Internet to solve hydro-engineering tasks in distributed teams. The HydroWeb aims at students as well as at professionals who want to learn ways of cooperation which are required for working in international and interdisciplinary projects.

In the following, the course concept is briefly explained. The course is given at different locations at the same time (see fig. 4.25); at each location, a group with several participants is involved. The course covers lectures on communication and documentation in the WWW, project management and teamwork as well as modeling of hydrosystems for engineering or management purposes. A few introductory lectures are given by a local advisor or by videoconference to ensure the same educational level. The participants only require a normal PC with Internet access and standard tools belonging to the WINDOWS or UNIX / LINUX operating system. One central WWW server is installed, providing file and document services. A modeling system is available for the solution of a defined hydro-engineering task; up to now, a river-engineering modeling system, which uses an Internet-based license service, has been applied to design a flood-protection system. The course participants are divided into distributed groups, i.e. participants from different locations are in each group which has to organize itself, i.e. subdivision of the task, coordination, communication and documentation. The supervision is carried out partially by local advisors and partially via the Internet.

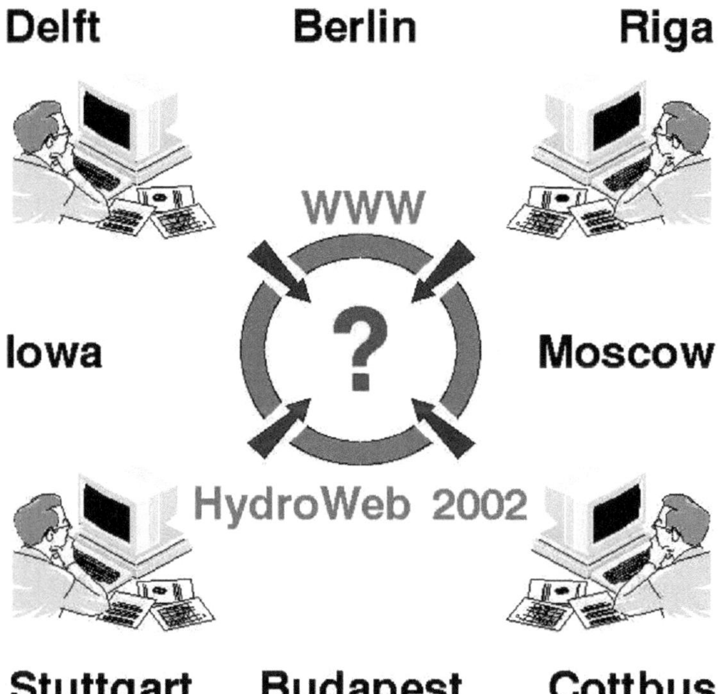

Fig. 4.25. HydroWeb: WWW-based Collaborative Engineering in Hydroscience, after MOLKENTHIN (2000 [181])

Overall, the course concept is considered very successful. In the future, a few extensions are conceivable. The HydroWeb can serve as a platform for advanced courses in the modeling of hydrosystems. A few common lectures and events (same location, same time) could improve the social interaction between the participants and lecturers. Finally, the HydroWeb can serve as a professional platform for WWW-based Collaborative Engineering.

5

Applications

In this chapter, several applications from hydro- and environmental engineering are discussed. Groundwater flow and transport processes, two-phase flow processes, two-phase / multicomponent flow and transport processes as well as free-surface flow and transport processes are dealt with. Special emphasis is laid on the interaction of the efficient numerical methods (see chap. 3) with the efficient information-processing techniques (see chap. 4) for the numerical simulations of different complex processes occurring in hydro- and environmental systems.

5.1 Groundwater flow and transport processes

This section contains numerical simulations of groundwater flow and transport processes in fractured-porous media (see sec. 2.1.2). The development of a so-called *equidimensional modeling* approach for fracture-matrix systems (see sec. 2.1.2) is in the foreground. In the context of efficient numerical methods, *adaptive* (see sec. 3.3) and *Multigrid Methods* are treated. An extension of a mesh generator (see sec. 4.1.7) for the equidimensional modeling approach is explained with respect to efficient information processing techniques. The modeling system MUFTE-UG (see sec. 4.2.4) served as the basis for the developments and applications. The content of this section is also published in NEUNHÄUSERER et al. (2001 [188]), NEUNHÄUSERER et al. (2002 [189]), GEBAUER et al. (2002 [89]), OCHS et al. (2002 [193]), or NEUNHÄUSERER (2003 [186]).

5.1.1 Motivation of the equidimensional modeling approach for fracture-matrix systems

In recent years, the numerical simulation of flow and transport processes in fractured-porous media has become important, for example, when investigating the long-term safety of subsurface waste-disposal sites (see fig. 5.1). Rock is considered to be a geological barrier and to prevent the transport of contaminants. However, due to tectonic stresses, these formations can be pervaded by fractures acting as major flow paths (see fig. 1.1). Transport processes in fractured-porous systems are generally characterized by high advection in the fractures and dominant diffusion and dispersion processes in the matrix (see sec. 2.3.2). When rock, for example granite, is considered, mechanical dispersion in the matrix can be neglected. The length scales of fractures and the matrix as well as the hydraulic properties differ vastly. Furthermore, the geometry can be very complex if several intersecting fractures or fracture plains are dealt with. All these circumstances make great demands on the numerical simulation.

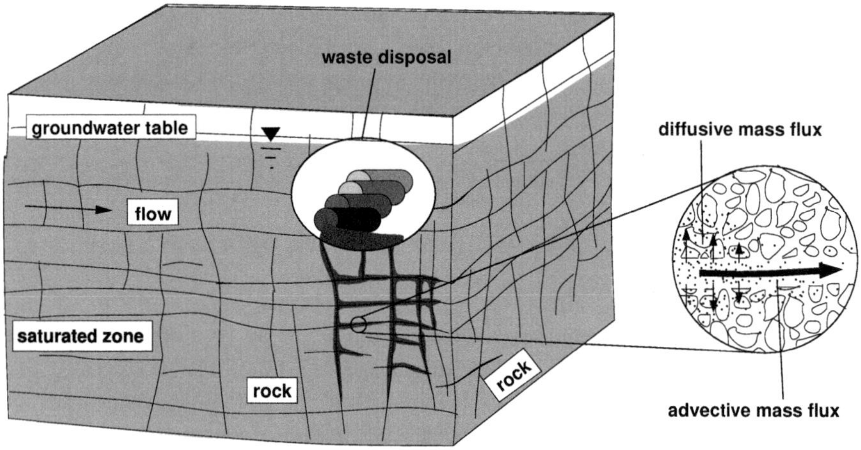

Fig. 5.1. Leaking of subsurface waste into a fractured rock, after NEUNHÄUSERER (2003 [186])

In the *combined approach* of the discrete model concept (see sec. 2.1.2), fractures or fault zones are generally modeled with elements which have a lower dimension than the surrounding matrix, i.e. fractures are discretized as one-dimensional elements in a two-dimensional porous medium and as one- or two-dimensional elements in a three-dimensional porous medium (see figs. 5.2, left, 3.46). However, this low-dimensional modeling approach has some disadvantages which are expected to be avoided by a so-called *equidimensional* modeling approach. If a fracture is treated as an element of a lower dimension, the flow velocity at the fracture-matrix interface is *discontinuous*, the flow within the fracture is forced to be parallel to the fracture axis and the physical processes can be smoothed, if parts of the local stiffness matrix are added up in the FEM (see sec. 3.1.4) or fluxes are added up in the FVM (see sec. 3.1.5). When the fractures are discretized equidimensionally, a continuous flow velocity at the fracture-matrix interface can be achieved by *Mixed* or *Mixed-Hybrid Methods* (see sec. 3.1.6). This is highly desirable for transport simulations as the flow velocity is its most important input parameter. Moreover, two- or three-dimensional flows can be simulated in the fractures. This is important for intersecting fractures and for better resolving sharp gradients perpendicular to the fracture axis, for example sharp gradients of the hydraulic head and of the concentration. The equidimensional modeling approach is also advantageous for *Particle Methods* (see sec. 3.1.6) because it enables any inflow and outflow of particles over the fracture-matrix interface as well as the determination of unique stream lines.

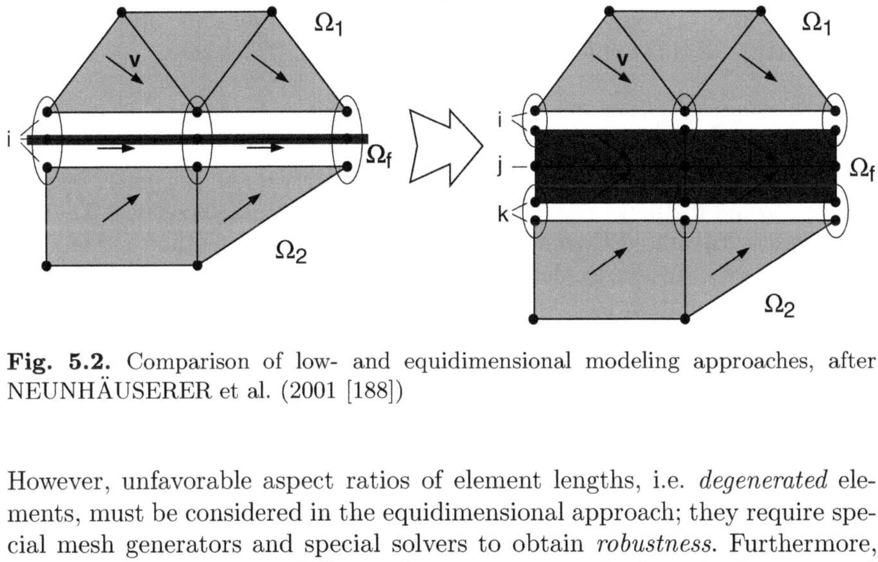

Fig. 5.2. Comparison of low- and equidimensional modeling approaches, after NEUNHÄUSERER et al. (2001 [188])

However, unfavorable aspect ratios of element lengths, i.e. *degenerated* elements, must be considered in the equidimensional approach; they require special mesh generators and special solvers to obtain *robustness*. Furthermore, it is advantageous to use different discretization methods in the fracture and the matrix depending on the dominant physical proccesses. Finally, it is desirable to apply and develop *adaptive* methods (see sec. 3.3), which adapt to the physical processes, and *parallel* methods (see sec. 3.2) to obtain an overall efficient numerical algorithm for modeling flow and transport processes in fracture-matrix systems. These points are discussed further in this section.

5.1.2 The numerical algorithm

The governing equations for groundwater flow are explained in section 2.3.1. Here, stationary flow problems are considered. The Standard-Galerkin FEM (see sec. 3.1.4) is chosen for the discretization. A two-level Multigrid Method based on a *hierarchical decomposition* into a matrix and a fracture space is developed. It leads to a decoupled treatment of the matrix and the fracture problem which are then combined in an iteration scheme. The matrix problem covers the whole computational domain, i.e. the matrix and fracture space, whereas the fracture problem is only defined on the fracture space. For the matrix problem, a $V-$Multigrid cycle (see sec. 3.4.4) with different grid levels is chosen with a direct solver (see sec. 3.4.5) for the coarse grid and a *Gauss-Seidel* smoother (see sec. 3.4.2) with 1 presmoothing step (see sec. 3.4.4). In order to achieve robustness for vanishing fracture width, an additional, geometrically adapted coarse-grid correction at the fracture-matrix interface has been introduced (see GEBAUER et al. (2002 [89])). As the fracture problem is small, it is solved directly. The solver is robust with respect to vanishing fracture width, strongly varying hydraulic conductivities and vanishing mesh size. Further details are found in GEBAUER et al. (2002 [89]).

The governing equations for transport processes in groundwater are given in section 2.3.2. Here, ideal tracers are treated. For the discretization in time, the fully implicit *Euler* Method (see sec. 3.1.2) is used and, in space, two different upwind methods, the Fully Upwind Box Method (see sec. 3.1.5) and a Streamline-Orientated Upwind Box Method. The latter method reduces unwanted cross diffusion by determining upwind coefficients according to the streamline orientation (see NEUNHÄUSERER et al. (2002 [189]), sec. 3.1.4). As a solver, the BiCGSTAB Method (see sec. 3.4.2) with Multigrid preconditioning is used (see sec. 3.4.3). A $V-$Multigrid cycle (see sec. 3.4.4) with different grid levels is chosen using a direct coarse-grid solver (see sec. 3.4.5), an ILU smoother and 2 pre- and postsmoothing steps (see sec. 3.4.4). Further details are given in NEUNHÄUSERER et al. (2002 [189]).

H-adaptive methods, which avoid hanging nodes, are applied to the transport simulations in section 5.1.6 and steered by heuristic error indicators (see sec. 3.3). Adaptive and Multigrid Methods are treated together with Local Multigrid Methods (see sec. 3.4.4).

5.1.3 Extension of the mesh generator ART

The mesh generator ART (see sec. 4.1.7) is able to deal with fracture-matrix systems according to the low-dimensional approach (see fig. 5.3, left). Its 2D-version has been extended to generate meshes for the equidimensional approach (see fig. 5.3, middle). To do this, the fractures are cut off, and the fracture nodes are duplicated. The original and the copied fracture nodes are shifted half a fracture width apart perpendicular to the fracture axis (see 5.3, right). Thus, the fractures are discretized with quadrilaterals, while the matrix is generated with triangles. As shown in figure 5.6, right, the fracture width can be discretized with several elements, during the extension of the mesh generator or during uniform or adaptive refinement within MUFTE-UG (see sec. 4.2.4); very small elements occur in the intersection area of the fractures. It must be mentioned that the fractures in figure 5.3 are relatively wide with $b = 1cm$ compared to element lengths of the matrix. In many cases, this aspect ratio is considerably less favorable, and the fracture elements degenerate. Therefore, the number of elements for discretizing the fracture width should be rather limited. Further measures for improving the mesh quality in the intersection areas of fractures and for dealig with fracture ends within the domain are available. These and further details are explained in NEUNHÄUSERER et al. (2001 [188]) and NEUNHÄUSERER (2003 [186]).

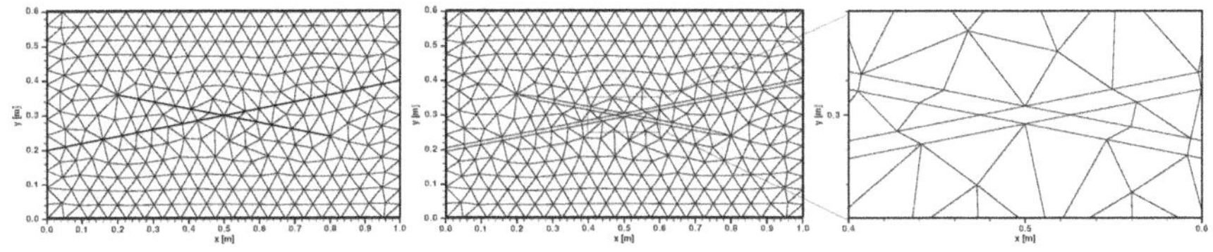

Fig. 5.3. Mesh with fractures; 1D-fractures, left; 2D-fractures, middle; zoom at the cross-over point of the 2D-fractures, right; after NEUNHÄUSERER et al. (2001 [188])

5.1.4 Comparison of the low- and equidimensional modeling approaches

The low- and the equidimensional modeling approaches are compared using the Streamline-Orientated Upwind Box Method and two similar systems, one with a single fracture (see fig. 5.4, left) and the other with a fracture network (see fig. 5.6, left). The systems have the same physical parameters as well as initial and boundary conditions. They only differ in the fractures. The systems have a length of $1m$ and a height of $0.6m$. The hydraulic conductivities are $K_f = 6.3 \cdot 10^{-4} m/s$ for the matrix and $K_f = 6.3 \cdot 10^{-1} m/s$ for the fracture(s). The porosities are $\phi = 0.2$ for the matrix and $\phi = 0.3$ for the fracture. The molecular diffusion is set to $D_{mol} = 1.0 \cdot 10^{-9} m^2/s$, and the dispersion lengths are $\alpha_L = \alpha_T = 0.01m$ in the matrix and $\alpha_L = 0.1m, \alpha_T = 0.001m$ in the fracture. The coarse mesh (Multigrid level 0) in figure 5.4 has about 300 nodes for both approaches, while the coarse mesh in figure 5.6 has 377 nodes for the low-dimensional modeling approach and 511 nodes for the equidimensional one. The grids are refined up to Multigrid level 4. The time-step size varies with the grid level. For level 3, a time step of $\Delta t = 30s$ is chosen for the mesh with the single fracture (see fig. 5.4) and a time step of $\Delta t = 1s$ for the mesh with the fracture network (see fig. 5.6).

The system is closed along the upper and lower boundaries and open along the left and right boundaries. For the flow simulation, the pressure is prescribed to $p = 100000 Pa$ on the left boundary and $p = 99900 Pa$ on the right boundary. For the transport simulation, the concentration is initially zero in the whole domain. Along the left boundary, the concentration is fixed at $c = 1 kg/m^3$ for the first $9960s$ and at $c = 0$ afterwards and, along the right boundary, the concentration is set to zero.

In figure 5.4, left, the coarse mesh is shown with an almost vertical fracture with a width of $b = 0.005m$. The pressure distribution is nearly linear from the left to the right boundary, some deviations occur around the fracture (see fig. 5.4, right). No visible differences between both modeling approaches are observed. If the concentration distributions are considered after $24000s$, see figure 5.5, the center of gravity of the front is propagated about $60cm$ into the system. The low-dimensional modeling approach shows smoother concentrations around the fracture.

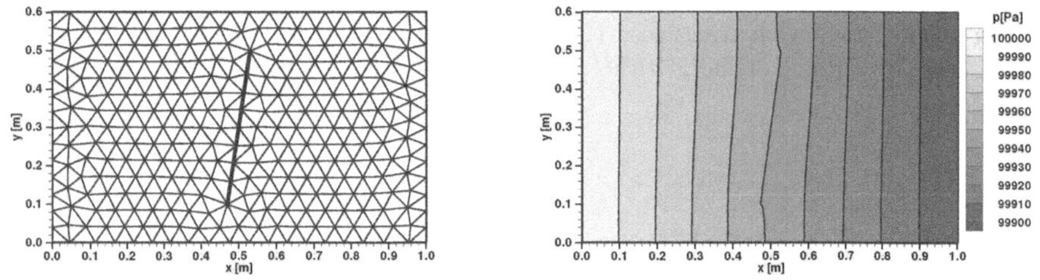

Fig. 5.4. Coarse mesh of the system with one almost vertical fracture, left; pressure distribution, right; after NEUNHÄUSERER et al. (2001 [188])

Figure 5.6, left, shows the system with the fracture network. The fracture width is $b = 0.005m$. The fracture network causes a larger deviation from a linear pressure distribution than the single fracture system (see 5.6, left). If the velocity field in the upper fracture intersection is considered (see 5.6, right), a two-dimensional flow field is observed. There are not only fluxes from the lower left to the lower right fracture, but also into both upper fractures. Such flow conditions cannot be simulated with one-dimensional fractures. Consequently, the tracer distributions differ considerably for both approaches in the intersection area (see fig. 5.7). The equidimensional modeling approach shows a larger smearing of the concentration gradients.

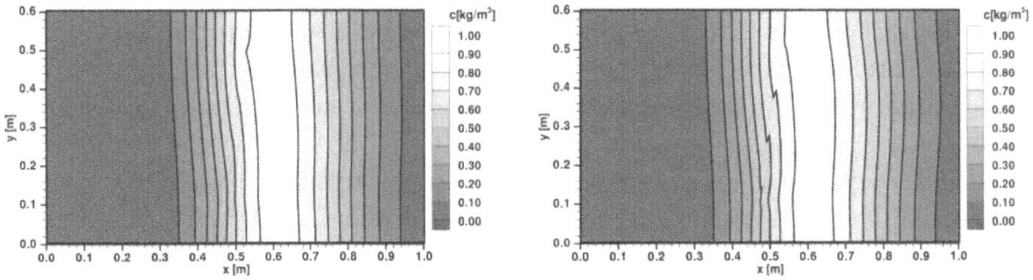

Fig. 5.5. Concentration distributions after 24000s; low-dimensional, left; equidimensional, right; after NEUNHÄUSERER et al. (2001 [188])

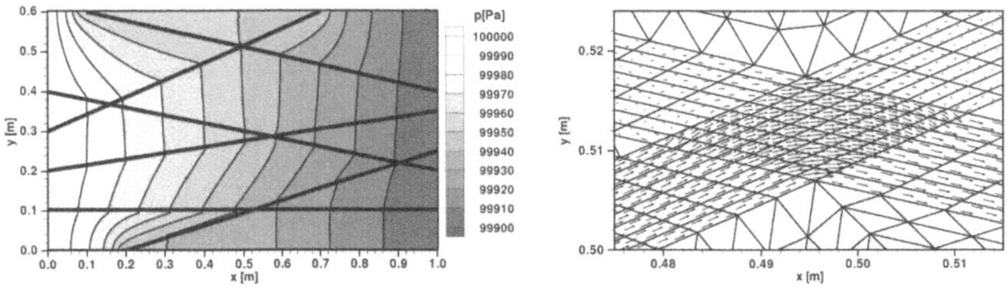

Fig. 5.6. Pressure distribution and fracture network, left; velocity distribution in the zoom into the upper fracture intersection, right; after NEUNHÄUSERER et al. (2001 [188])

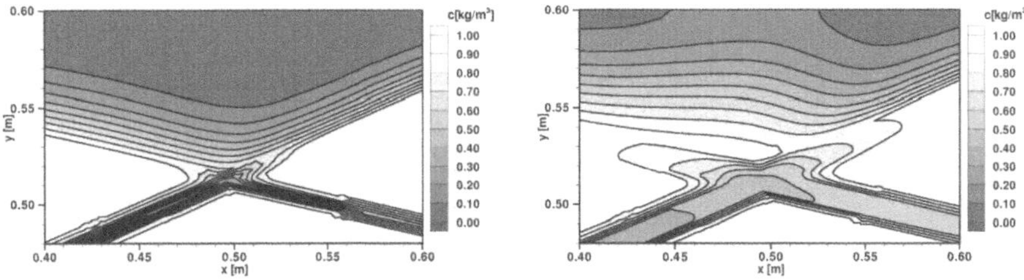

Fig. 5.7. Tracer distribution after 10000s; low-dimensional, left; equidimensional, right; after NEUNHÄUSERER et al. (2001 [188])

Similar results are obtained by OCHS et al. (2002 [193]). Finally, it is mentioned that the low- and equidimensional approaches lead to different results in locally limited areas around the fractures. However, if integral simulation results, such as breakthrough curves, are considered, differences are hardly visible with the methods developed up to now.

5.1.5 Comparison of upwind methods

The Fully Upwind Box Method and the Streamline-Orientated Upwind Box Method are compared using the system with one fracture (see fig. 5.4, left), the equidimensional modeling approach and different multigrid levels. In figure 5.8, the results for both methods, which are obtained on the grid of figure 5.4, left, refined twice, are shown. The fully upwinding leads to a stronger smearing of the solution. In figure 5.9, tracer profiles and breakthrough curves are presented. The steepness of the profile increases with increasing multigrid levels. Moreover, the streamline orientation on level 2 leads to a steeper front than the fully upwinding on level 4. Finally, it is assumed that the streamline-orientated upwinding produces better results than the fully upwinding.

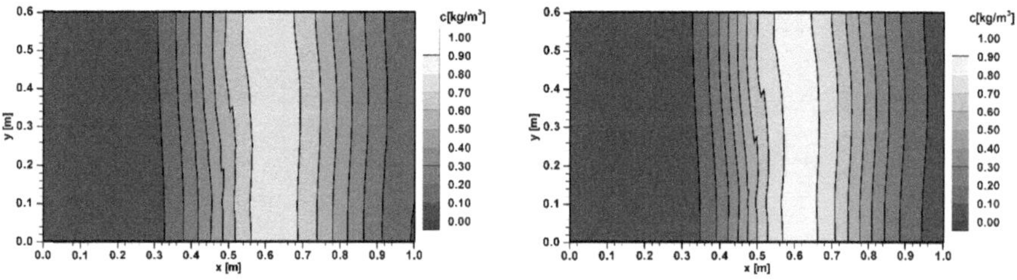

Fig. 5.8. Tracer distribution after 24000s; fully upwinding (fu), left; streamline-orientated upwinding (so), right; after NEUNHÄUSERER et al. (2001 [188])

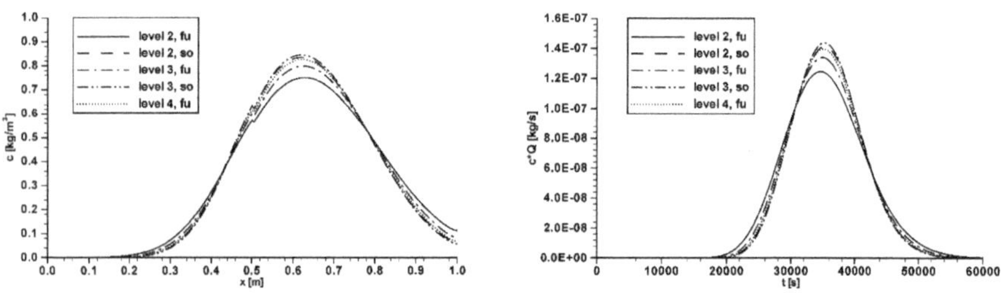

Fig. 5.9. Tracer profile at $y = 0.3m$, left; breakthrough curves, right; after NEUNHÄUSERER et al. (2001 [188])

5.1.6 Adaptive methods

Test-case problem: System with a single horizontal fracture

The investigations are carried out with a system similar to that in figure 5.4 and with the parameters similar to those in section 5.1.4. The differences are mentioned in the following. The grid in figure 5.10, left, has 349 nodes and one horizontal fracture of $b = 0.005m$ width in the middle of the system. The fracture porosity is set to $\phi = 1.0$ to obtain a better comparison of the numerical results with an analytical solution of TANG et al. (1981 [244]). Dispersion is neglected. Along the left boundary, a stationary concentration boundary condition is imposed with $c = 1.0mg/l$. The equidimensional modeling approach with the Fully Upwind Box Method and a difference indicator (see sec. 3.3.2) are chosen.

In figure 5.10, right, the adaptively refined grid up to level 3 is given. The mesh is refined around the concentration front in the fracture as well as along the concentration front in the matrix at the inflow boundary (see fig. 5.10, left).

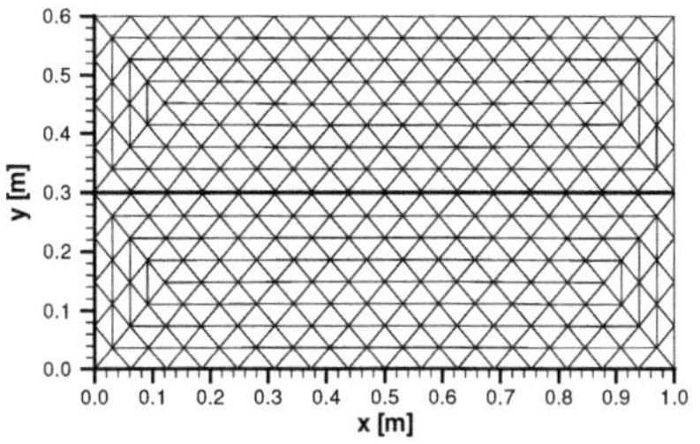

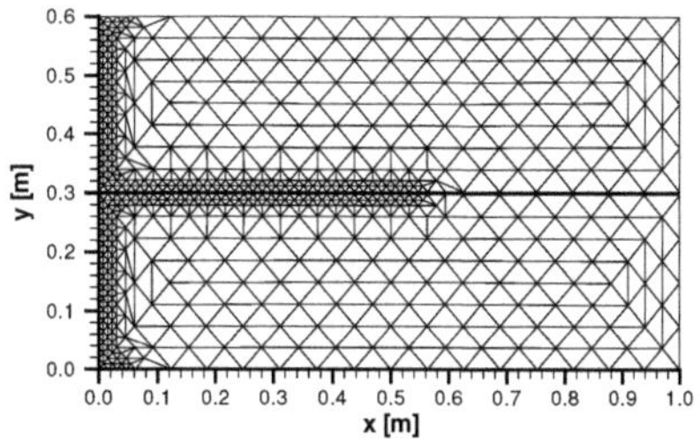

Fig. 5.10. Mesh (level 2) left, and adaptively refined mesh (level 4) right, of the system with one horizontal fracture; after OCHS et al. (2002 [193])

In figure 5.11, left, the concentration distribution after $800s$ is shown. The front propagates about $50cm$ into the fracture, mainly driven by advection, and about $1cm$ into the matrix, mainly driven by diffusion. In figure 5.11, right, the concentration front along the fracture axis after $800s$ is presented for different variants. For level 4, the adaptively and uniformly refined solutions coincide well. However, they differ considerably from the analytical solution because the Fully Upwind Box Method in space and the fully implicit *Euler* Method in time lead to high numerical dispersion. The front becomes steeper and propagates faster when the refinement level increases up to 6. However, the adaptively refined solution up to level 6 does not agree very well with the analytical solution of TANG et al. (1981 [244]). One the one hand, further refinement in space and time further improve the results. On the other hand, assumptions in the analytical solution do not fully agree with the model assumptions (see OCHS et al. (2002 [193])).

In figure 5.12, different heuristic indicators are compared. If similar tolerances are given, the numerical results as well as the adaptive meshes are very similar. Differences only occur at the front in the fracture. The fracture width

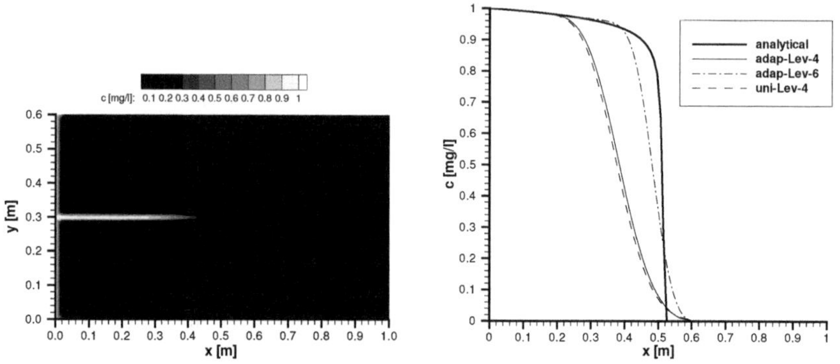

Fig. 5.11. Concentration distribution in the domain, left, and concentration distribution in the fracture for different variants, right, after 800s; after OCHS et al. (2002 [193])

is initially discretized with 4 elements (see fig. 5.12). The difference indicator refines the whole fracture width (see fig. 5.12, left), while the gradient indicator does this only along the fracture-matrix interface (see fig. 5.12, right). If many adaptive refinement levels are used, the error indicators should be formulated anisotropically (see sec. 3.3.3), for example by restricting a refinement perpendicular to the fracture axis.

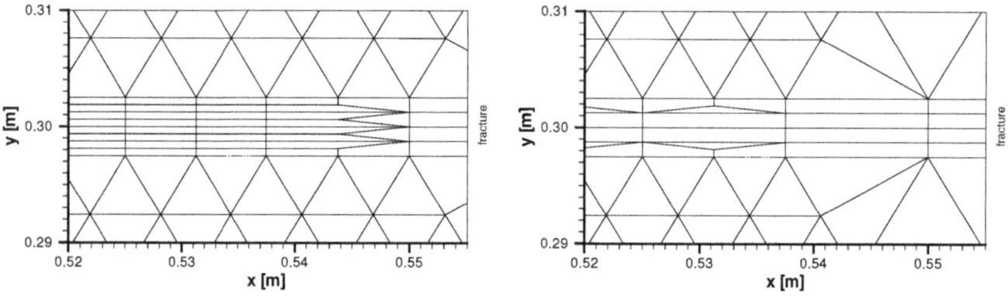

Fig. 5.12. Comparison of a difference indicator, left, with a gradient indicator, right; after OCHS et al. (2002 [193])

In figure 5.13, the manner of the Local Multigrid Method (see sec. 3.4.4) is visualized. The mesh is uniformly refined once and adaptively refined twice. On grid levels where adaptive refinement is carried out, only the refined elements and some transitional elements are stored; not, however, further elements to cover the whole domain.

Table 5.1 gives run times and averaged numbers of unknowns per time step for the simulation up to 800s. The adaptively refined mesh up to level 4 only

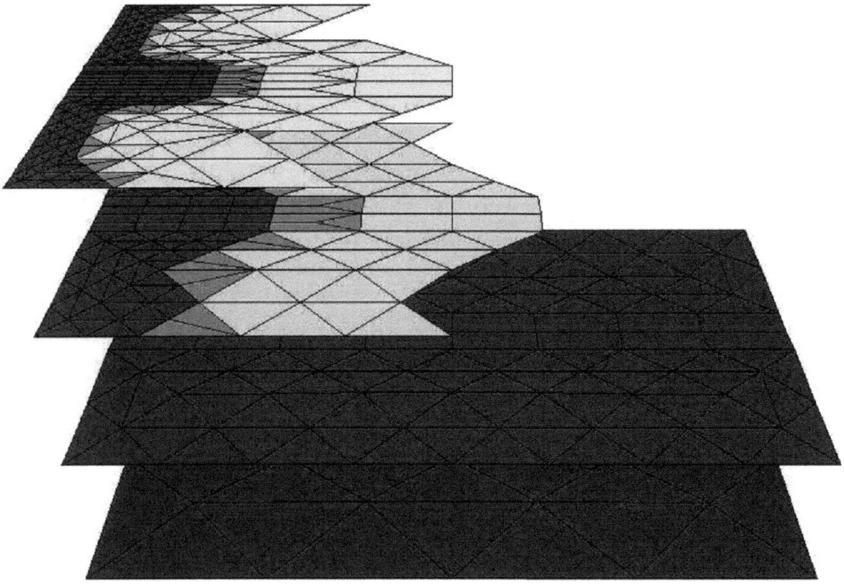

Fig. 5.13. Local Multigrid Method

requires about 22% of the CPU and 17% of the storage compared to the uniformly refined solution. On level 6, the advantages of the adaptive solution will be even larger. Overall, the superiorities of adaptive methods are clearly demonstrated.

refinement	CPU time [s]	related to uniform [%]	number of unknows [-]	related to uniform [%]
Lev-2	12	-	349	-
adap-Lev-4	790	21.7	906	17.3
uni-Lev-4	3630	-	5223	-
adap-Lev-6	2440		7266	

Table 5.1. Comparison of run times and unknowns

Real-case problem: System with several fractures

A real-case problem taken from SILBERHORN-HEMMINGER (2002 [237]) is chosen to investigate the adaptive methods on a complex geometry. The system represents the cross section between the boreholes B31 and B21 of an sandstone outcrop (see fig. 5.14). An extensive survey has been undertaken and is underway to obtain field data, especially on the complex fracture systems. The fractures have a width of $b = 1cm$. The hydraulic conductivity is set to $K_f = 3.69 \cdot 10^{-7} m/s$ for the matrix and to $K_f = 1.36 \cdot 10^{-2} m/s$ for the fractures. The porosities are $\phi = 0.3$ for the matrix and $\phi = 1.0$ for the fractures. The molecular diffusion is set to $D_m = 6.12 \cdot 10^{-5} m^2/s$ in the matrix as well as in the fracture. Dispersion lengths of $\alpha_L = 0.001m$ and $\alpha_T = 0.001m$ in the matrix and in the fracture are chosen. The initial grid has 432 nodes. The equidimensional modeling approach with the Fully Upwind Box Method and a difference indicator (see sec. 3.3.2) are applied.

Initially, the concentrations are set to zero. The system is closed along the top and bottom boundaries. On the left boundary, a pressure of $p = 103410Pa$ is imposed and, on the right boundary, a pressure of $p = 102900Pa$. Constant concentration boundary conditions are given with $c = 1.0mg/l$ along the left boundary and $c = 0.0mg/l$ along the right boundary.

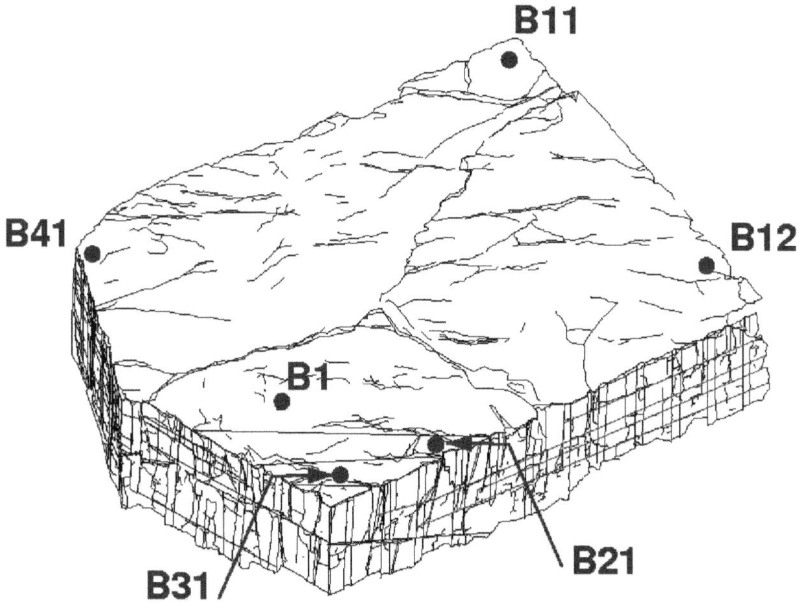

Fig. 5.14. Sandstone outcrop, after SILBERHORN-HEMMINGER (2002 [237])

In figure 5.15, left, the adaptively refined grid up to level 3 is presented while in figure 5.15, right, the concentration distribution after $40s$ is shown. The tracer migrates comparatively fast within the fracture network and slowly in the matrix. In areas of steep concentration gradients, a mesh refinement is carried out. The performance of the adaptive methods depends slightly on the complexity of the mesh. Similar performance results can be expected as shown in previous example because the initial meshes have approximately the same number of nodes (see tab. 5.1).

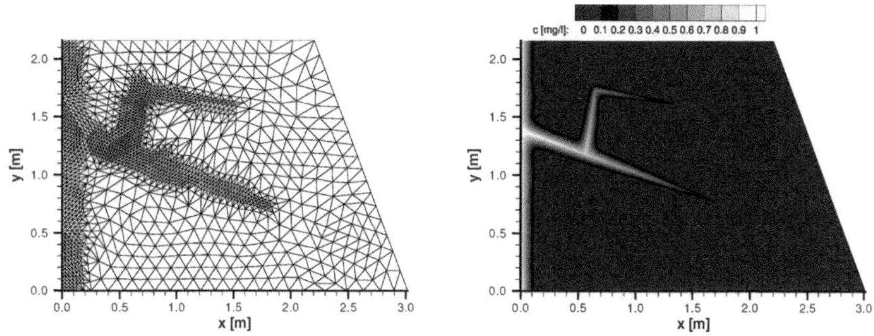

Fig. 5.15. Adaptively refined grid, left, and concentration distribution, right; after OCHS et al. (2002 [193])

5.1.7 Evaluation and future work

In space, the flow simulation method is of second order consistency for the piezometric head and of first order consistency for the flow velocity. It is globally mass conservative. The transport simulation method represents the correct physical solution, is monotonous and locally mass conservative. Due to the implicit formulation, there is no stability constraint for the time step. The disadvantages are given by, generally, only first order consistency in space and time, i.e. possibly high numerical dispersion. The coupled flow and transport solver is applicable to unstructured grids. Moreover, it uses adaptive Multigrid Methods and can easily be extended for parallelization. However, vectorization is hardly possible.

With the equidimensional approach, the basis for Mixed-Hybrid Methods is provided yielding the possibility to obtain a continuous flow field at the fracture-matrix interface and to increase the order of consistency of the flow velocity. This is expected to lead to a considerable improvement of the transport simulation and to have advantages for Particle Methods. If the hierarchical decomposition is extended to the transport simulation, it is possible to chose different discretization methods in fracture and matrix depending on the dominant physical processes and moreover, to couple different model concepts, for example to couple the flow in a porous medium with the flow in a pipe. Overall, such methods are very promising coupling methods which can be used in the field of *multiphysics*.

5.2 Two-phase flow processes in the subsurface

This section deals with the numerical modeling of two-phase flow processes in the subsurface. *Parallel* and *adaptive methods* together with *fast non-linear Multigrid-based solvers* are applied as efficient numerical methods (see secs. 3.2 - 3.4). Nearly all the methods described in secs. 4.1 - 4.3 are used in the context of efficient information-processing techniques. MUFTE-UG (see sec. 4.2.4) serves as the modeling system. The model concept was developed by HELMIG (1997 [98]).

5.2.1 Numerical algorithm

The governing equations as well as the chosen pressure-saturation formulation are discussed in section 2.5. The discrete model concept (see sec. 2.1.2) is applied to fractures which are modeled as elements of one dimension lower than that of the matrix. For the discretization, the Fully Upwind Box Method (see sec. 3.1.5) is used in space and the fully implicit *Euler* Method in time (see sec. 3.1.2). For heterogenous systems, the use of a special *interface condition* is recommended (see BASTIAN et al. (2000 [22])). The non-linearities are treated with the *Newton-Raphson* Method (see sec. 3.4.5), and the linearized equations are solved with the BiCGSTAB Method (see sec. 3.4.2) and Multigrid preconditioning (see sec. 3.4.3). A $V-$Multigrid cycle (see sec. 3.4.4) with different grid levels is chosen using a direct coarse-grid solver (see sec. 3.4.5), an ILU smoother and 2 pre- and postsmoothing steps (see sec. 3.4.4).

The parallel methods used in section 5.2.4 are based on the message-passing programming model and an algebraic parallelization (see secs. 3.2.2, 4.2.4). A dynamic load balancing can be chosen, and the domain can be split up into any number of subdomains (see secs. 3.2.4, 3.2.4). H-adaptive methods, which avoid hanging nodes, are applied in section 5.2.3 and steered by heuristic error indicators (see sec. 3.3). Adaptive and Multigrid Methods are treated together with Local Multigrid Methods (see sec. 3.4.4). The time-step adaptation is controlled by the number of non-linear iterations (see sec. 3.3.2).

In the following, the numerical algorithm is briefly evaluated. The advantages are the representation of the correct physical solution, the applicability to unstructured grids, the local mass conservation and the monotonicity. Due to the implicit formulation, there is no stability constraint; the time-step size is controlled by the number of non-linear iterations. Furthermore, the algorithm uses parallel, adaptive Multigrid Methods and is thus computationally highly efficient. The disadvantages are only first-order consistency in space and time as well as possibly high numerical dispersion caused by the fully upwinding and the fully implicit methods. Moreover, vectorization is hardly possibly. Finally, it is mentioned that nowadays, no algorithm can fulfill all criteria satisfactorily. The philosophy of the numerical algorithm consists of exploiting the advantages of 'simple' methods and compensating for their disadvantages with high mesh resolutions which can be efficiently solved with parallel adaptive Multigrid Methods. Consequently, the whole numerical algorithm, in spite of some drawbacks, can be evaluated as very good.

5.2.2 Example: Methane migration processes in coal mining areas

Problem description

For the past couple of decades, coal has been exploited as an energy resource from a decreasing number of mines at many locations across the world. Methane gas adsorbed by a coal seam is released while mines are in operation, although this release is generally carefully controlled by means of ventilation (see fig. 5.16). In recent years, the closure of such mines has led to the termination of the gas ventilation and the rise of the water table after water pumping ceased. However, the closed mines continue to produce methane gas which may reach the surface of the earth via old mine workings, shafts, permeable strata, faults and mining-induced fractures. Methane-gas emission, for example close to residential areas, is dangerous because methane gas is toxic for humans, and is a fire and explosion risk; thus, the gas which accumulates in buildings can jeopardize human life. However, it can be used to supply energy if the flow is large enough and if a controlled suction is possible.

A number of simplified test cases have been carried out to determine the dominant parameters and processes for simplified geological structures; some of them are shown in the following. The test cases deal with different permeabilities (see 'Multi-layered systems') and different fault zones (see 'System with different fault zones'), as well as with the influence of small-scale heterogeneities (see sec. 'System with small-scale heterogeneities'). These features have an important influence on the gas-water flow as well as on the amounts and the locations of gas escaping to the atmosphere. A number of real case data which stem from an abandoned coal mine (see 'Setting up a 3D-model of an existing coal mine') are available. For these test cases, dissolution processes are neglected (see 'Future work'). The overall aim consists of building up a three-dimensional model of an existing coal mine, calibrating a numerical model in the range of available data and making predictions for possible remediation measures. The numerical simulations can detect endangered areas or be used for the optimal positioning of wells which draw off the gas in a controlled way.

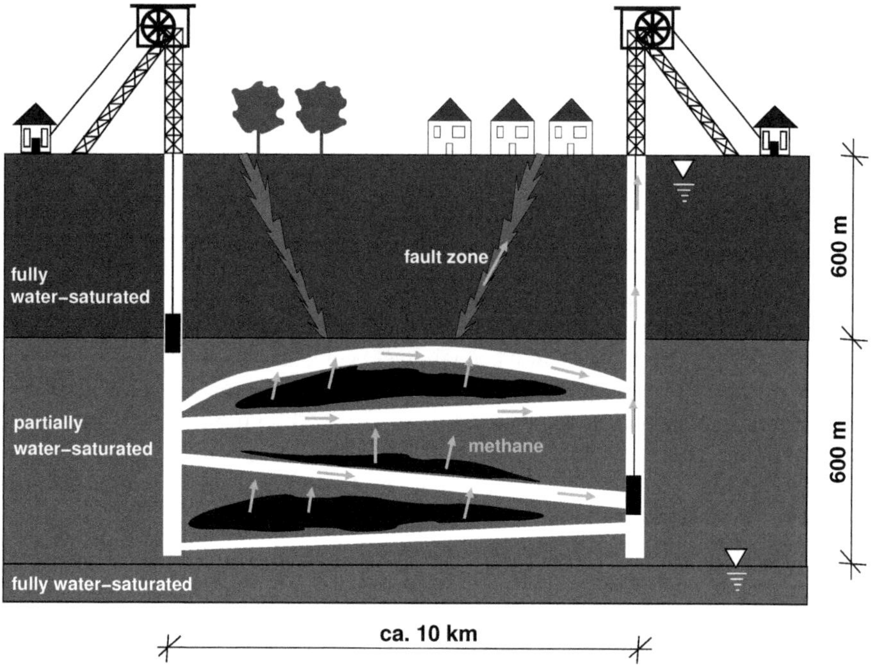

Fig. 5.16. Sketch of a coal mine, after BREITING et al. (2000 [41])

Multi-layered systems

The content of this section is also discussed in BREITING et al. (2000 [43]) and BREITING et al. (2000 [41]). The two-layered system in figure 5.17 and the four-layered system in figure 5.19 are chosen to investigate the influence of different permeabilities on the gas-water flow processes. Both systems differ only in the permeability distribution. They have a length of $500m$ and a height of $10m$. The porosity is set to $\phi = 0.48$, the density of water to $\rho_w = 10^3 kg/m^3$ and the dynamic viscosity of water to $\mu_w = 1.1 \cdot 10^{-3} Pas$. The gas phase is assumed to be a mixture of 50% methane and 50% air, i.e. the density of the gas is $\rho_w = 0.9 kg/m^3$ and the dynamic viscosity of the gas $\mu_w = 1.4 \cdot 10^{-5} Pas$; the pressure dependency of the gas is taken into account. The constitutive relationships of BROOKS, COREY (1964 [46]) are chosen with $p_d = 5000 Pa$, $\lambda = 2.0$, $S_{wr} = 0$ and $S_{nr} = 0$. The grid consists of 2816 elements; in the central area of $100m$ length around the methane-gas source, quadratic elements with a length of $0.625m$ are chosen and, in the outer areas, rectangular elements with a length of $12.5m$ and a height of $0.625m$. To be independent of the boundary conditions, the system is enlarged in the horizontal direction. The main interest focuses on the inner area.

As initial conditions, the system is fully saturated with water, and a hydrostatic pressure distribution is given. The system is closed on the left, right and lower boundaries. At the open upper boundary, a $50m$ water column, i.e. $p_w = 5 \cdot 10^5 Pa$, and no gas saturation, i.e. $S_n = 0.0$, are imposed as *Dirichlet*-boundary conditions. A coal seam is idealized as a gas-source term with the magnitude $q_n = 5 \cdot 10^{-7} kg/s$.

In figure 5.17 and 5.18, the permeability and the simulation results after 100 days in a $26m$ long region around the gas source are shown for the two-layered system. Due to the lower density, the gas migrates upwards to the layer with the lower permeability. It spreads horizontally, and it penetrates into the top layer in areas where the gas pressure exceeds the entry pressure of the lower permeability layer. The methane source causes a pressure increase in the central area of the system, whereas the pressure tends to be hydrostatic in the other parts. The flow velocities of the gas phase are higher than the ones of the water phase. The flow velocities of the gas phase are comparatively high above the methane source and at the layer boundary.

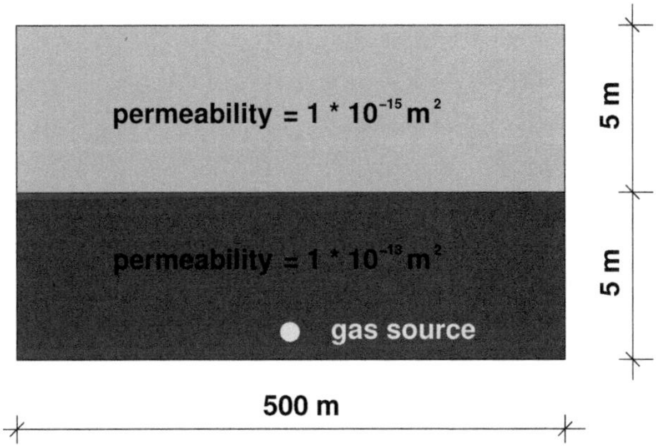

Fig. 5.17. Permeability distribution of the two-layered system, after BREITING et al. (2000 [41])

In figure 5.19, the distributions of the permeabilities and gas saturations are given for different time steps. The permeability decreases from the bottom to the top, four fault zones with the permeability of the bottom layer are included, and there is an inclination between the lower two layers. Again, the gas migrates upwards and spreads horizontally. However, it does not reach the left fault zone because of the layer inclination. The gas migrates further upwards along the fault zones and the layer boundaries. In areas where the entry pressure is exceeded, it also migrates through layers of lower permeability.

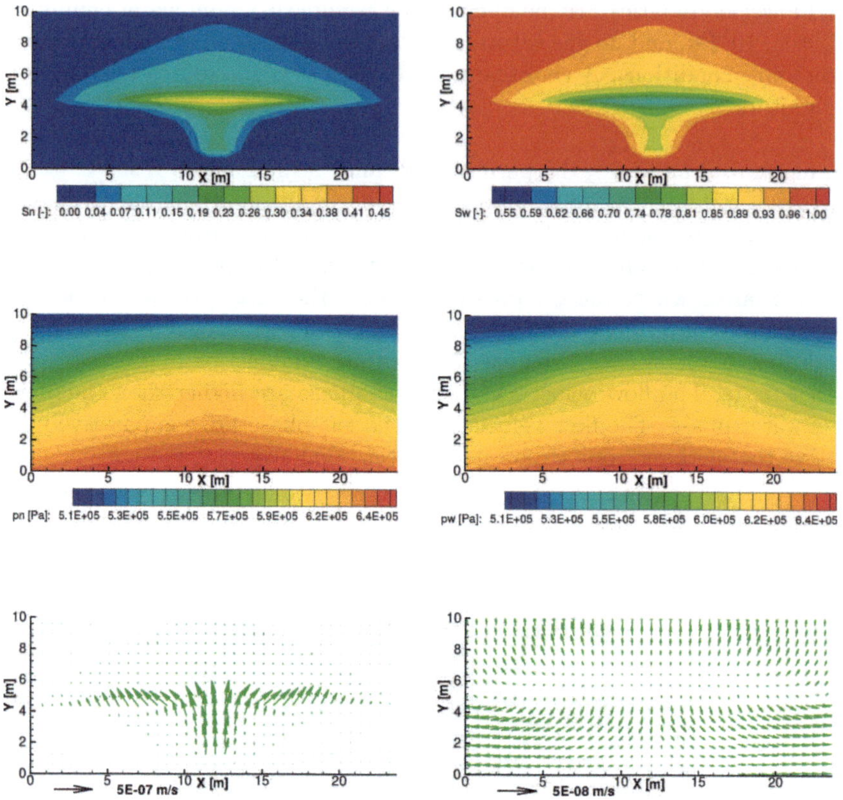

Fig. 5.18. Simulation results after 100 days; left: gas phase, right: water phase; top: saturations, middle: pressures, bottom: flow velocities, after BREITING et al. (2000 [41])

The simulations indicate the strong influence of permeability differences, fault zones and layer inclinations. They help to explain observations in nature suggesting that the locations where the gas reaches the surface of the earth need not be directly above the coal seam.

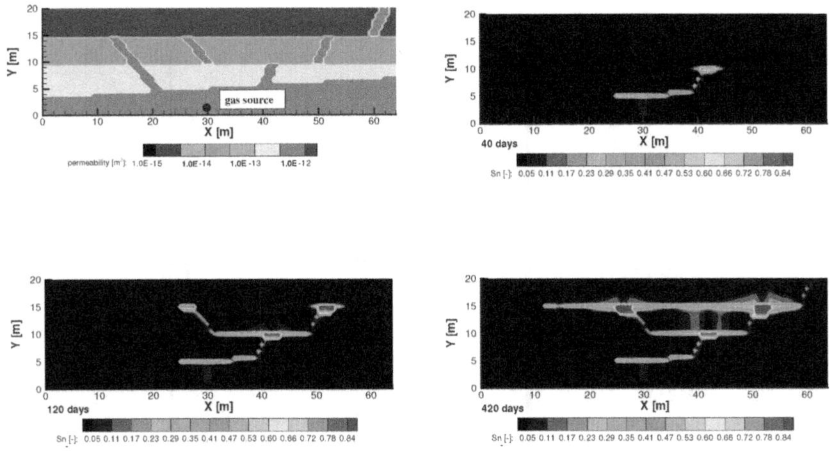

Fig. 5.19. Distributions of the permeabilities and gas saturations after 40, 120 and 420 days, after BREITING et al. (2000 [41])

System with different fault zones

The work described in the following is partially discussed in HINKELMANN et al. (2002 [112]). The system shown in figure 5.20 is used to make a study of how a shaft, a tunnel and fractures affect the gas-water flow processes. The system is $400m$ long and wide and its set-up is shown in figure 5.20. The permeability of the matrix is $K = 10^{-16}m^2$ and the porosity $\phi = 0.04$. For simplicity reasons, the permeabilities of the shaft, the tunnel and the fractures are all set to $K = 10^{-8}m^2$, and the porosities to $\phi = 0.90$. Shaft and tunnel have a width of $b = 5.0m$, and the fracture is $b = 0.3m$ wide. All the fault zones are modeled as one-dimensional elements. In these simulations, the flows in the tunnel and shaft are very slow and take place in the range of the *Darcy* law; this can be crucical in other cases. Capillarity is neglected. The other parameters are similar to those of the test case 'Multi-layered systems'. The grid in figure 5.21, left, consists of 2572 triangular elements.

As initial conditions, the system is fully water-saturated, and the pressure is atmospheric. The system is closed on the left, right and lower boundaries; at the top, the water pressure is set to $1bar$ and the gas saturation to zero. In the coalbed, a methane-gas source term is given with $q_n = 3 \cdot 10^{-7}kg/s$.

In figure 5.21, right, the distribution of the water pressure is presented. The pressure is hydrostatic in left part of the domain, and a pressure increase caused by the methane source is observed in the right part. In figure 5.22,

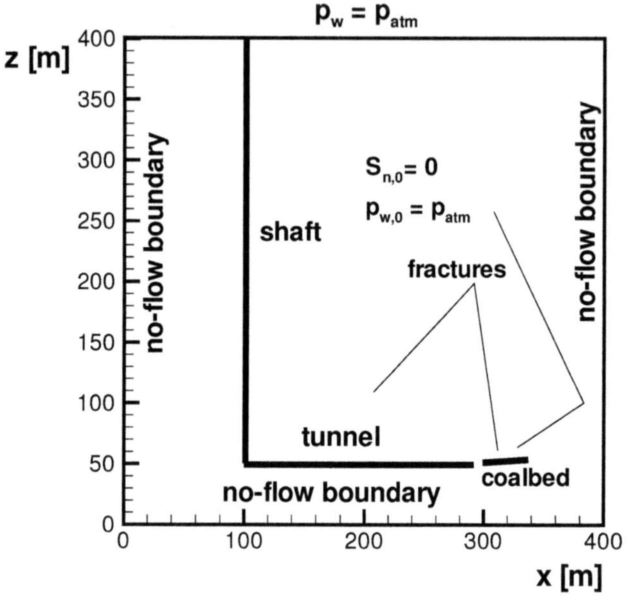

Fig. 5.20. Model set-up

the gas saturations are shown. The methane source leads to an increase of the saturations around the coalbed. The gas moves upwards and passes the different fault zones where it has low saturations. After 960 days, the gas migrates out of the top end of the fractures into the matrix (see fig. 5.22, left). After 8530 days, this process continues, and a further methane migration from the tunnel into the matrix is observed (see fig. 5.22, right). In such a case, methane gas reaches the surface of the earth via the shaft as well as via the matrix. Finally, this simulation indicates the strong influence of the different fault zones on the amount as well as on the location of gas escaping to the atmosphere. //

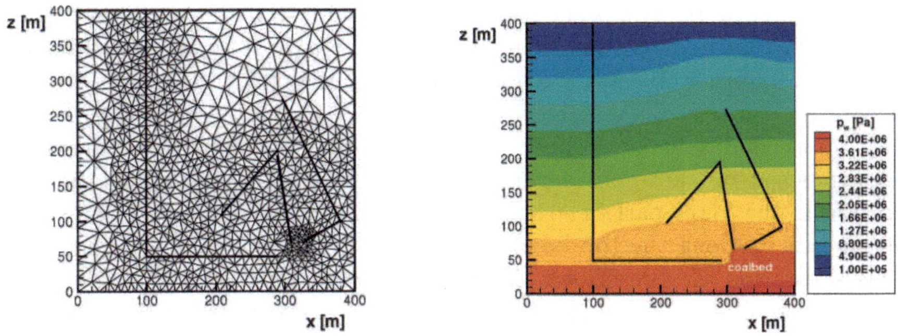

Fig. 5.21. Mesh, left; distribution of the water pressure, right

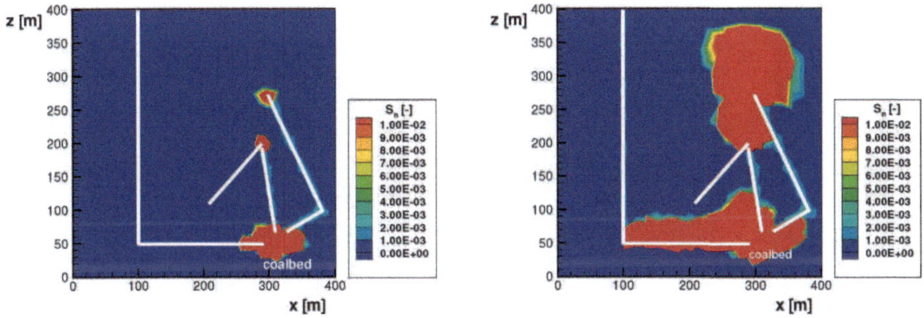

Fig. 5.22. Distribution of the gas saturation after 960 days, left, and 8530 days, right

System with small-scale heterogeneities

The content of this section is also discussed in KOBAYASHI et al. (2002 [146]). Small-scale heterogeneities have a significant influence on the gas-water flow processes as well, depending on the scales considered. To analyze this, the geostatistical model SIMSET (see BARDOSSY (1992 [15]), sec. 4.1.6) is applied. Its parameters are systematically varied for systems of different scales focusing on the amount, the location and the distribution of gas escaping to the surface of earth.

In the following, two similar systems which only differ in the length scale are considered. The system of $10m$ width and $20m$ height is called the small system (SS), and the one of $100m$ width and $200m$ height is called the large system (see fig. 5.23). The average permeability is $K = 10^{-13}m^2$ and the porosity $\phi = 0.57$. The other parameters are similar to the those of the test case 'Multi-layered systems'. The domain is discretized by 800 quadratic elements of $0.5m$ for the small and $5m$ for the large system.

As initial conditions, atmospheric water pressure and complete gas saturation are imposed in the upper part and hydrostatic pressure and zero gas saturation in the lower part. The system is closed along the left, right and lower boundaries. Along the top boundary, atmospheric pressure and (nearly) full gas saturation are prescribed. A gas line source is given at $Z = 1m$ in the small system and at $Z = 10m$ in the large one. The results are analyzed at steady-state conditions.

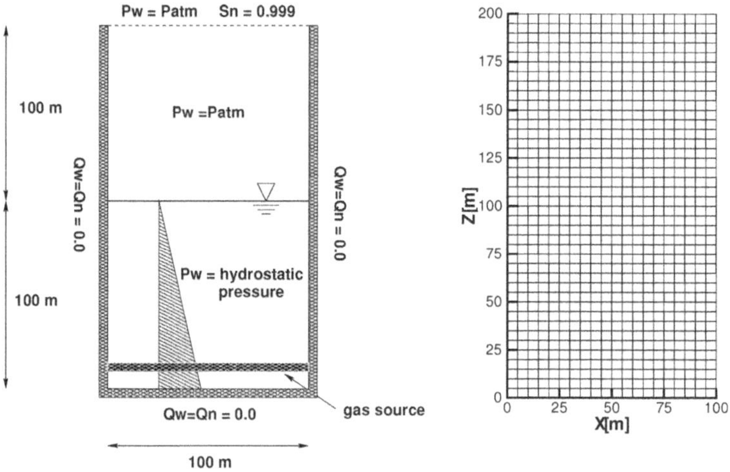

Fig. 5.23. Model set-up, left; grid, right; after KOBAYASHI et al. (2002 [146])

The notation, for instance c2v1a0.2 as shown in table 5.2, means that the variogram (see sec. 4.1.6) has a **c**orrelation length of $2m$, a **v**ariance of 1 and an **a**nisotropy ratio of 0.2. The anisotropy reflects the ratio of the vertical to the horizontal correlation length. Figure 5.24 shows the permeability fields, left, the distributions of the gas saturation, middle, and the gas velocity, right, with the variograms c20v0.25a0.2 in lower row and c20v1a0.2 in the upper row.

In order to quantify the influences due to the changes of the variance and the correlation length, the variances of the distribution of the gas velocity at the cross section $Z = 18.5m$ for the small system and $Z = 185m$ for the large system are determined using the following formula:

$$vv = \frac{1}{n} \sum_{i=1}^{n} \left(\frac{v_{z,i} - v_{z,ave}}{v_{z,ave}}\right)^2 \qquad (5.1)$$

SS(10m × 20m)	homogeneous	c2v1a0.2	c2v0.25a0.2	c0.5v1a0.2	c0.5v0.25a0.2
vv [-]	0	2.94	0.863	2.79	0.681
LS(100m × 200m)	homogeneous	c20v1a0.2	c20v0.25a0.2	c5v1a0.2	c5v0.25a0.2
vv[-]	0	1.02	0.190	4.54	1.86

Table 5.2. Variances of the velocity distribution for different variograms and scales, after KOBAYASHI et al. (2002 [146])

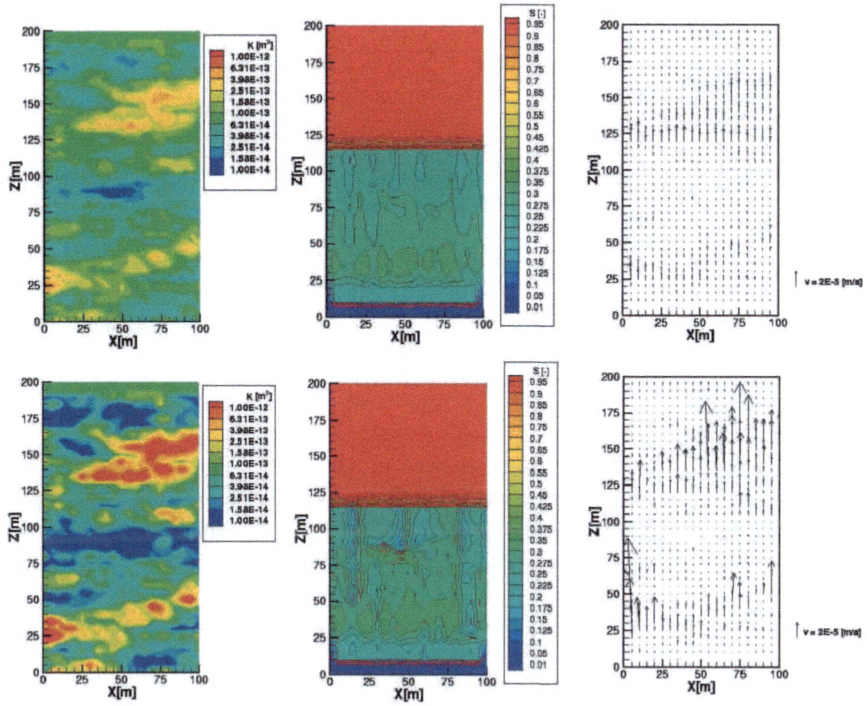

Fig. 5.24. Geostatistical permeability fields, left; gas saturations, middle; gas velocities, right; variogram c20v0.25a0.2, bottom, and c20v1a0.2, top; after KOBAYASHI et al. (2002 [146])

In this equation, $v_{z,i}$ denotes the velocity of the gas phase at node i and $v_{z,ave}$ the average gas velocity, both determined at the cross section.

Several tendencies are deduced from figures 5.24 and 5.25 and table 5.2. The methane saturation and the velocity distribution become more heterogeneous if the variance of the variogram increases, here from $v = 0.25$ to $v = 1$. It can be quantitatively argued from table 5.2 that, if the variance of the variogram increases, the variance of the velocity distribution at the cross section also increases in both systems. In the small system, the influences due to the changes of the correlation length are not as high as the influences due to the

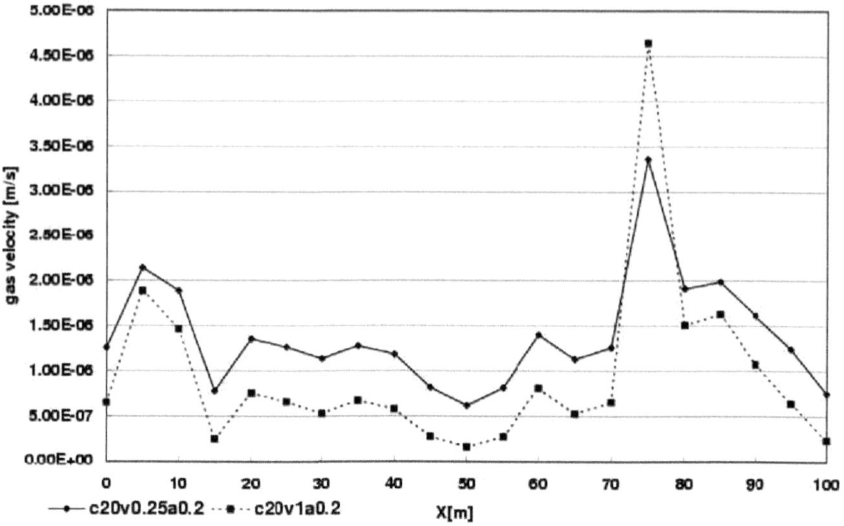

Fig. 5.25. Vertical gas velocity at the cross section $Z = 185m$ with the variograms c20v0.25a0.2 and c20v1a0.2, after KOBAYASHI et al. (2002 [146])

changes of the variance. In the large system, the influences due to the changes of the correlation length are relatively high compared to the small system. Furthermore, the variance of the velocity distribution increases as the correlation length decreases. However, since the ratio of the correlation length to the system size is relatively large with regard to the small system, it is questionable whether the simulation results obtained from one geostatistical realisation of the permeability field can really be representative for the realisations, especially when the system is relatively small.

Overall, the results can be summarized as follows: If the variance in the variogram increases, this leads to an increase of the variance of the velocity distributions in the upper region of the systems. Other scale-dependent tendencies cannot be clearly seen. Therefore, further investigations should deal with the *scaling of variogram parameters* and the development of *upscaling methods.*

Setting up a 3D-model of an existing coal mine

The work described in the following is also discussed in HINKELMANN et al. (2002 [112]). A three-dimensional model of the abandoned coal mine *Westfalen* is set up (see fig. 5.26, bottom). It is located close to the town Ahlen in the *Ruhrgebiet* in the western part of Germany (see fig. 5.26, top). Information about the geometry of the geological structures is provided by several exploration boreholes and cross sections which are processed with AutoCAD (see sec. 4.1.2). One cross section is shown in figure 4.3. A particular difficulty for setting up the model and the numerical simulations consists of integrating shafts, tunnels and coal seams (see fig. 5.27). A number of physical parameters are available. They are assigned to the geological structures by the database-management system MySQL (see sec. 4.1.4) as shown in figure 4.5. The geometrical and physical data serve as the input for the mesh generator ART (see sec. 4.1.7). A computational mesh is given in figure 4.13. In this context, attention is drawn to the overview of the modeling system MUFTE-UG in figures 4.20 and 4.19.

Future work

Firstly, the database of the 3D-domain will be extended to contain more information. Secondly, 3D numerical simulations will be carried out. As the computational domain, consisting of multiple layers, fractures, void spaces etc. (see fig. 5.28), is rather complex, 3D-subdomains which contain only 1 or 2 layers will be modeled first, starting in the lower regions at the coal seams and then moving upwards. Moreover, investigations will consider whether dissolution processes must be taken into account in certain areas; this is the case when such methane-migration processes contribute significantly to the overall amount of methane escaping to the atmosphere. Furthermore, optimization methods for gas-water flow and transport processes will be developed, for example to optimize methane suction.

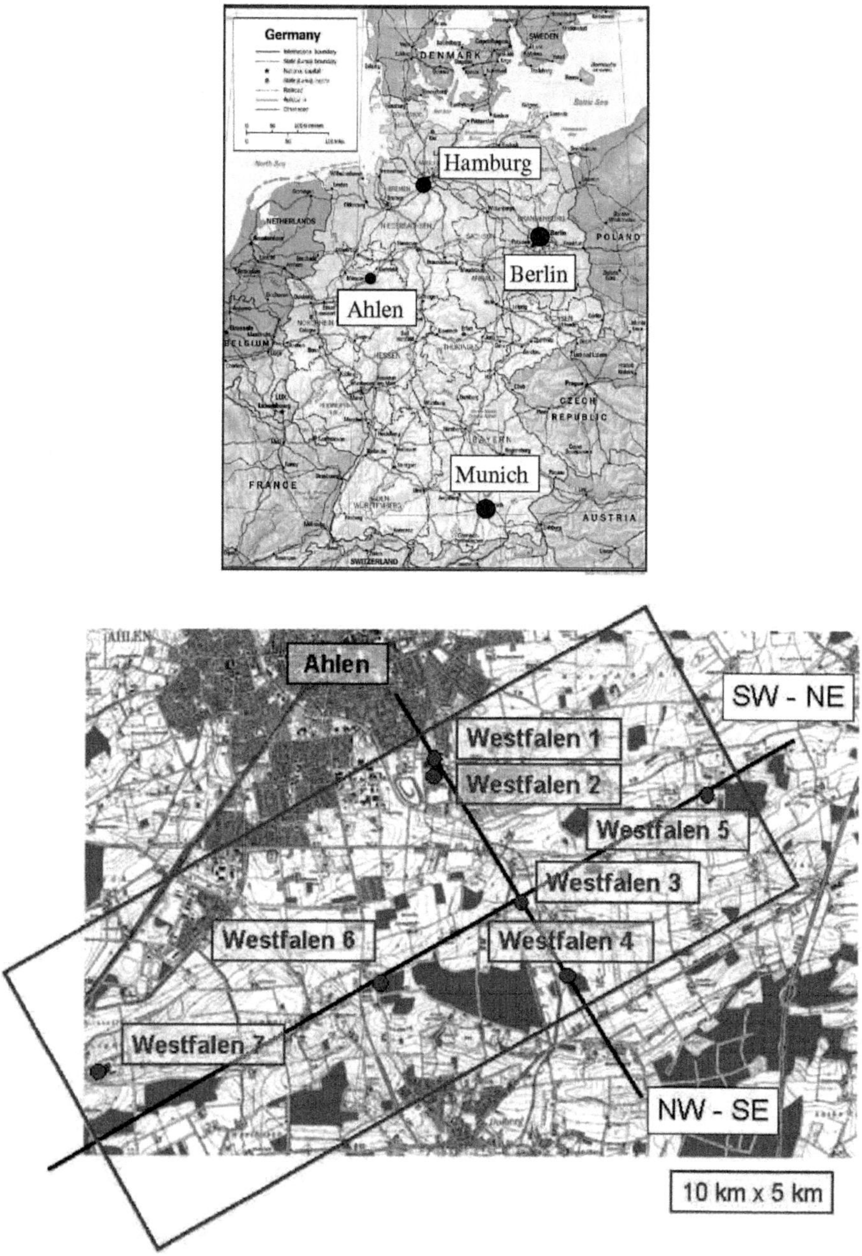

Fig. 5.26. Location of Ahlen, top, and of the Westfalen mine, shafts and cross sections, bottom; after HINKELMANN et al. (2002 [112])

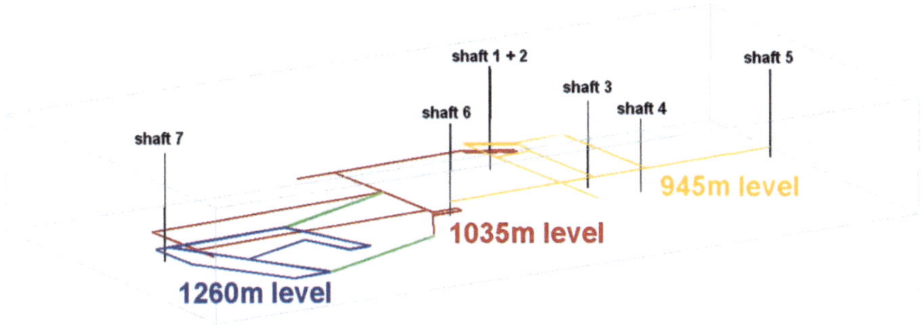

Fig. 5.27. CAD plan of the tube system in the Westfalen mine, after HINKELMANN et al. (2002 [112])

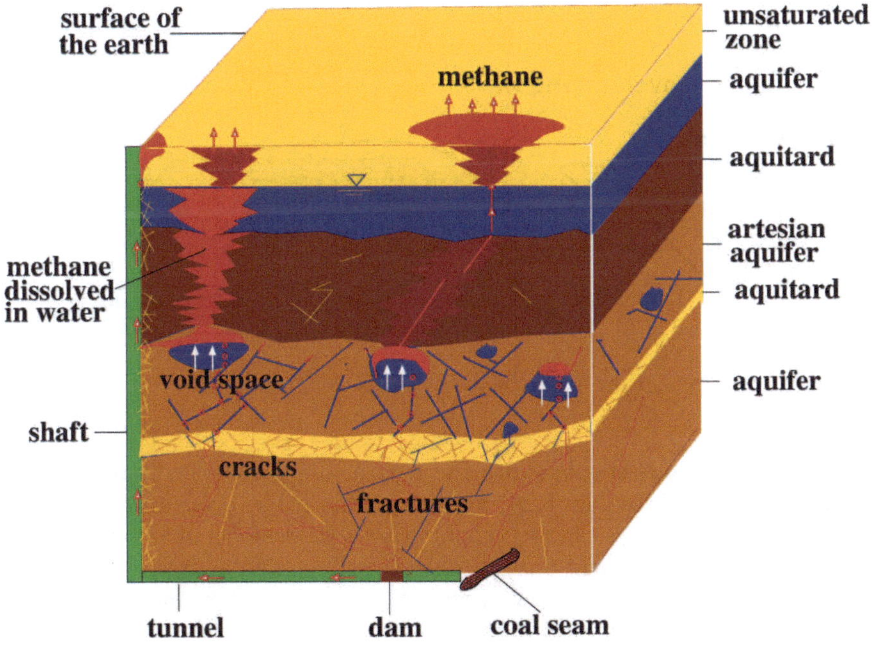

Fig. 5.28. Sketch of a multi-layered system over coal seams

5.2.3 Example: Gas-water flow processes in dike systems

Problem description

The content of this section is discussed in PAUL et al. (2000 [205]), PAUL et al. (1999 [204]) and PAUL (2003 [203]). Dikes play an important role in the protection of human life and goods, especially in the field of coastal engineering (see fig. 5.29). Investigations of damaged dikes in the nature as well as in the laboratory have shown that the failure often starts on the land side in areas where air is pressed out of the system by overtopping and breaking waves. As the height and occurence of storm surge levels are expected to increase in the near future, the loads on such coastal defense structures as well as the probability of their overtopping will increase as well. These effects underline the need for tools for simulating current and future loads in order to understand the dominant processes and to improve such coastal defense structures. The overall task requires a coupling of a fluid mechanics model, which describes the surface waves and the flow in the porous medium, with a structural mechanics model, which determines the stresses and the strains.

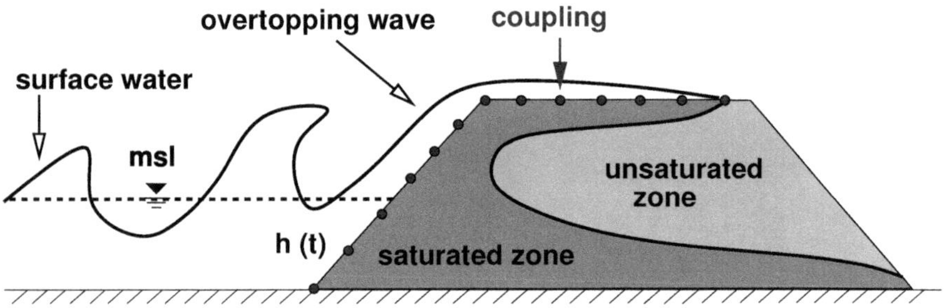

Fig. 5.29. Sketch of an overtopping dike

The surface wave is coupled with the flow in the porous medium along the common interface (see fig. 5.29) via the pressure. The pressure field coming from the surface waves can be computed by a wave model or it can be taken from experimental or field data. Problems with a 'simple' free-water surface in the dike system can be treated with a groundwater flow model which determines the free water surface iteratively. This is, for exemple, the case for seepage problems when the water level is higher on the sea side than on the land side. If overtopping waves occur, the shape of the free surface is complex, as shown in figure 5.29; air can be trapped and processes of the pressed-out air must be taken into account. In such cases, a two-phase flow model consisting of a water and a gas phase must be applied. Naturally, this model concept includes the determination of a free-water surface in a dike system.

Simulation results using adaptive methods

The dike system shown in figure 5.30 is chosen for experimental and numerical simulations. It consists of 4 different zones. Zone 1 and 3 are of the same soil; the permeability of zone 2 is about 4 times lower than that of zones 1 and 3, and zone 4 has a comparatively high permeability for experimental reasons. The soil properties are given in table 5.3. The constitutive relationships after VAN GENUCHTEN (1980 [92]) are used. In the numerical simulations, capillarity is neglected. The other parameters are similar to those from section 5.2.2.

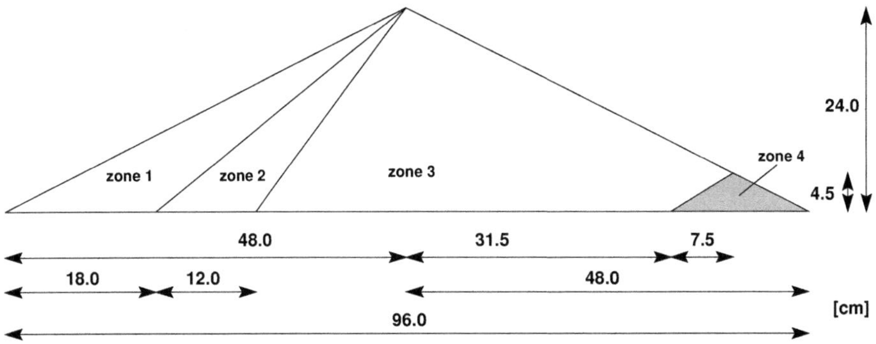

Fig. 5.30. Geometry of the dike system, after PAUL et al. (2000 [205])

zone	K $[m^2]$	ϕ [-]	α [-]	n [-]	S_{wr} [-]	S_{gr} [-]
1	$9.8 \cdot 10^{-10}$	0.37	0.0037	4.70	0.05	0.00
2	$3.6 \cdot 10^{-11}$	0.35	0.0011	4.70	0.10	0.00
3	$9.8 \cdot 10^{-10}$	0.37	0.0037	4.70	0.05	0.00
4	$3.6 \cdot 10^{-8}$	0.20	0.1000	4.70	0.15	0.00

Table 5.3. Soil parameters, after PAUL et al. (2000 [205])

As initial conditions, zone 1 and 2 are fully water-saturated up to the height of the left boundary condition, and the water pressure is hydrostatic. The other areas of the domain have the residual saturations, and the pressure is atmospheric. The system is closed along the lower boundary. On the left side, a rising water table is given, i.e. the water pressure is prescribed and the gas saturation is set to zero. Along the other boundaries, atmospheric pressure and full gas saturation are imposed.

In figure 5.31, the numerical results are shown for steady-state conditions. An overhanging water front develops in zone 2 and a free-water surface in zone 3. The numerical results agree very well with the experimental results given in figure 5.32.

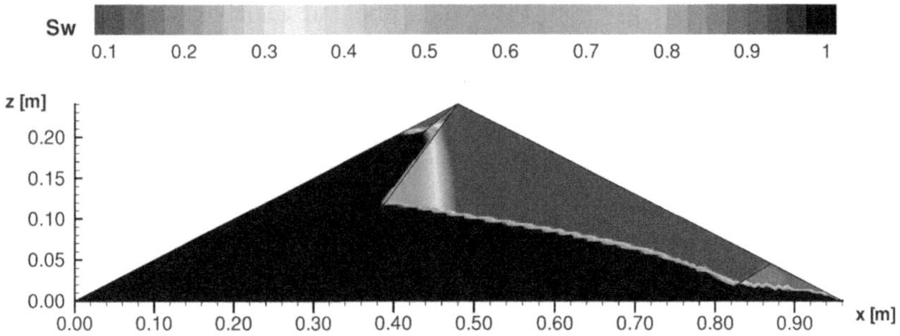

Fig. 5.31. Distribution of the water saturation for steady state, after PAUL et al. (2000 [205])

Fig. 5.32. Experimental results for steady state, after PAUL et al. (2000 [205])

For the adaptive simulations, a gradient indicator (see sec. 3.3.2) applied to the water saturation is used for the refinement and coarsening in space. An adaptive refinement is restricted up to level 3. The initial grid on level 1 consists of 3274 triangular elements (see fig. 5.33) and is uniformly refined once. In figure 5.34, the adaptively refined and coarsened grid with 2528 elements is shown for the steady state. In large areas, the grid is coarsened to level 0; in the area along the sharp front, the grid is refined up to level 3. The adaptive solution agrees with the uniformly refined solution. It requires about 5% of the number of elements and about 10% of the CPU time needed to compute the uniformly refined solution. Again, the excellent performance of adaptive methods is demonstrated.

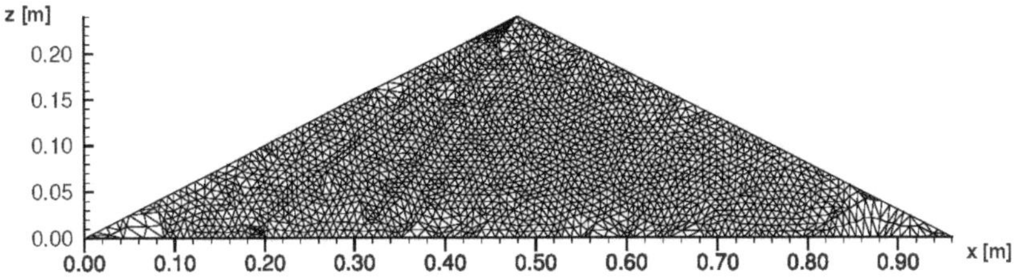

Fig. 5.33. Initial grid for the dike system, after PAUL et al. (2003 [203])

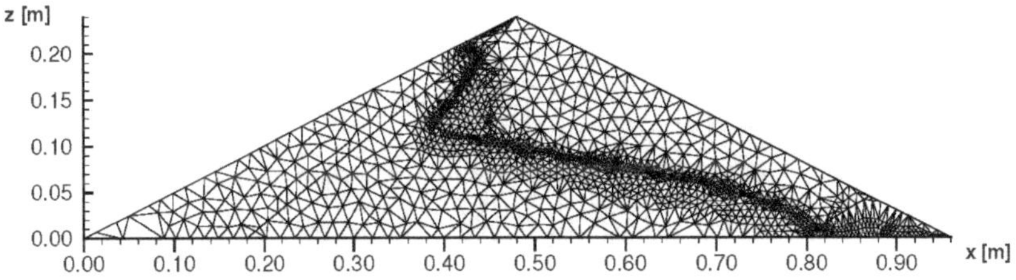

Fig. 5.34. Adaptively refined and coarsened grid for the dike system, after PAUL et al. (2003 [203])

5.2.4 Example: Multi-step outflow experiment

The work described in the following is also discussed in ÖLMANN et al. (2002 [194]). The *multi-step outflow experiment* is a laboratory-scale experiment to determine soil parameters. The system in figure 5.35, left, is $6cm$ long and $12cm$ high. It consists of a medium sand with an embedded coarse lense in the upper part and an embedded fine lense in the lower part. The soil parameters are given in table 5.4. The other parameters are similar to the those chosen in section 5.2.2. The coarse grid consists of 1054 quadratic elements. A grid partitioning for 7 processors is shown in figure 5.35, right. For a better visualization, the subdomains are shrunk. The influence of different multigrid levels as well as the parallel run-time behavior are already discussed in section 3.4.6.

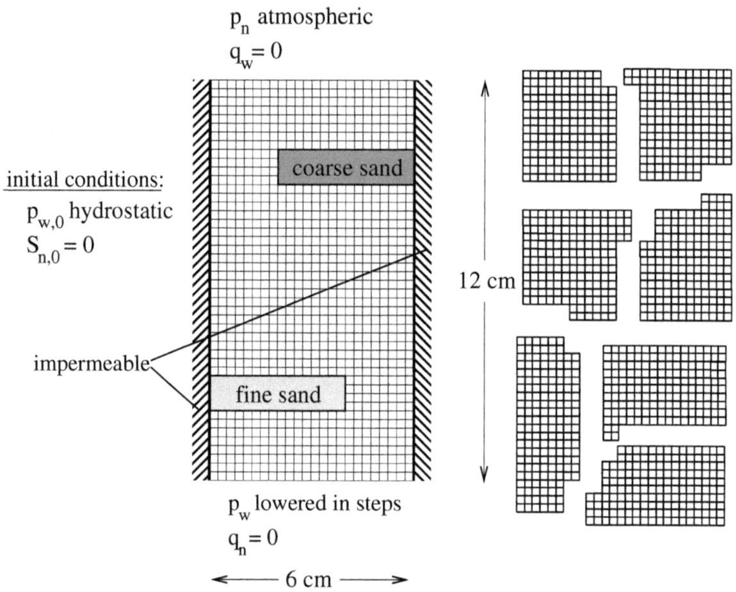

Fig. 5.35. Model set-up, left, and grid partitioning, right; after ÖLMANN et al. (2002 [194])

The system is initially fully saturated with water. It is closed on the left and right boundaries, and for the water flux on the upper boundary and for the gas flux on the lower boundary. Along the upper boundary, gas can enter the system; along the lower boundary, a time-dependent pressure, which is lowered stepwise, is prescribed (see fig. 5.36). As a consequence, water flows out of the system through the open bottom and air enters the system through the open top. The cumulative outflow of the water phase reacts directly on the pressure boundary condition. If the pressure gradient is large, so is the gradient of the cumulative water outflow; if small, the gradient of the cumulative water

parameter	fine sand	medium sand	coarse sand
porosity ϕ [-]	0.5	0.4	0.35
permeability K $[m^2]$	$1.427 \cdot 10^{-12}$	$1.426 \cdot 10^{-11}$	$1.66 \cdot 10^{-10}$
Brooks-Corey p_d $[Pa]$	1668.3	606.0	166.8
Brooks-Corey λ [-]	1.29	1.29	1.29
residual saturation S_{nr} [-]	0	0	0
residual saturation S_{wr} [-]	0.5	0.1	0.05

Table 5.4. Soil parameters, after ÖLMANN et al. (2002 [194])

outflow is also small (see fig. 5.36). As the amount of water in the system is limited, the course of the cumulative outflow of water asymptotically reaches a limes. With the measured water outflow as a response of the system, soil parameters can be deduced with, for example, *inverse modeling.* Moreover, a comparison of computations and measurements can be used to check the validity of certain model assumptions, such as the time-independence of the capillary pressure-saturation relationship.

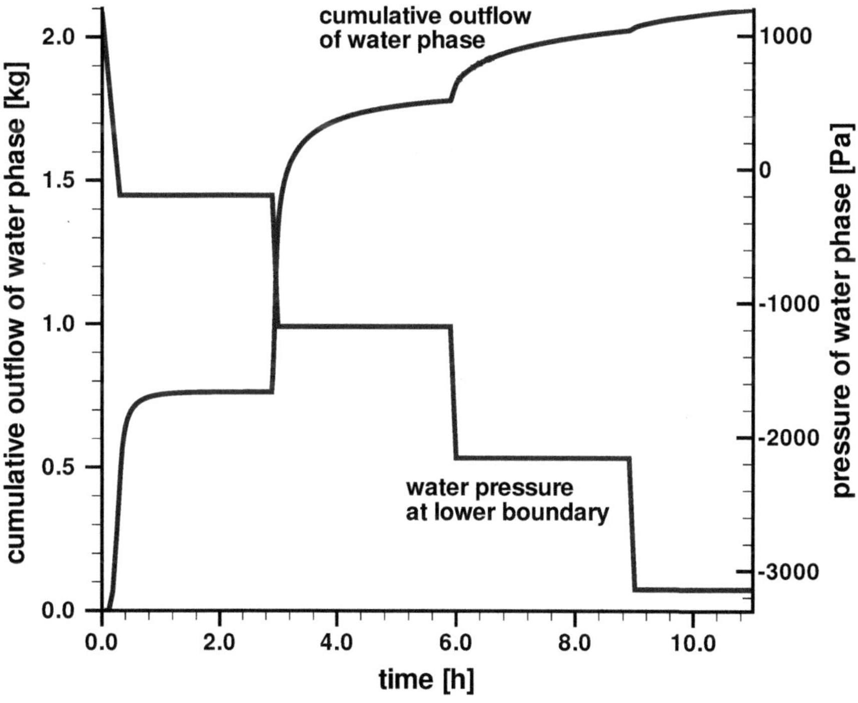

Fig. 5.36. Pressure boundary condition and cumulative outflow of the water phase, after ÖLMANN et al. (2002 [194])

5.3 Two-phase / multicomponent flow and transport processes in the subsurface

This section contains numerical simulations of two-phase / three-component flow and transport processes in the subsurface (see sec. 2.5). Here, *complex physical processes* are in the foreground. Moreover, *fast non-linear solvers* (see sec. 3.4.5) are applied with respect to efficient numerical methods, and a *geostatistical method* (see sec. 4.1.6) as an efficient information-processing technique. In order to ensure a relative independence of this section, a few repetitions of parts of sections 5.1 and 5.2 occur. The modeling system MUFTE-UG (see sec. 4.2.4) is applied to the numerical simulations. The development of the model concept is based on the work of HELMIG (1997 [98]) and CLASS (2000 [62]) and is described in HINKELMANN et al. (2002 [115]).

5.3.1 Numerical algorithm

The governing equations and the choice of the primary variables depending on the phase state are described in section 2.5. For the discretization, the Fully Upwind Box Method (see sec. 3.1.5) is applied in space and the fully implicit *Euler* Method in time (see sec. 3.1.2). The non-linearities are handled with the *Newton-Raphson* Method (see sec. 3.4.5), and the linearized equations are solved with the BiCGSTAB Method (see sec. 3.4.2). The time-step adaptation depending on the number of non-linear iterations (see sec. 3.3.2) is carried out. The evaluation of the numerical algorithm is comparable to that in section 5.2.1.

5.3.2 Example: Two-phase / multicomponent flow and transport processes in the interaction area groundwater - surface water

The content discussed in the following is also described in HINKELMANN et al. (2002 [115]) and KALERIS et al. (2001 [141]).

Problem description

In recent years, the phenomenon of *submarine groundwater springs* (see figs. 1.1, 1.4), also called *vents*, which are groundwater fluxes into sea water, has been detected by field measurements in several coastal zones all around the world - e.g. the Baltic Sea, the North Sea, the Mediterranean Sea, the Gulf of Mexico or the Chinese Sea. It has been assumed that the transport of contaminants or nutrients involved has a considerably larger influence on the water quality in coastal areas than expected before. In order to improve the understanding of these complex processes, several field measurement and monitoring campaigns were carried out in methane-rich coastal sedimentary environments of the Baltic Sea and numerical models have been developed. The numerical simulation approach is twofold. On the regional scale, the large-scale water balances and flow patterns taking into account (single-phase) groundwater flow coupled with salinity transport were investigated in order to determine the large-scale conditions for submarine groundwater fluxes. These processes are not investigated here; see KALERIS et al. (2002 [142]), KALERIS et al. (2001 [141]). Around locations where the submarine groundwater flows into the sea water, methane gas bubbles which are formed by microorganisms occur. Therefore, small-scale considerations are concerned with the interaction of the phases water and gas, also taking into account the component transport of freshwater, salt water and methane. The numerical simulation of such processes requires a two-phase / three-component model concept for a coastal aquifer or a porous medium. Here, special emphasis is placed on the so-called phase phase switch of methane, i.e. the transition of methane dissolved in water to the gas phase which did not exist before. This effect is responsible for the formation of the gas bubbles (see fig. 5.37).

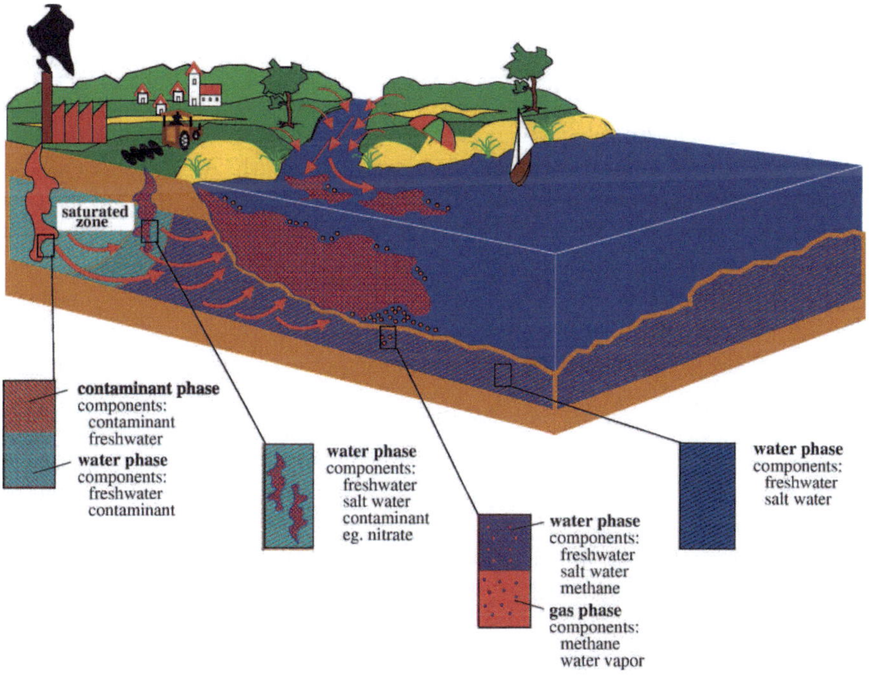

Fig. 5.37. Flow and transport processes in a coastal aquifer, after HINKELMANN, HELMIG (2002 [111])

Different flow and transport processes, which occur in a coastal aquifer, as well as the corresponding model concepts are briefly discussed (see fig. 5.37). In order to model groundwater flow coupled with salinity transport, a one-phase / two-component model concept (phase: water; components: water, salt; see sec. 2.3) should be applied in most cases (see OLDENBURG & PRUESS (1995 [199]), HINKELMANN et al. (2000 [114])). If methane is integrated in the simulations, low concentrations where methane is dissolved in water are first considered. Then, no further phase occurs, but a third component. Such situations require a one-phase / three-component model approach (phase: water; components: water, salt, methane; see sec. 2.3, HINKELMANN et al. (2000 [116])). The flow of the two phases water and gas (see sec. 2.4), assuming immiscibility of the phases, has been investigated intensively (see SHETA (2000 [230]), BASTIAN (1999 [18])). This model concept must be applied if dissolution processes are of minor importance and high methane concentrations are considered, because then the maximum solubility of methane in water is exceeded and a gas phase occurs. If dissolution processes must be taken into account additionally, a two-phase / three-component model approach (phases: water, gas; components: water, salt, methane, see sec. 2.5) is required.

Here, the major aim of the two-phase / three-component model is to simulate the methane switch numerically. A volume-averaged methane saturation is determined with a continuum approach. It is not possible to simulate the 'fate' of single gas bubbles which are even smaller than the element or REV (representative elementary volume, see sec. 2.1.2) length. As the bottom layer of the submarine aquifer under investigation has a very high porosity, the pore space can be disturbed by upward flowing gas bubbles, and thus preferential flow paths are formed. This effect is taken into account qualitatively using geostatistical distributions of the permeability field; this is explained later.

Model set-up

The system under investigation is located in the *Eckernförde Bay*, about $25km$ north-east of the town Kiel (see fig. 5.46). In figure 5.38, the measured bathymetry of a typical vent location is shown. This figure also indicates sediment flushed out by the vents (see zoom bottom left) as well as methane gas measured in the sediment (see zoom bottom right). Special measurement techniques were developed and a number of measurement campaigns were carried out. One interesting result is the fact that the vertical flow velocities from the groundwater into the sea water are comparatively high with up to $1cm/s$, and thus are in the lower range of the horizontal flow velocities in the sea water.

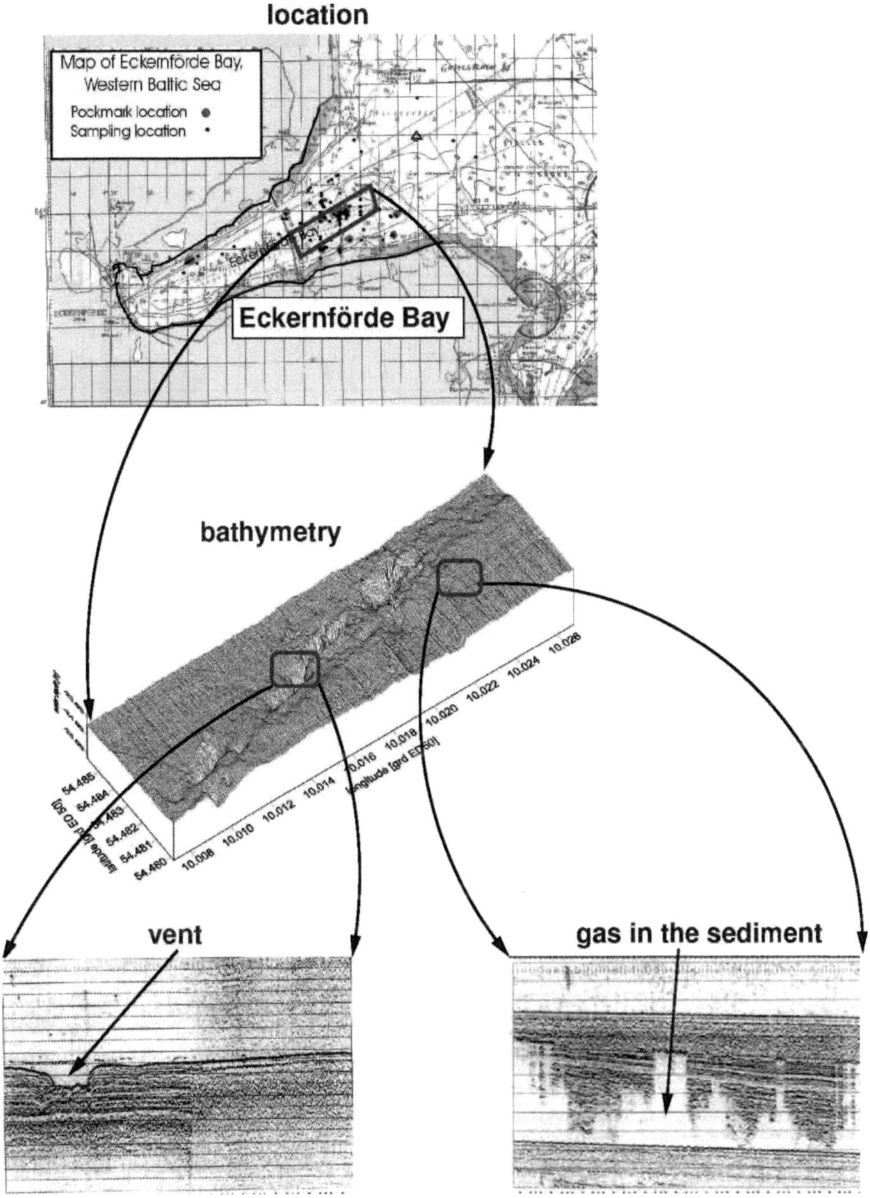

Fig. 5.38. Location and bathymetry of a vent, after SAUTER (2001 [229])

For the numerical simulations, an idealized two-dimensional system consisting of a $1m$-deep and $6m$-wide mud layer (see fig. 5.39) is chosen to investigate the influence of different submarine groundwater discharges which are called *vent*, *partial vent* and *non-vent* in the following. The lower boundary of the system is given by the aquifer layer where the inflow occurred. The upper boundary is given by the sea bottom.

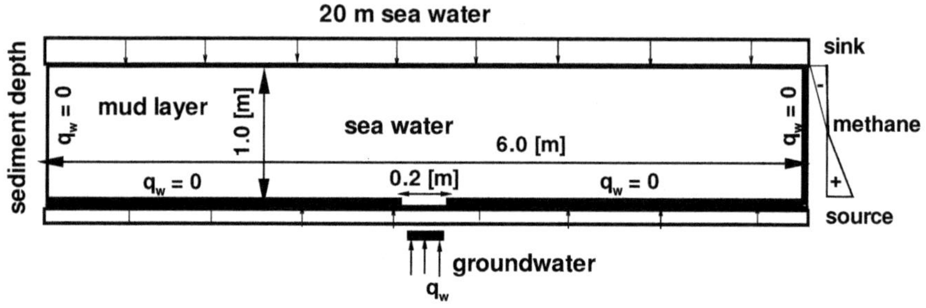

Fig. 5.39. System, initial and boundary conditions, after HINKELMANN et al. (2002 [115])

The porosity is set to $\phi = 0.9$, the permeability to $K = 10^{-9}m^2$, the molecular weights of water, salt and methane to $M^w = 0.018kg/mol, M^s = 0.058kg/mol, M^m = 0.016kg/mol$ respectively, the density of fresh water to $\rho_w = 1000kg/m^3$ and the dynamic viscosities of the water and the gas phases to $\mu_w = 0.001Pas$ and $\mu_g = 1.24 \cdot 10^{-5}Pas$, respectively. The dependence of the water density on the salinity is taken into account, whereas the viscosities are assumed to be constant. For the determination of the dispersion tensors, the molecular diffusion coefficients are set to $D_w^s = 1.0 \cdot 10^{-9}m^2/s$, $D_w^g = 9.0 \cdot 10^{-10}m^2/s$ and $D_g^w = 10^{-6}m^2/s$, the tortuosity to $\tau = 1.0$ and the longitudinal and transversal dispersivities to $\alpha_L = 0.05m$ and $\alpha_T = 0.01m$. The capillary pressure and the relative permeability after BROOKS, COREY (1964 [46]) are used with $p_d = 500Pa$, $\lambda = 2.0$, $S_{wr} = 0.0$ and $S_{gr} = 0.0$.

As initial conditions, a hydrostatic pressure distribution of a $20m$ sea-water column with a salinity of $23^o/_{oo}$ and complete water saturation without a methane component are imposed. The left and the right boundaries are closed, whereas the upper boundary is open. Along the upper boundary, the salt-water pressure, the salinity of 23 $^o/_{oo}$, which corresponds to a chloride concentration of $390mmol/l$ (see figs. 5.41 - 5.44, right), and zero methane are used to determine the boundary conditions. The lower boundary is also closed with the exception of the $20cm$ wide vent. Freshwater discharges are given as boundary conditions according to measurements of $q_w^f = 10.8l/(m^2h)$ for the vent, $q_w^f = 4.0l/(m^2h)$ for the partial vent, and $q_w^f = 0.1l/(m^2h)$ for the non-vent location. In the mud layer, methane formation and oxidation caused by microorganisms occurr. However, no data for corresponding sink and source terms are available. The distribution of the methane sink and source is linearly imposed with $q_w^m = 0.015 mol/(m^2 h)$ at the lower boundary and with $q_w^m = -0.005 mol/(m^2 h)$ at the upper boundary, because these values lead to the best agreement between computations and measurements, and they remain constant in the three simulations for the vent, the partial vent and the non-vent location. The freshwater discharge is the only value that changes in the following three simulations which are run until steady-state conditions. The system is discretized with 55 elements in the horizontal direction and 13 elements in the vertical directions, i.e. 715 elements. Further information can be found in HINKELMANN et al. (2002 [115]) and SAUTER (2001 [229]).

Vent location

Figure 5.40, left, shows the pressure, which is nearly hydrostatic. In figure 5.40, right, the flow velocity of the water phase represents the equilibrium between advective fluxes caused by the upward-flowing freshwater and diffusive fluxes caused by the different salinities.

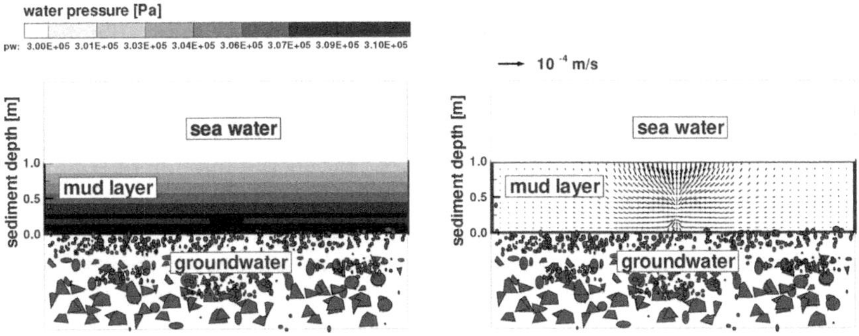

Fig. 5.40. Distribution of the pressure and flow velocity in the water phase at the vent location

Figure 5.41, left, shows the equilibrium between advective and dispersive salt fluxes. It is mentioned that the influence of the diffusion is much stronger than that of the mechanical dispersion. In figure 5.41, right, a comparison between simulations and measurements at the vent location is presented. Chloride corresponds to the salt component or mole fraction in the water phase; it falls to about a third of the prescribed sea-water concentration due to the comparatively high freshwater inflow.

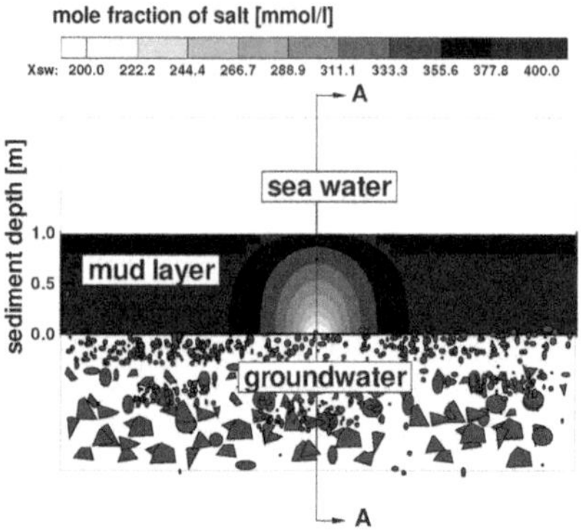

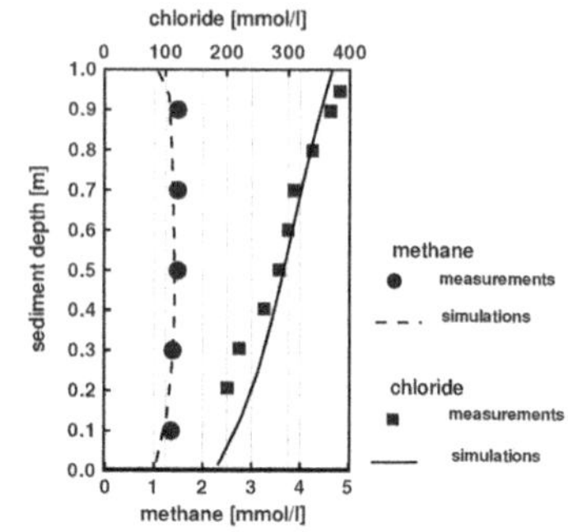

Fig. 5.41. Vent location; left: distribution of the mole fraction of salt in the water phase, right: simulated and measured chloride and methane profiles in section A-A, right: after HINKELMANN et al. (2002 [115])

In figure 5.42, left, the distribution of the methane phase is given, indicating the equilibrium between gas saturation, methane formation and oxidation. For the isothermal conditions in these simulations, the gas phase consists of about 95% methane and 5% water vapor with minor variations. With the exception of the inflow area, the methane production causes the maximum solubility of methane in water to be exceeded. Consequently, a gas phase occurs. Methane gas is held in the lower part of the mud layer due to capillary forces. At the vent location, the methane dissolved in water is vertically advected in such a strong way that the concentrations do not exceed the maximum solubility and, consequently, no gas phase occurs. The profile of the methane component in the water phase is nearly constant (see fig. 5.42, right). It moves as a component in the water through the mud layer and leave the system through the upper boundary. Overall, a reasonable agreement of computations and measurements is obtained.

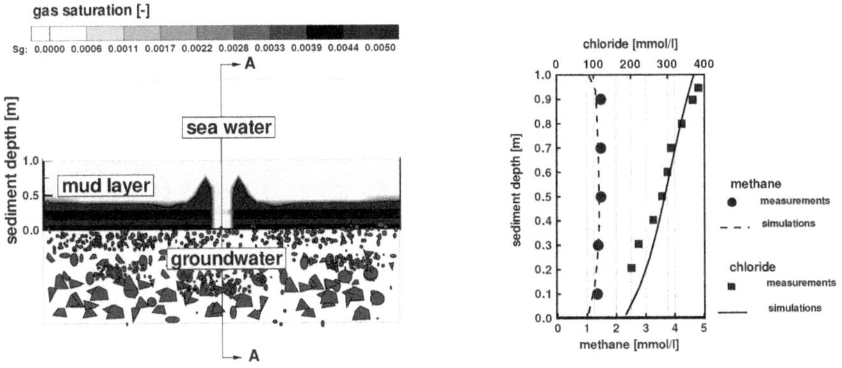

Fig. 5.42. Vent location; left: distribution of the gas saturation, right: simulated and measured chloride and methane profiles in section A-A, after HINKELMANN et al. (2002 [115])

Partial vent location

As already mentioned, the only change in the simulations is a different freshwater discharge at the lower boundary. Figure 5.43, left, shows the distribution of the gas saturation at the partial vent. When compared to figure 5.42, left, the results strongly resemble one another. Again, a gas phase only occurs outside the vent location. In section A-A, the methane concentration appears to be zero in the upper part, caused by the methane sink, and it increases towards the bottom boundary (see fig. 5.43, right). The chloride gradient in figure 5.43, right, is much smaller when compared to the vent location (see fig. 5.41, right) due to the smaller advection. Again, simulations and measurements agreed well.

Non-vent location

As the magnitude of the freshwater discharge is negligible here, a gas phase occurs in the entire bottom of the system (see fig. 5.44, left). In section A-A, methane increases drastically from zero at the sea bottom to the maximum solubility close to the bottom of the system (see fig. 5.44, right). The chloride profile is more or less constant with the water concentration of the sea bottom (see fig. 5.44, right). Overall, simulations and measurements agree reasonably.

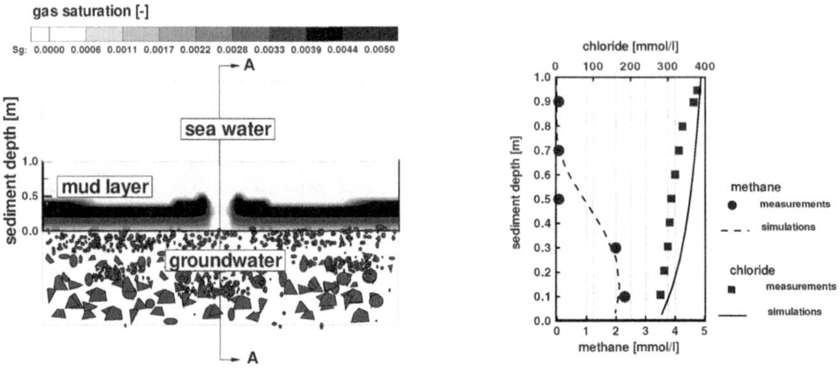

Fig. 5.43. Partial vent location; left: distribution of the gas saturation, right: simulated and measured chloride and methane profiles in section A-A, after HINKELMANN et al. (2002 [115])

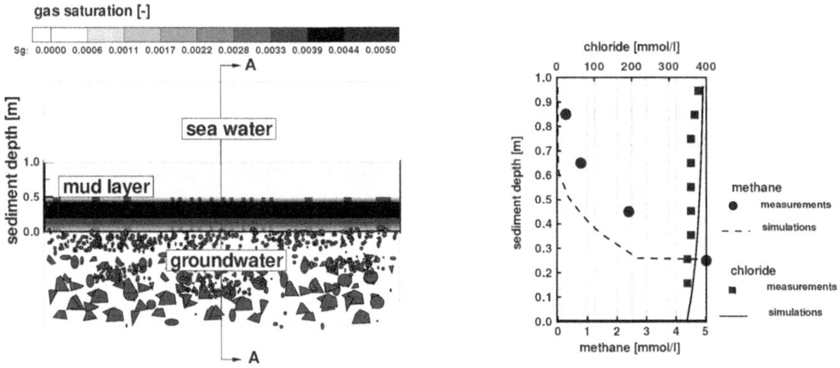

Fig. 5.44. Non-vent location; left: distribution of the gas saturation, right: simulated and measured chloride and methane profiles in section A-A, after HINKELMANN et al. (2002 [115])

Release of gas bubbles

In the field, it was observed that methane is released as gas bubbles into the sea water. The numerical simulations indicate that the gas is trapped in the lower part of the mud layer. One reason could be that the methane production, which is assumed to be constant in the horizontal direction, increases locally. Another reason is given by the influence of small-scale heterogeneities which can be taken into account by geostatistical methods (see sec. 4.1.6). It is well known that there are different dominant forces affecting the movement of gas. One of these forces is the buoyancy, which acts on the gas adue to the large difference between the densities of gas and water. The force acting in the opposite direction is the capillary force or the entry pressure p_d, which inhibits the upward movement of the gas. However, the gas can be released into sea water at locations where the entry pressure is not high enough to prevent its upward movement. The variation of the entry pressure can be determined directly with the help of the geostatistical variation of the permeability K using the *Leverett* function (see sec. 4.1.6).

Therefore, a geostatistical variation of the permeability field ($K_m = 10^{-9} m^2$) is determined with the SIMSET model (see BARDOSSY (1992 [15])), using an exponential variogram with a horizontal correlation length of $c_h = 0.1m$, a vertical correlation length of $c_v = 0.5m$ and a variance of $v = 0.05$ (see fig. 5.45, left). Thus, the small-scale heterogeneities are of an order which can be expected in the field. As the principle effect is to be investigated, just a $1m$-wide part of the mud layer for the non-vent case is analyzed. In figure 5.45, right, the gas saturation is shown. Because the correlation length is higher vertically than horizontally, a fingering is observed in the areas with higher permeability. This effect together with the methane production are the most important reasons for the occurrence of methane gas bubbles.

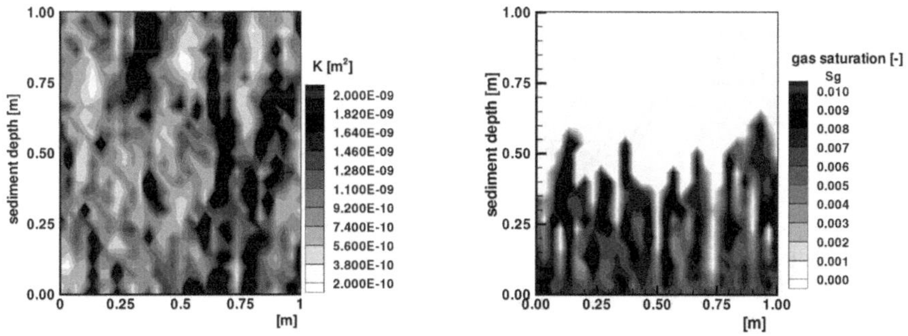

Fig. 5.45. Non-vent location: Geostatistical permeability field and distribution of the gas saturation, after HINKELMANN et al. (2002 [115])

5.4 Flow and transport processes in surface water

In this section, numerical simulations of vertically-integrated and three-dimensional flow and transport processes in shallow-water systems are presented (see sec. 2.6). *Parallel Operator-Splitting* (see secs. 3.1.2, 3.1.6) and *Conjugate Gradient Methods* (see secs. 3.4.2) are chosen as efficient numerical methods. Special emphasis is put on a *grid partitioning* tool (see sec. 4.1.7) with respect to efficient information-processing techniques. In order to guarantee a relative independence of sections 5.4.1 and 5.4.2, some repetitions are unavoidable.

The numerical models are applied to the Baltic Sea estuary *Darß-Zingster Boddenkette* and the North Sea estuary *Weser*. The location of these estuaries in the northern part of Germany is shown in figure 5.46.

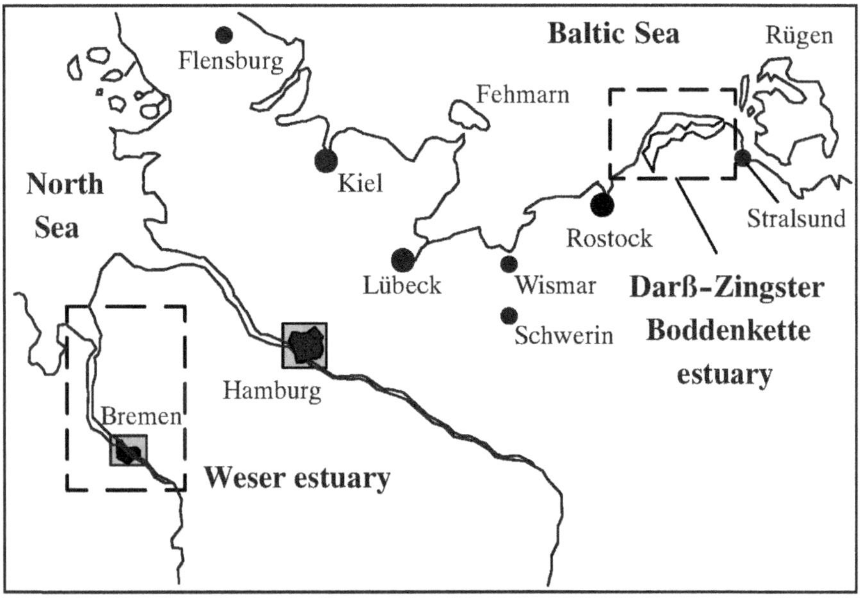

Fig. 5.46. Location of the estuaries Darß-Zingster Boddenkette and Weser, after HINKELMANN (1997 [108])

5.4.1 Vertically-integrated flow and transport processes in shallow water

The numerical simulations are carried out with the *TELEMAC-2D modeling system* ([246]). The sequential algorithm was developed by GALLAND et al.

(1991 [86]), and is further described, for example, in JANIN (1992 [131]). The algorithm is parallelized by HINKELMANN (1997 [108]), and is further discussed in HINKELMANN, ZIELLE (1996 [118]), or HINKELMANN (2000 [109]).

Numerical algorithm

Discretization methods and solvers

The numerical algorithm is based on a parallel Operator-Splitting Method consisting of an *advection (adv)* and a *diffusion step (dif)*. For a clearer comprehension, the governing equations, which are the *Saint-Venant* (eqs. 2.69 - 2.71) and the transport equations (eq. 2.75), are recalled:

continuity equation:

$$\underbrace{\frac{\partial h}{\partial t}}_{adv/dif} \underbrace{+ v_x \frac{\partial h}{\partial x} + v_y \frac{\partial h}{\partial y}}_{adv} \underbrace{+ h \frac{\partial v_x}{\partial x} + h \frac{\partial v_y}{\partial y} = q_w/\rho_w}_{dif} \tag{5.2}$$

momentum equation in x-direction:

$$\underbrace{\frac{\partial v_x}{\partial t}}_{adv/dif} \underbrace{+ v_x \frac{\partial v_x}{\partial x} + v_y \frac{\partial v_x}{\partial y}}_{adv}$$
$$\underbrace{- \frac{\partial}{\partial x}(\nu_{wh} \frac{\partial v_x}{\partial x}) - \frac{\partial}{\partial y}(\nu_{wh} \frac{\partial v_x}{\partial y}) = \frac{f_x}{\rho_{w0}} - g \frac{\partial (h + z_b)}{\partial x}}_{dif} \tag{5.3}$$

momentum equation in y-direction:

$$\underbrace{\frac{\partial v_y}{\partial t}}_{adv/dif} \underbrace{+ v_x \frac{\partial v_y}{\partial x} + v_y \frac{\partial v_y}{\partial y}}_{adv}$$
$$\underbrace{- \frac{\partial}{\partial x}(\nu_{wh} \frac{\partial v_y}{\partial x}) - \frac{\partial}{\partial y}(\nu_{wh} \frac{\partial v_y}{\partial y}) = \frac{f_y}{\rho_{w0}} - g \frac{\partial (h + z_b)}{\partial y}}_{dif} \tag{5.4}$$

salinity transport equation:

$$\underbrace{\frac{\partial S}{\partial t}}_{adv/dif} \underbrace{+ v_x \frac{\partial S}{\partial x} + v_y \frac{\partial S}{\partial y}}_{adv} \underbrace{- \frac{\partial}{\partial x}(\nu_{th} \frac{\partial S}{\partial x}) - \frac{\partial}{\partial y}(\nu_{th} \frac{\partial S}{\partial y}) = \frac{q_S - q_w S}{\rho_w}}_{dif} \tag{5.5}$$

The variables are denoted in section 2.6. The salinity S is chosen as the tracer in the transport equation.

For the generation of a grid, triangular or quadrilateral elements can be chosen. The hyperbolic and parabolic parts in the governing equations are split up and treated separately with methods which are 'optimal' for the particular problem class. First, the hyperbolic terms are dealt with in the advection step (adv). Second, the parabolic and further remaining terms are treated in the diffusion step (dif). The corresponding parts are marked by underbraces in equations 5.2 - 5.5. The temporal derivative terms are split up as follows and used in both steps:

$$\frac{\partial f}{\partial t} = \frac{f^{n+1} - f^n}{\Delta t} = \frac{f^{n+1} - f^{adv}}{\Delta t} + \frac{f^{adv} - f^n}{\Delta t} \tag{5.6}$$

f stands for the quantities water level h, flow velocity v_x, v_y, or salinity S; the 'exponent' ^{adv} specifies the result of the advection step.

The advection step is solved with the Method of Characteristics (see sec. 3.1.6), based on a *Lagrangian* point of view. With a first-order *Runge-Kutta* Method and the velocity field of the old time step, the pathline of each node is traced back to the base point of the characteristic. In order to contribute to the solution at the current time step, the base-point value is interpolated to the nodal values of the element in which the base point is located. This approach is explicit.

With the results of the advection step as initial conditions, the diffusion step is carried out with a semidiscrete Finite-Element Method (see sec. 3.1.4) based on an *Eulerian* point of view. Linear shape functions are applied. An implicit formulation leads to a linear non-symmetric system of equations for the shallow-water equations, which has three unknowns per node and which is solved with the BiCGSTAB Method (see sec. 3.4.2), as well as a linear symmetric system of equations for the salinity transport, which has one unknown per node equation and which is solved with the PCG Method (see sec. 3.4.2). Both solvers can apply different preconditioners (see sec. 3.4.3).

Parallel methods

The Operator-Splitting Method is parallelized with the message-passing programming model (see sec. 3.2.2). An algebraic parallelization with inconsistent storing for the interprocessor nodes is used (see secs. 3.2.2, 3.2.3). The domain is split up into 2^n subdomains which are assigned to the processors. The data and load distribution is static (see sec. 3.2.4), and it is the same for the advection and diffusion step. The parallelization of the Method of Characteristics is described in section 3.2.5. The tracking of the pathlines is initially done completely in parallel. An interprocessor node is treated on each processor / subdomain it belongs to. However, depending on the flow field, the corresponding base point is found on one processor only. Therefore, a local communication is performed for the interprocessor nodes after each advection step (see sec. 3.2.3). Since the base points are almost always located in the accompanying node patch, the parallelization of the pathline tracking is also practicable for distributed-memory systems (see sec. 4.2.1). However, additional steps, which are described in HINKELMANN (1997 [108]), are necessary. The treatment of the semidiscrete Finite-Element Method consists of two parts, assembling the system matrices and solving the systems of equations. The first part is performed as far as possible without communication. The parallelization of the BiCGSTAB and PCG Methods for inconsistent storing is based on the techniques described in sections 3.2.3 and 3.4.2.

Evaluation of the numerical algorithm

Even though the numerical algorithm contains an explicit part, it is not constrained by a stability criterion. Therefore, time steps which are several times larger than the *Courant* number can be chosen (see HINKELMANN (1997 [108])). The advection step has to be performed only once for the shallow-water equations and any number of transport equations. Unstructured triangular or quadrilateral grids can be used. Although the algorithm is not monotonous, it is rather well suited to the modeling of *sharp-front problems*, even though the result can be somewhat damped due to the interpolation during the advection step. Conservativity problems occur, e.g. of mass, which result from the Method of Characteristics and which require additional correction algorithms. Furthermore, a first-order accuracy with respect to time of the Operator-Splitting Method is disadvantageous. Second-order accuracy with respect to space is given. The combined explicit and implicit method, as presented above, can be parallelized well including the Conjugate Gradient solvers and to a larger extent also vectorized. When adaptive and Multigrid Methods are considered, the algorithm must be enlarged by dynamic data structures. In conclusion, this numerical algorithm, even with some drawbacks, can be evaluated as very good.

The presented parallel numerical algorithm is used at, for example, the *Federal Waterways Engineering and Research Institute* in Germany for large-scale simulations (see KOPMANN, JANKOWSKI (2000 [151])).

Example: Flow and salinity transport processes in the Darß-Zingster Boddenkette

Model set-up and simulations

The performance of the implemented algorithms is examined using a practical engineering example which deals with the numerical modeling of the Baltic-Sea estuary Darß-Zingster Boddenkette (see fig. 5.46). This estuary consists of four lakes which are connected by narrow passages and which are connected to the Baltic Sea at the eastern end. It covers an area of about 200/cra^2 and is very shallow with an average water depth of about 2m. The goal of this modeling work is to calibrate a numerical model in order to make predictions regarding future interference with the system, e.g. an artificial connection with the Baltic Sea (see HINKELMANN et al. [117]), STÜCKRAD, HINKELMANN (1995 [243])). For the model calibration, a simulation period is chosen for which a huge data set from a specially performed measurement program is available. A map as well as a coarse grid are shown in figure 5.47, a sattelite photo is given in figure 1.2.

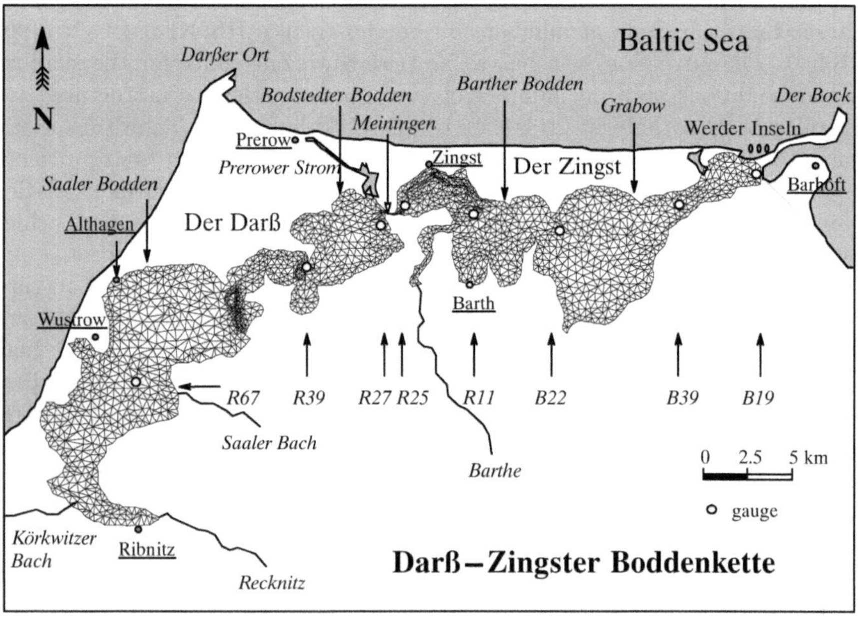

Fig. 5.47. Map, coarse grid and position of measurement devices, after HINKELMANN (1997 [108])

The model is mainly calibrated with the *Taylor*-friction coefficient. A comparatively high friction coefficient $\lambda = 0.03$ leads to the best agreement between computations and measurements; this is caused by very dense sea-bottom vegetation and low water levels. The turbulent viscosity and diffusivity are less sensitive during the model calibration; the turbulent viscosity is set to $\nu_{wh} = 1m^2/s$, and the turbulent diffusivity to $\nu_{th} = 0.1m^2/s$ to stabilize the solutions. The *Crank-Nicholson* factor is $\theta = 0.6$. A diagonal preconditioner is applied. As initial conditions, the water level and the salinity distributions are assumed to be linearly distributed over the domain, and they are constructed from measured data; the flow velocities are set to zero. Measured water levels and salinity concentrations at the open boundary to the Baltic Sea (gauge B19) as well as fresh-water inflows and wind velocities serve as boundary conditions. Along the closed boundaries, the flow velocities are set to zero. Further information can be found in HINKELMANN (1997 [108]).

The water levels in figure 5.48 show an incoming wave from the Baltic Sea which is damped more and more with progressing propagation. The comparison between measurements and computations here is very good. However, a satisfactory numerical simulation of the complex flow patterns in the estuary is not achieved (see fig. 5.49, left); this is discussed in detail in HINKELMANN (1997 [108]). A comparison of the measured and computed salinity concentration (see fig. 5.49, right) shows only small differences, in general less than 1%. In figure 5.50, the distributions of the discharge per meter width and the salinity concentrations are given. Large discharge vectors can be found at the inflow area close to the Baltic Sea and at the Meiningen narrow, while only small discharge vectors occur within the estuary. The salinity concentration increases from west to east. The mixing with incoming fresh water from rivers and creeks can be seen in regions with very small salinity concentrations.

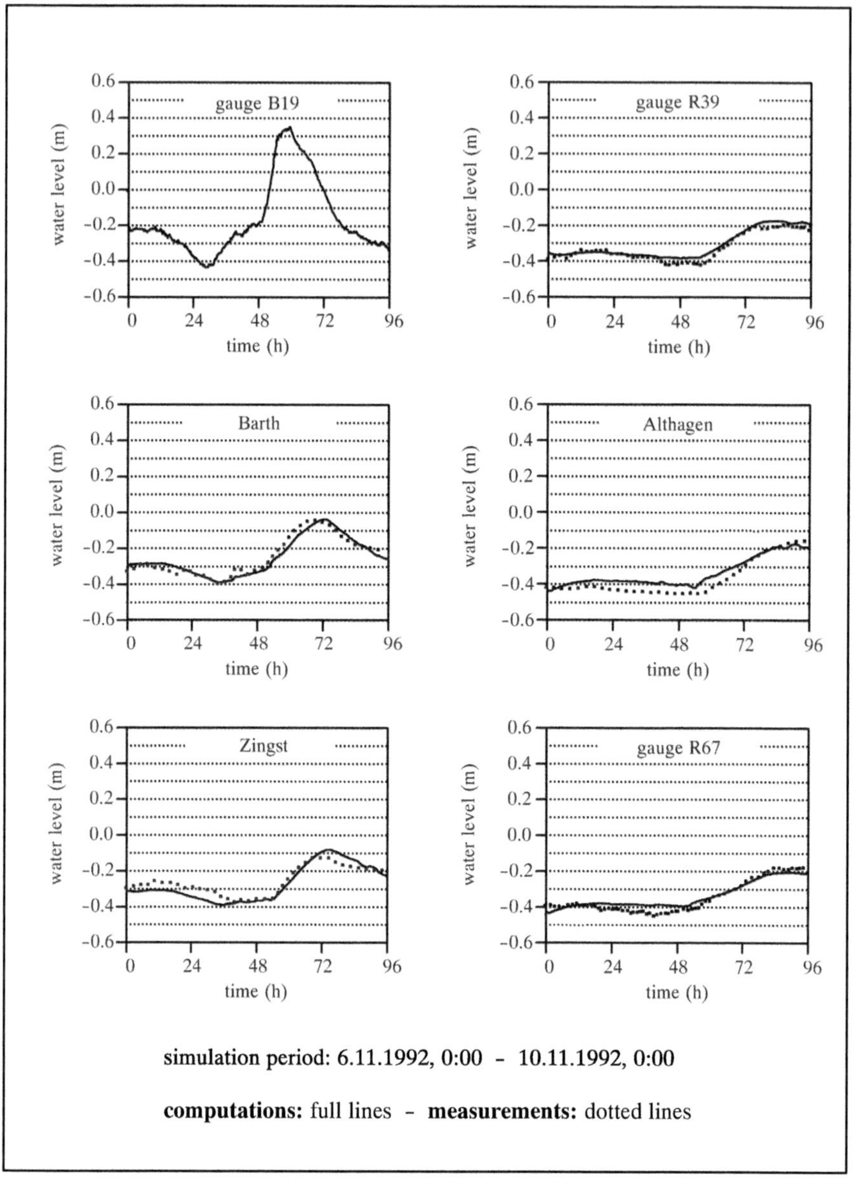

Fig. 5.48. Time series of water levels, after HINKELMANN (1997 [108])

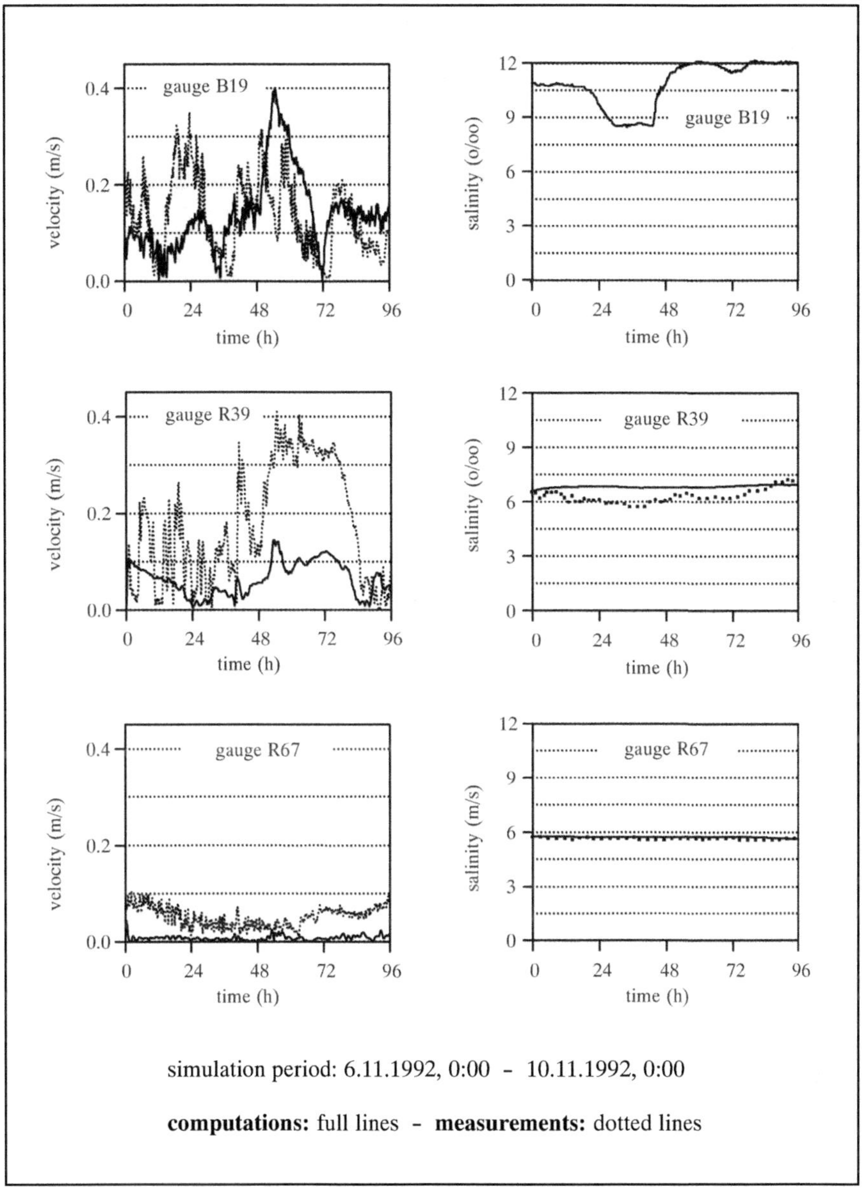

Fig. 5.49. Time series of flow velocities and salinity concentration, after HINKELMANN (1997 [108])

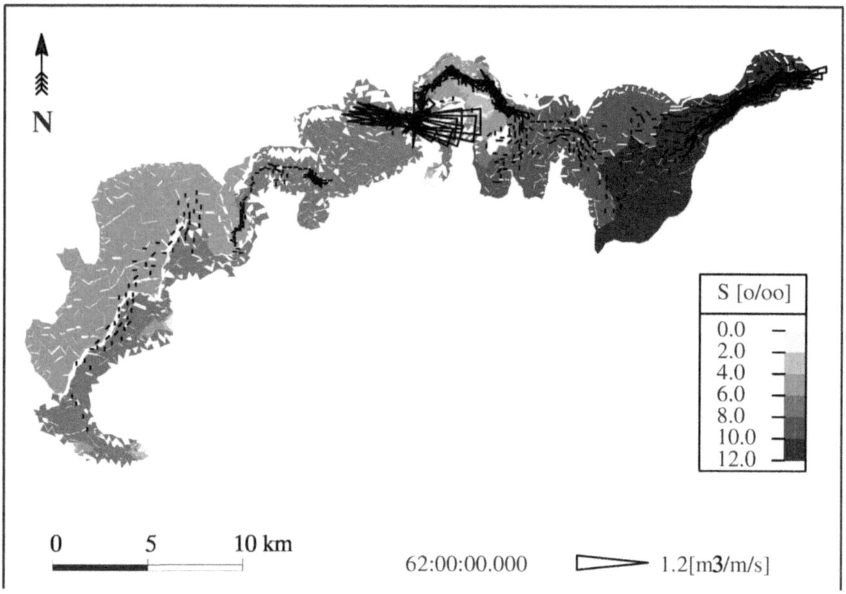

Fig. 5.50. Discharge and salinity concentration distributions after $62h$, after HINKELMANN (1997 [108])

Partitioning of the domain and load balancing

Some of the load balancing methods described in section 3.2.4 are investigated on two triangular grids which differ in the number of nodes and elements by a factor of approximately 4. The coarse grid in figure 5.47 consists of about 2200 nodes and 3900 elements, while the fine grid in figure 5.53 is made up of approximately 8900 nodes and 16300 elements. Further, a small creek in the northern part of the Darß-Zinster Boddenkette is included in the fine grid. This creek is important for some of the questions under investigation, e.g. the artifical connection with the Baltic Sea (see STÜCKRAD, HINKELMANN (1995 [243])).

Figures 5.51 and 5.52 show the partitioning of the domain on a coarse grid for 8 and 32 processors. The interprocessor nodes and a small surrounding are not displayed. For the variants with up to 8 processors, the domain is split up in 'stripes, i.e. interior subdomains adjoin 2 other subdomains and boundary subdomains just 1. Further, the mapping described in section 3.2.4 is applied. It can be seen in figure 5.51 that the modified Recursive-Coordinate Bisection combined with the *Kernighan-Lin* heuristic (RCB+KL) does not yield a connected subdomain for processor $P4$, even though only 8 processors are used. This is not the case if the Recursive-Spectral Bisection combined with the *Kernighan-Lin* heuristic (RSB+KL) is applied. The partitioning of the domain for 32 processors (see fig. 5.52) looks like a chess board for RCB+KL, but it is irregular for RSB+KL. For complex grids, it is barely recognizable that RSB+KL is better than RCB+KL. Figure 5.53 contains a partitioning of the fine grid for 128 processors using RSB+KL. For the area around the Meiningen narrow, it has to be taken into account that, due to the extremely complex topography, no connected subdomains occur for more than 16 processors. Therefore, the numbers of elements in the subgrids in the zoom of figure 5.53 are not the same.

Table 5.5 contains the parallel efficiencies (see sec. 4.2.3) for the simulation on the coarse grid. They are achieved using the very simple - supposedly worse - RCB and the high-quality - supposedly very good - RSK+KL on a nCUBE2S parallel computer (see sec. 3.4.6). The interesting result is the very small difference between the RCB and the RSB+KL. This is caused by the grid partitioning for the dual graph (see sec. 3.2.4) and the comparatively small parallel overhead for such a problem size; more detailed explanations are given in HINKELMANN (1997 [108]). For increasing problem size, the differences between the two load-balancing methods become even smaller. In summary, it is stated that, on a parallel computer suitable for fine-granular problems (see sec. 4.2.2), different static load balancing methods applied to FDM, FEM or FEM differ only slightly.

processors	1	2	4	8	16	32
RSB + KL [%]	100.0	97.6	93.8	87.5	75.6	58.8
RCB [%	100.0	97.4	93.3	87.3	75.3	58.1

Table 5.5. Parallel efficiency for RSB+KL and RCB on 1 to 32 processors, after HINKELMANN (1997 [108])

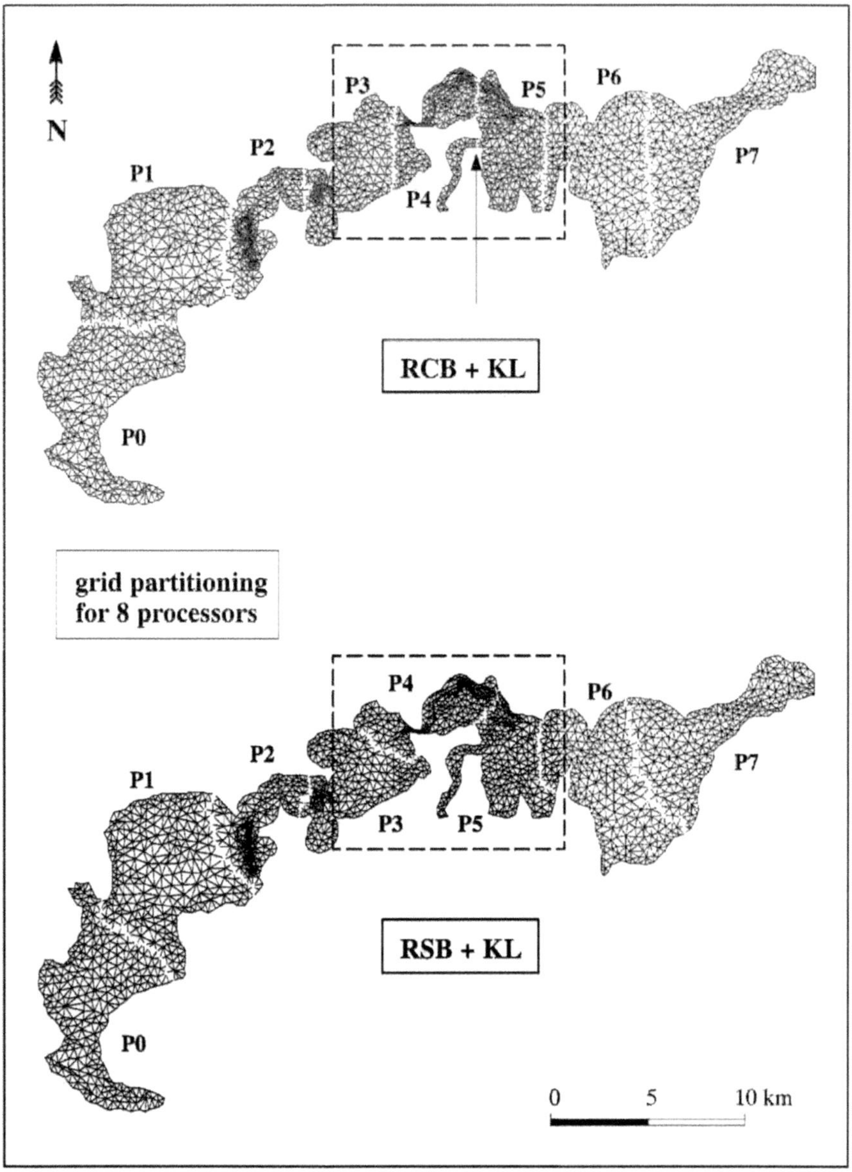

Fig. 5.51. RCB+KL and RSB+KL for 8 processors on the coarse grid, after HINKELMANN (1997 [108])

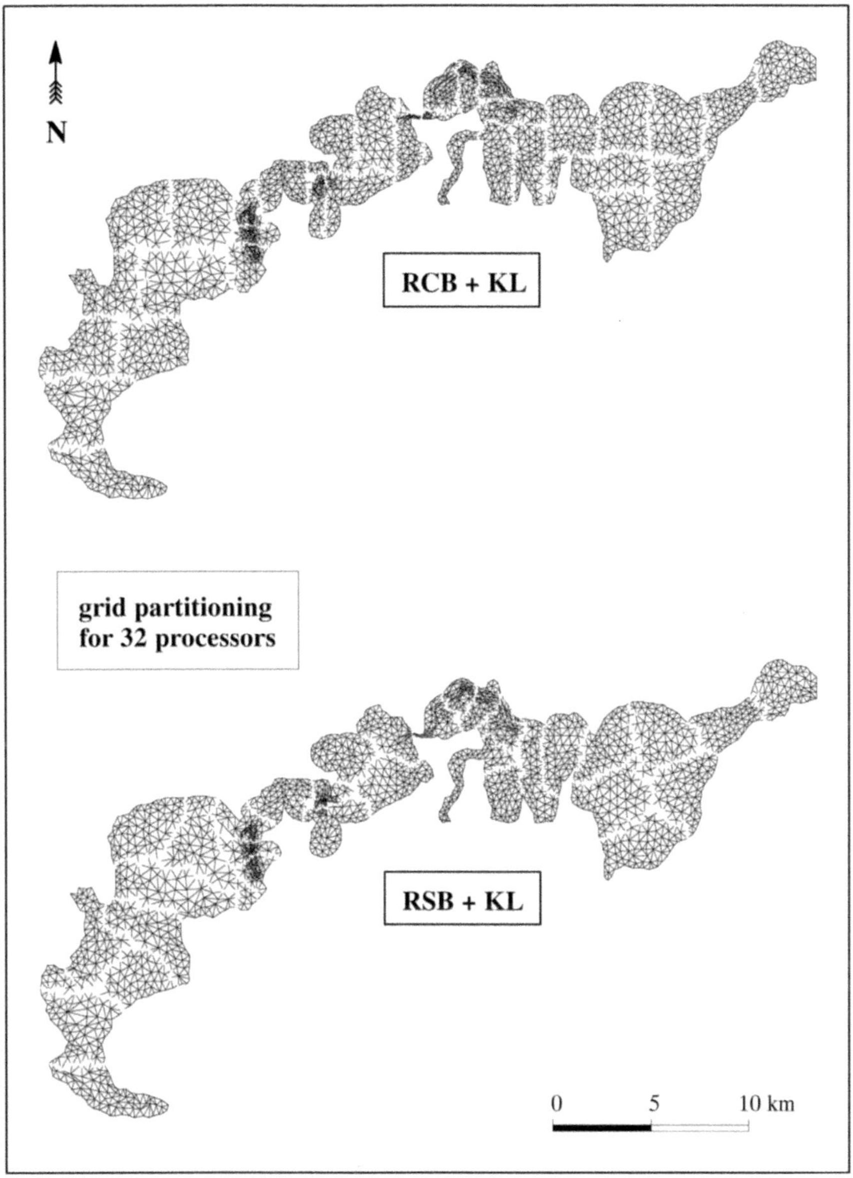

Fig. 5.52. RCB+KL and RSB+KL for 32 processors on the coarse grid, after HINKELMANN (1997 [108])

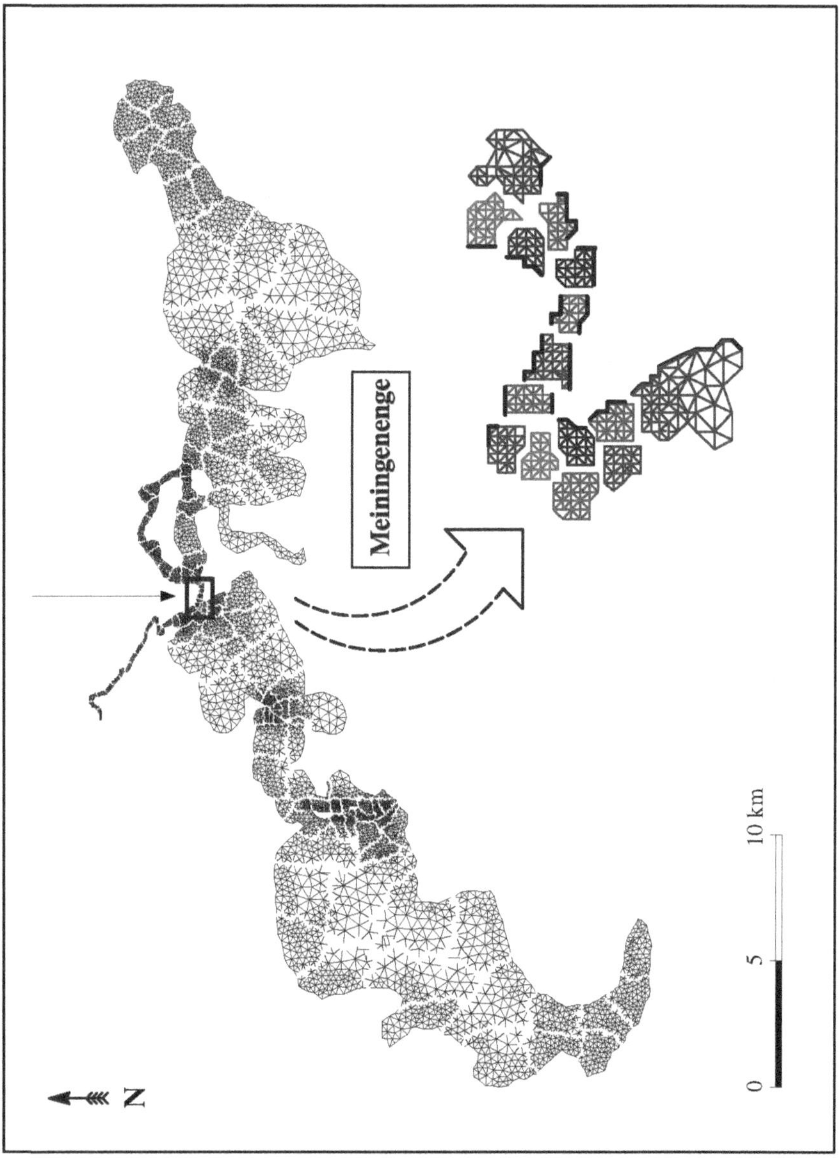

Fig. 5.53. RSB+KL for 128 processors on the fine grid, after HINKELMANN (1997 [108])

<u>Parallel run-time behavior</u>

For the evaluation of the parallel run time behavior, 100 time steps are investigated on the nCUBE2S parallel computer (see sec. 3.4.6); a time step of 15s is chosen for the coarse grid (approximately 2200 nodes and 3900 elements), and a time step of 6s for the fine grid (approximately 8900 nodes and 16300 elements). This corresponds to a *Courant* number of approximately 6 for both grids.

Figure 5.54 shows the parallel efficiency and the speedup on 1 to 128 processors. As a consequence of the parallel overhead (see sec. 3.2.3), the parallel efficiency and the inclination of the parallel speedup decrease with increasing problem size, while the parallel speedup increases. Overall, an efficient parallelization of the entire numerical algorithm is achieved. On the coarse grid, only a partitioning of the domain up to 32 processors is carried out since, with 32 processors, there are already fewer than 100 nodes per processor left, and the parallel overhead for even fewer nodes per processor increases dramatically. The following recommendation can be made for such two-dimensional numericals simulations: for a given problem size, the number of processors should be limited in such a way that at least 100 nodes or even better 100 elements are assigned to each processor. This rule of thumb limits the number of processors to 32 and 128 for the coarse grid and the fine grid respectively. In this range of application, the parallel efficiency and speedup are always larger than 57% and 18 for the coarse grid and always larger than 49% and 63 for the fine grid; these numbers are still reasonable lower limits for unstructured grids.

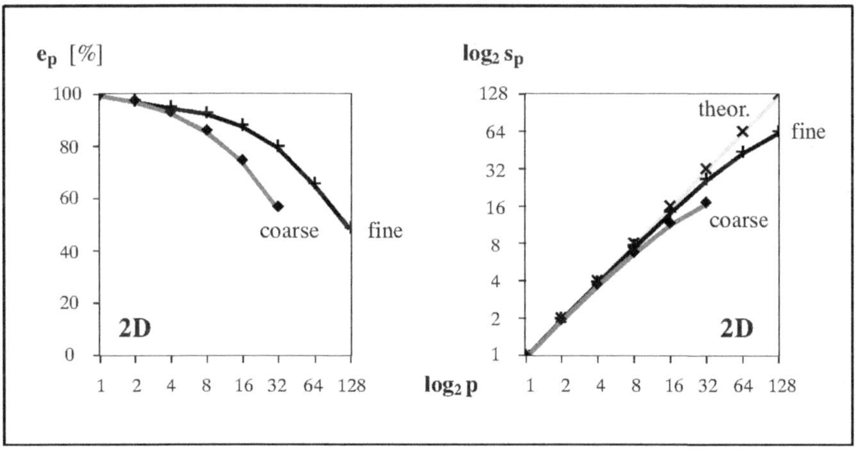

Fig. 5.54. Parallel efficiency and speedup for the coarse and fine grid, after HINKELMANN (1997 [108])

The variation of the number of time steps, time-step sizes, simulation periods, physical parameters (e.g. turbulent viscosity) as well as numerical parameters (e.g. *Crank-Nicholson* factor) does not have a remarkable influence on the parallel run-time behavior. The use of different storage techniques for the system matrices, different preconditioners (see sec. 3.4.3) as well as an automated time-step control are discussed in HINKELMANN (1997 [108]).

Vector computer and parallel vector computer

The Operator-Splitting method described was vectorized based on an Element-By-Element storage technique (see sec. 3.4.3) by LEPEINTRE (1992 [166]). As a prerequisite for using the vectorized algorithm, a special element numbering must be generated in order to avoid recursions and data dependencies (see HINKELMANN (1997 [108])). A 70% usage of the vectorization unit is achieved on the vector computer *S400/40* from *Siemens-Nixdorf / Fujitsu*; this is a reasonable number for an algorithm based on an unstructured grid. The S400/40 has a substantially higher processor performance than 32 nCUBE2S processors. Therefore, the S400/40 is several orders of magnitude faster than 32 nCUBE2S processors.

The simulations on the coarse grid are also performed on the parallel vector computer NEX SX-4. For this purpose, a special storage technique for the system matrices is developed (see HINKELMANN (1997 [108])). The simulations on 1 to 32 parallel vector computers demonstrated the collaboration of parallelization and vectorization on principle. Far better computation times are achieved than on other parallel and vector computers, but the results by no means fulfilled the expectations based on the performance characteristics of the NEX SX-4. The main reason for this is that the problem size is much too small with approximately 2200 nodes and 3900 elements. An efficient usage of parallel vector computers, even with a moderate number of processors, can be achieved only for large-scale-problem sizes, i.e. more than 10^6 nodes. Further information can be found in HINKELMANN (1997 [108]).

5.4.2 Three-dimensional flow and transport processes in shallow water

The numerical simulations are undertaken with the *TELEMAC-3D modeling system* ([246]). The sequential algorithm was developed by JANIN et al. (1992 [132]) and is further described in, for example, MALCHEREK (1995 [170]). The algorithm is parallelized by HINKELMANN (1997 [108]) and is further discussed in HINKELMANN et al. (1998 [113]), HINKELMANN, ZIELLE (2000 [119]), or HINKELMANN (2000 [109]).

Numerical algorithm

Discretization methods and solvers

The numerical algorithm is based on a parallel Operator-Splitting Method consisting of an *advection (adv)*, a *diffusion (dif)* and a *free-surface-continuity-pressure step (fcp)*. For a better understanding, the governing equations, which are the three-dimensional shallow-water (eqs. 2.60, 2.63 - 2.65) and the transport equations (eq. 2.75) together with an equation of state for the density of water (eq. 2.76), are recalled:

continuity equation:

$$\underbrace{\frac{\partial v_x}{\partial x} + \frac{\partial v_y}{\partial y} + \frac{\partial v_z}{\partial z} = q_w/\rho_w}_{fcp} \tag{5.7}$$

momentum equation in x-direction:

$$\underbrace{\frac{\partial v_x}{\partial t}}_{adv/dif/fcp} \underbrace{+ v_x \frac{\partial v_x}{\partial x} + v_y \frac{\partial v_x}{\partial y} + v_z \frac{\partial v_x}{\partial z}}_{adv}$$

$$\underbrace{- \frac{\partial}{\partial x}(\nu_{wh} \frac{\partial v_x}{\partial x}) - \frac{\partial}{\partial y}(\nu_{wh} \frac{\partial v_x}{\partial y}) - \frac{\partial}{\partial z}(\nu_{wv} \frac{\partial v_x}{\partial z})}_{dif} = \underbrace{\frac{1}{\rho_{w0}} f_x}_{dif} \underbrace{- \frac{1}{\rho_{w0}} \frac{\partial p}{\partial x}}_{fcp} \tag{5.8}$$

momentum equation in y-direction:

$$\underbrace{\frac{\partial v_y}{\partial t}}_{adv/dif/fcp} \underbrace{+ v_x \frac{\partial v_y}{\partial x} + v_y \frac{\partial v_y}{\partial y} + v_z \frac{\partial v_y}{\partial z}}_{adv}$$

$$\underbrace{- \frac{\partial}{\partial x}(\nu_{wh} \frac{\partial v_y}{\partial x}) - \frac{\partial}{\partial y}(\nu_{wh} \frac{\partial v_y}{\partial y}) - \frac{\partial}{\partial z}(\nu_{wv} \frac{\partial v_y}{\partial z})}_{dif} = \underbrace{\frac{1}{\rho_{w0}} f_y}_{dif} \underbrace{- \frac{1}{\rho_{w0}} \frac{\partial p}{\partial y}}_{fcp} \tag{5.9}$$

momentum equation in z-direction:

$$\underbrace{\frac{1}{\partial\rho_w}\frac{\partial p}{\partial z} + g = 0}_{fcp} \qquad \Leftrightarrow \qquad \underbrace{p = \rho_{w0}g(z_s - z) + \rho_{w0}g\int_z^{z_s}\frac{\Delta\rho_w}{\rho_{w0}}dz}_{fcp} \quad (5.10)$$

salinity transport equation:

$$\underbrace{\frac{\partial S}{\partial t}}_{adv/dif} \underbrace{+ v_x\frac{\partial S}{\partial x} + v_y\frac{\partial S}{\partial y} + v_z\frac{\partial S}{\partial z}}_{adv}$$

$$\underbrace{-\frac{\partial}{\partial x}(\nu_{th}\frac{\partial S}{\partial x}) - \frac{\partial}{\partial y}(\nu_{th}\frac{\partial S}{\partial y}) - \frac{\partial}{\partial z}(\nu_{tv}\frac{\partial S}{\partial z}) = \frac{q_S - q_w S}{\rho_w}}_{dif} \quad (5.11)$$

equation of state:

$$\rho_w(S) = \rho_{w0}(1 + 0.00075S) \quad (5.12)$$

The variables are denoted in section 2.6. The salinity S is chosen as the tracer in the transport equation.

The computational domain is discretized with prismatic elements consisting of triangles in the horizontal projection which are duplicated along the vertical (see fig. 5.55). This special mesh topology causes simple node and element addressing and is very advantageous for the grid partitioning (see fig. 5.56). Due to the water-level elevation, the three-dimensional mesh moves with time. By applying the so-called *σ-transformation*, the real mesh is projected onto a stationary one, the so-called *σ-mesh*, with the unsteadiness being considered in the transformation relationship. With $z^* = 0$ representing the bottom z_b and $z^* = 1$ the free water surface z_s, the following holds:

$$z(x, y, t) = z_b(x, y) + z^*[z_s(x, y, t) - z_b(x, y)],$$
$$0 \leq z^* \leq 1 \quad \Leftrightarrow \quad z^* = \frac{z - z_b}{z_s - z_b} \quad (5.13)$$

As in the two-dimensional model (see sec. 5.4.1), the hyperbolic and parabolic terms in the basic equations are split up and treated separately with methods which are 'optimal' for the particular problem class. First, the hyperbolic terms are dealt with in the advection step (adv). Second, the parabolic and some right-hand-side terms are treated in the diffusion step (dif). In the last step, the free surface, continuity and pressure terms (fcp) are determined. The corresponding parts are marked by underbraces in equations 5.7 - 5.11. The temporal derivative terms are split up as follows and used in all three steps:

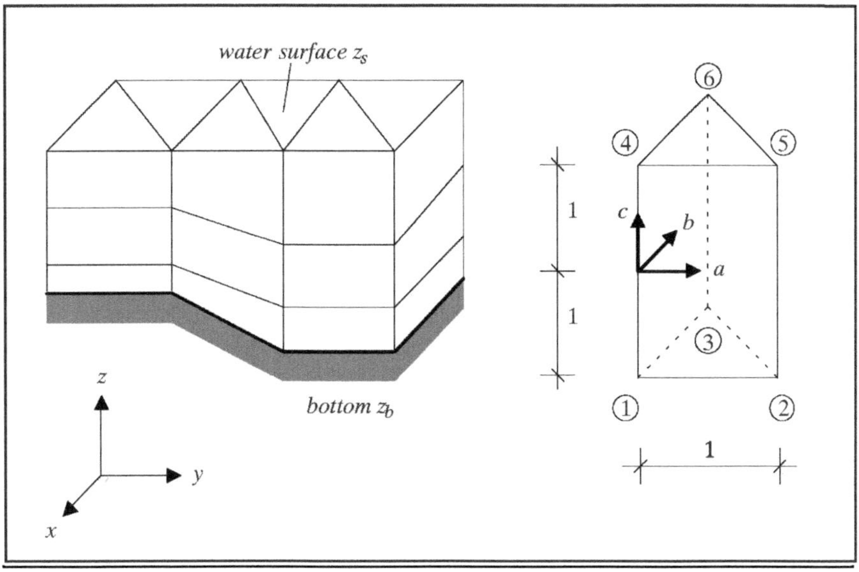

Fig. 5.55. Space discretization with prisms, after HINKELMANN (1997 [108])

$$\frac{\partial f}{\partial t} = \frac{f^{n+1} - f^n}{\Delta t} = \frac{f^{n+1} - f^{dif}}{\Delta t} + \frac{f^{dif} - f^{adv}}{\Delta t} + \frac{f^{adv} - f^n}{\Delta t} \tag{5.14}$$

Here, f represents the quantities flow velocity components v_x, v_y, v_z or salinity concentration S; the 'exponent' ^{adv} specifies the result from the advection step and ^{dif} the one from the diffusion step.

The advection step is solved with the *Method of Characteristics* (see sec. 3.1.6) in the σ-mesh based on a *Lagrangian* point of view. With a first-order *Runge-Kutta Method* and the velocity field of the old time step, the pathline of each node is traced back to the base point of the characteristic. In order to contribute to the solution at the current time step, the base-point value is interpolated to the nodal values of the element in which the base point is located. This approach is explicit.

With the results of the advection step as initial condition, the diffusion step is carried out with a *semidiscrete Finite-Element Method* (see sec. 3.1.4) in the real mesh based on an *Eulerian* point of view. Linear shape functions are applied. The horizontal flow velocities as well as the salt concentrations are decoupled and calculated subsequently. An implicit formulation leads to three linear symmetric systems of equations with one unknown per node, using a PCG solver (see sec. 3.4.2) with different preconditioners (see sec. 3.4.3) for their solutions. The integration of the system matrix is carried out analytically with the help of a mathematical expert system (see JANIN et al. (1992 [132])).

Here, it is advantageous that the analytical integration can be vectorized in contrast to the *Gauss-Point* integration (see sec. 3.1.4). The calculation of the salinity transport terminates after the diffusive step.

With the solution of the diffusive step as the initial conditions, the remaining terms are treated in the free-surface-pressure-continuity step in the real mesh based on an *Eulerian* point of view. For this purpose, the flow velocities are averaged vertically. The equations to be treated represent the *Saint-Venant* equations in a reduced form. They are solved as described in section 5.4.1. It is mentioned that the influence of density-induced flows is considered in this step.

Parallel methods

The Operator-Splitting Method is parallelized with the message-passing programming model (see sec. 3.2.2). An algebraic parallelization with inconsistent storing for the interprocessor nodes is used (see secs. 3.2.2), 3.2.3). The partitioning of the whole domain is carried out for the horizontal projection of the grid in the same way as for the two-dimensional model (see 5.4.1). The horizontally projected grid is split up into 2^n subdomains which are assigned to the processors with all prisms lying below. The data and load distribution is static (see sec. 3.2.4), and it is the same for the advection, diffusion and free-surface-continuity-pressure step. The parallelization of the Method of Characteristics is described in section 3.2.5. The tracking of the pathlines is initially carried out completely in parallel. An interprocessor node is treated on each processor / subdomain where it belongs to. However, depending on the flow field, the corresponding base point is found on one processor only. Therefore, a local communication is performed for the interprocessor nodes after each advective step (see sec. 3.2.3). Since the base points are almost always located in the accompanying node patch, the parallelization of the pathline tracking is also practicable for distributed-memory systems (see sec. 4.2.1). However, additional steps are necessary; these are described in HINKELMANN (1997 [108]). The treatment of the semidiscrete Finite-Element Method in the diffusion step consists of two parts: assembling system matrices and solving the systems of equations. Assembling the system matrices is performed as far as possible without communication. The parallelization of the PCG and BiCGSTAB Methods for inconsistent storing is based on the techniques described in sections 3.2.3 and 3.4.2. The free-surface-continuity-pressure step is carried out as described in section 5.4.1.

Evaluation of the numerical algorithm

Even though the numerical algorithm contains an explicit part, it is not constrained by a stability criterion. Therefore, time steps which are several times larger than the *Courant* number can be chosen (see HINKELMANN (1997 [108])). The advection step has to be performed only once for the shallow-water equations and any number of transport equations. The grid consisting of prismatic elements is unstructured in the horizontal projection and structured in the vertical. It is well suited for 'compact' water bodies; however, for complex vertical structures, such grids lead to an over-discretization in shallow regions and an under-discretization in deep regions. Although the algorithm is not monotonous, it is rather well suitable for the modeling of *sharp-front problems*, even though a certain numerical damping can occur due to the interpolation during the advection step. Above all, three points are disadvantageous: the vertical grid structure with the same number of layers in the whole domain together with the σ-transformation; global (mass) conservation problems which result from the Method of Characteristics and which require additional correction algorithms; only first-order accuracy in time. Second-order accuracy with respect to space is given. The combined explicit and implicit method, as presented above, is very suitable for parallelization including the Conjugate Gradient solvers and to a larger extent also for vectorization. When adaptive and Multigrid Methods are considered, an extension to dynamic data structures is required. In conclusion, this numerical algorithm, even with some drawbacks, can be evaluated as rather good.

Example: Flow and salinity transport processes in the Weser

Model set-up and simulations

The performance of the implemented algorithms is demonstrated using a field study of the North-Sea estuary Weser (see fig. 5.46). The investigated part is located between Nordenham and the Hemelingen weir; it is about $60km$ long, $400 - 2000m$ wide and $3 - 10m$ deep (see fig. 5.56). The model was calibrated by MALCHEREK (1995 [170]) with the help of a very large and coherent data set. The aim of this work consisted of developing a fundamental process understanding of the dynamics of suspended sediments in the Weser and its numerical simulation. For the investigations described in the following, the suspended sediment transport is not taken into account, and some small extensions are made in order to achieve a better calibration. A map and the horizontal projection of the grid with approximately 1500 2D-nodes and 2300 2D-elements is shown in figure 5.56. For the model calibration, 12 planes are distributed logarithmically in the vertical direction. With respect to the vertical structure, the domain can be characterized as rather compact; this is advantageous for the use of the σ-transformation.

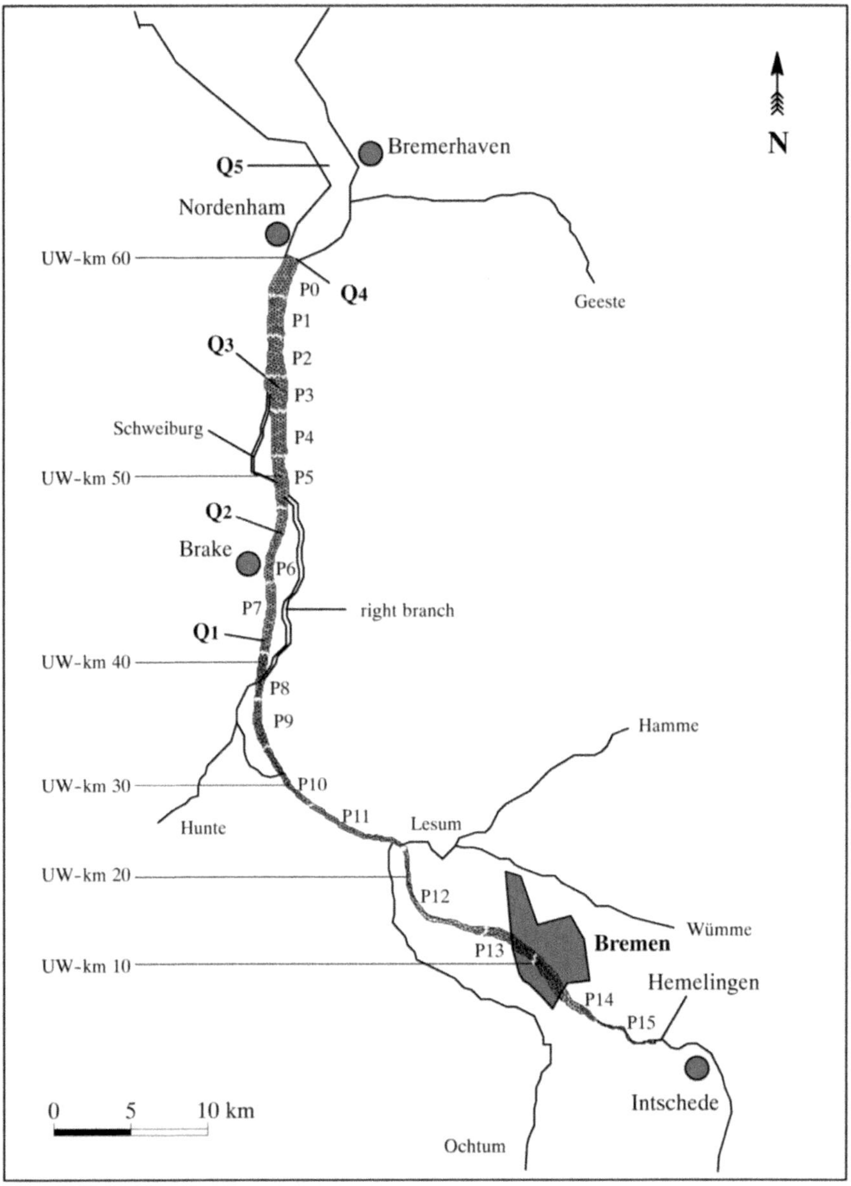

Fig. 5.56. Map, gauge locations, grid and partitioning for 16 processors with RSB. after HINKELMANN (1997 [108])

The model was calibrated with the friction coefficient and the turbulence parameters. A *Chezy* friction coefficient of 28.9 was determined from measurements (see MALCHEREK (1995 [170])). The horizontal turbulent viscosity was set to $\nu_{wh} = 0.1 m^2/s$ and the horizontal turbulent diffusivity to $\nu_{th} = 1.0 m^2/s$. In the vertical direction, the turbulence parameters ν_{wv}, ν_{tv} were computed with the mixing length model of LEHFELDT (1991 [165], see sec. 2.6.1). The *Crank-Nicholson* factor was $\theta = 0.55$. As initial conditions, a constant water level distribution with the value of position Q_4 was chosen for the whole domain, the flow velocities were set to zero, and the salinity distribution was assumed to be linearly distributed between position Q_4 and the Hemelingen weir. The measured water levels and salinity concentrations (during inflow only) were specified as boundary conditions at the open boundary to the sea at position Q_4. At the upstream boundary at the Hemelingen weir, the measured discharge and salinity concentration were prescribed as boundary conditions. Along the closed boundaries, the flow velocity is set to zero. Moreover, measured wind velocities were given. MALCHEREK (1995 [170]) developed the measured water levels and salinity concentrations at position Q_4 in *Fourier* series. They are parallized as described in section 3.2.5. Further information can be found in MALCHEREK (1995 [170]) and HINKELMANN (1997 [108]).

Figure 5.57 contains time series of calculated and measured flow velocities at position Q_2 and Q_3 (see fig. 5.56) which show a very good agreement. The typical influence of the tides can be observed in the periodic and sinus-like course of the results. The variability of the vertical flow profile is clearly seen at position Q_2 by comparing the flow velocities close to the bottom (1 meter above bottom, $1mab$) and close to the surface ($8mab$). Figure 5.58 shows simulated and measured salinity concentrations. At positions Q_3 and Q_4, the agreement is rather good. However, larger deviations occur at position Q_2 during the second half of the simulation. Reasons for this are discussed in detail in HINKELMANN (1997 [108]). Figure 5.59 contains the distribution of the vertically integrated salinity concentration and shows the maximum penetration of the salt tongue into the estuary during the eighth tide period after 90 hours simulation time.

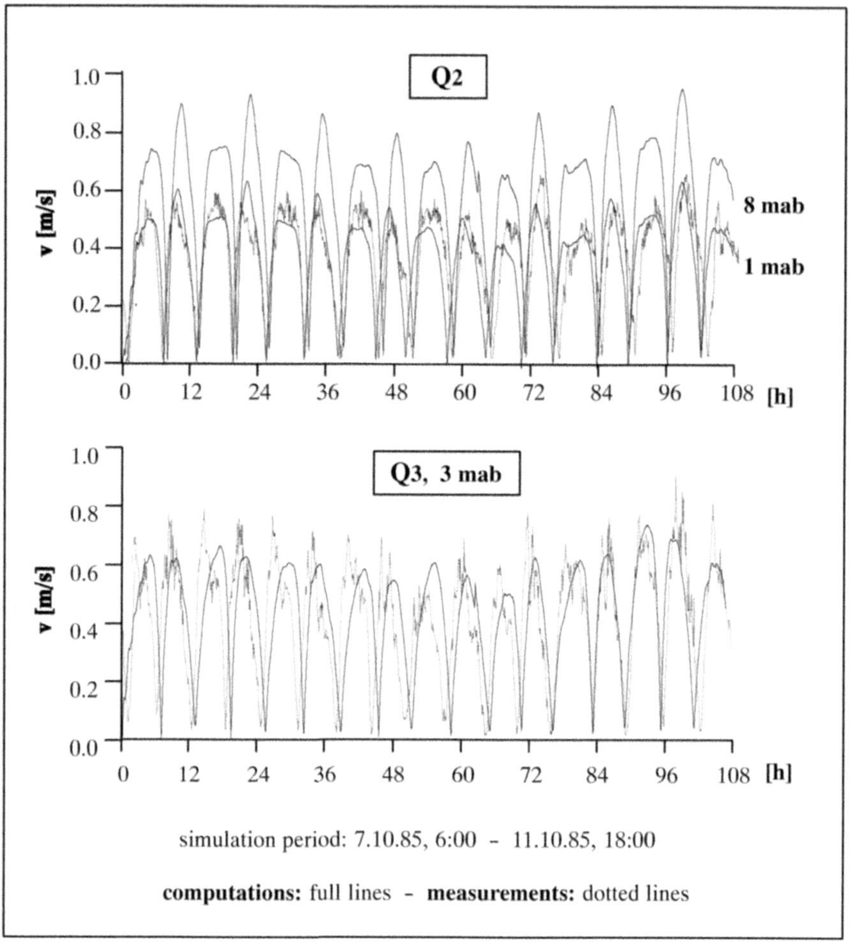

Fig. 5.57. Time series of flow velocities, after HINKELMANN (1997 [108])

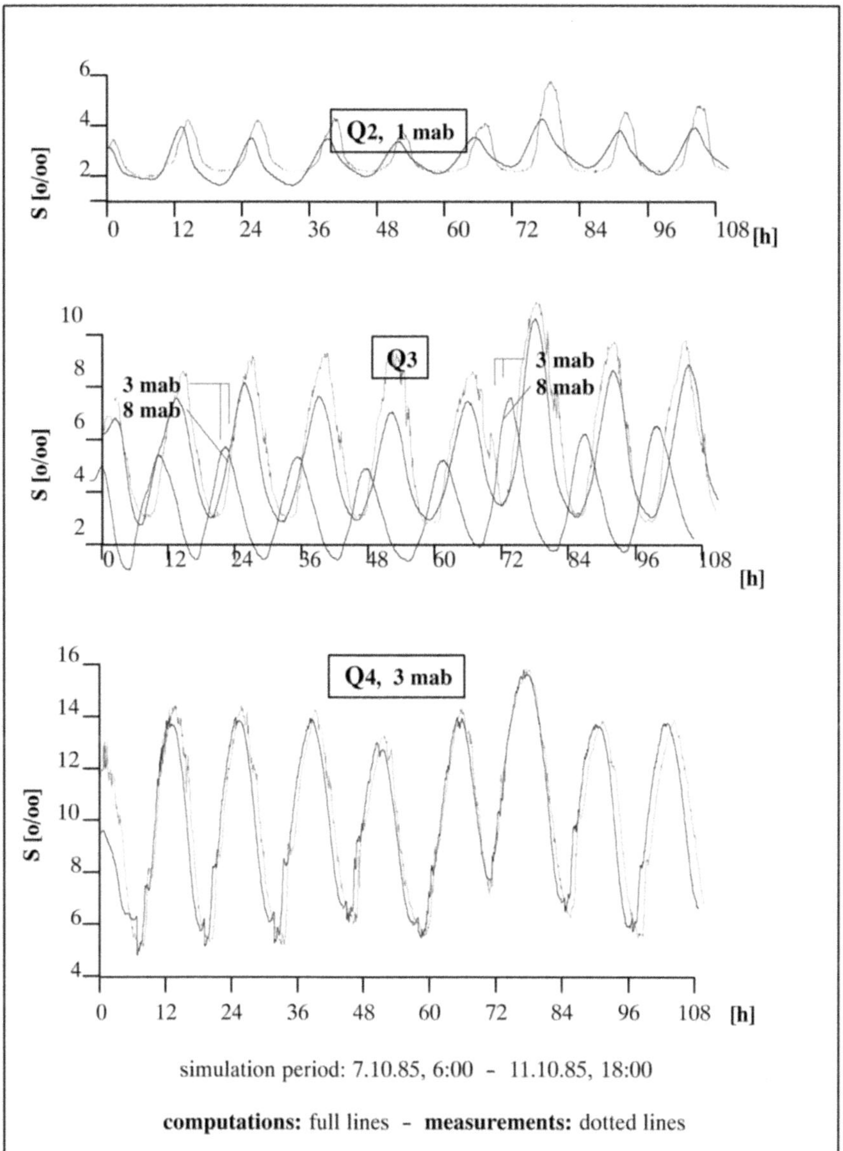

Fig. 5.58. Time series of salinity concentrations, after HINKELMANN (1997 [108])

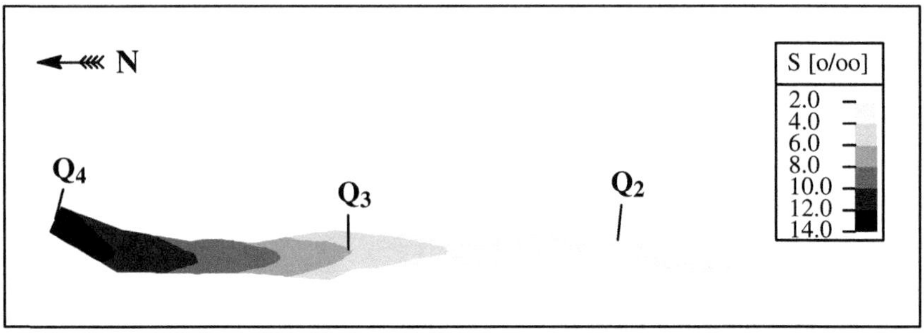

Fig. 5.59. Vertically integrated salinity concentration distribution after 90 hours, after HINKELMANN (1997 [108])

Partitioning of the domain and load balancing

The grid consists of about 1500 2D-nodes, 2300 2D-elements in the horizontal projection and 12 planes, i.e. 18000 3D-nodes and 25000 3D-elements. The partitioning of the grid is performed with the load-balancing method Recursive-Spectral Bisection as described in section 3.2.4. Here, the combination with the *Kernighan-Lin* heuristic has no influence worth mentioning. Due to the simple, line-like geometry of the horizontal grid, only very short interprocessor boundaries with few nodes occur. In figures 5.56 and 5.60, the partitionings for 16 and 64 processors are given; here, the interprocessor boundaries and a small surrounding region are not displayed. The partitioning for 16 processors leads to a 'stripe decomposition', i.e. each inner subdomain adjoins 2 other sudomains and the boundary subdomain just 1 other. Furthermore, the mapping as described in section 3.2.4 can be recognized. Figure 5.60 shows a zoomed part of the grid at the boundary to the North Sea. The partitioning of the grid is carried out for up to 256 processors, and each subdomain has a connected graph (see sec. 3.2.4).

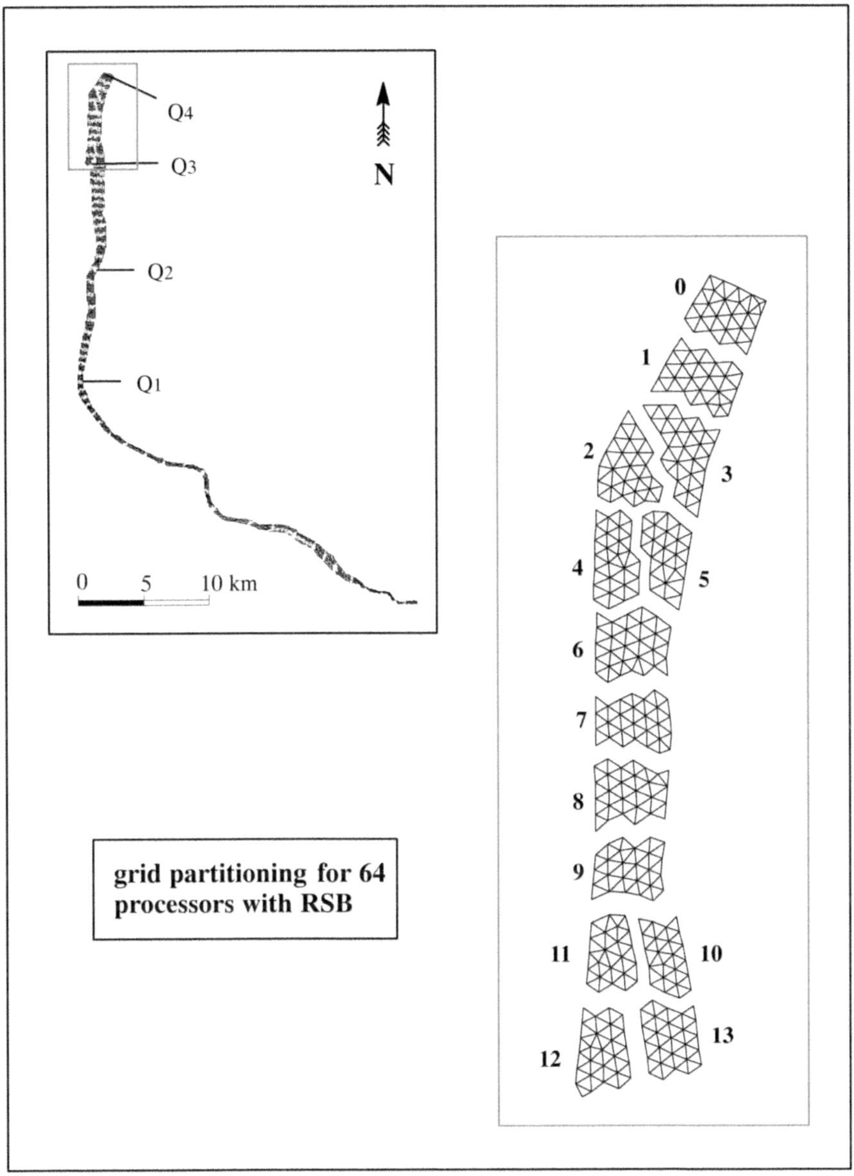

Fig. 5.60. Grid partitioning for 64 processors with RSB, after HINKELMANN (1997 [108])

Parallel run-time behavior

For the evaluation of the parallel run-time behavior, 120 time steps of 30s are investigated; this corresponds to a *Courant* number of approximately 10. The simulations are carried out on the parallel computer *CRAY T3D* using PVM for the communication (see sec. 3.2.2). The three-dimensional grid has about 18000 3D-nodes and 25000 3D-elements.

Figure 5.61 shows the parallel efficiency and the speedup on 1 to 256 processors. Values larger than 100% or 1 can be explained by cache effects (see sec. 4.2.3) which compensate for the parallel overhead up to 8 processors; on 16 processors, the speedup turns out to be extremely high (15). For more than 2 processors, the parallel overhead causes the parallel efficiency and the inclination of the parallel speedup to decrease with increasing problem size, while the parallel speedup increases. With 64 processors, the parallel speedup becomes 38 and the efficiency 60%. The simulations with more than 100 processors are above all supposed to demonstrate that the parallel algorithm works in principle. Overall, an efficient parallelization of the entire numerical algorithm is achieved. For the three-dimensional calculations, the number of processors and the problem size should be adjusted in a such way that a minimum of 300 3D-nodes (or, even better, 3D-elements) are assigned to each processor. The recommendation for the number of processors restricts the area of application for the Weser example to 64 processors.

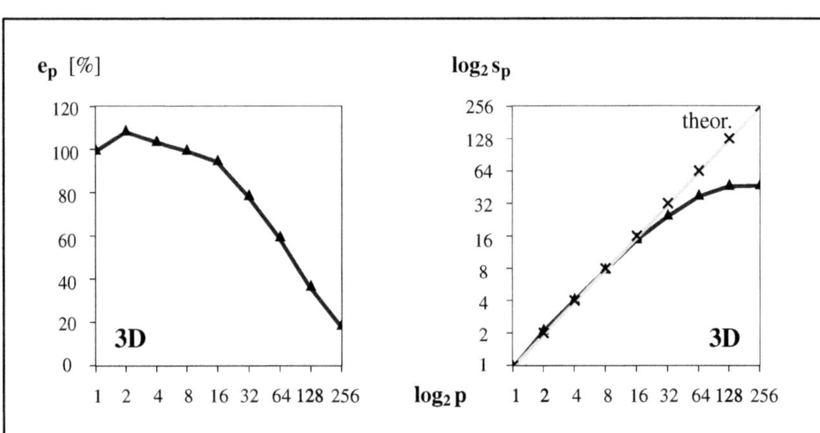

Fig. 5.61. Parallel efficiency and speedup on the T3D, after HINKELMANN (1997 [108])

The variation of the number of time steps, time-step sizes, simulation periods, physical parameters (e.g. turbulent viscosity) as well as numerical parameters (e.g. *Crank-Nicholson* factor) have no remarkable influence on the parallel run-time behavior. The use of different preconditioners (see sec. 3.4.3) is discussed in HINKELMANN (1997 [108]).

In order to examine the influence of the problem size on the parallel run-time behavior, the number of planes and thus the number of the 3D-degrees of freedom is varied (see fig. 5.62). The calculations are performed on the nCUBE2S parallel computer (see sec. 3.4.6) for 8 up to 128 processors. For the simulations with 24 planes, at least 8 processors are required for storage, i.e. RAM, reasons. Therefore, the simulations with 8 processors serve as reference value, i.e. exemplarily, the theoretical relative speedup on 128 processors is $s_{prel} = 128/8 = 16$. On 128 processors, the relative speedups for 3, 6, 12, and 24 planes are 5.3, 6.9, 8.5, and 9.5. For an increasing problem size, the relative parallel speedup as well as its inclination increase.

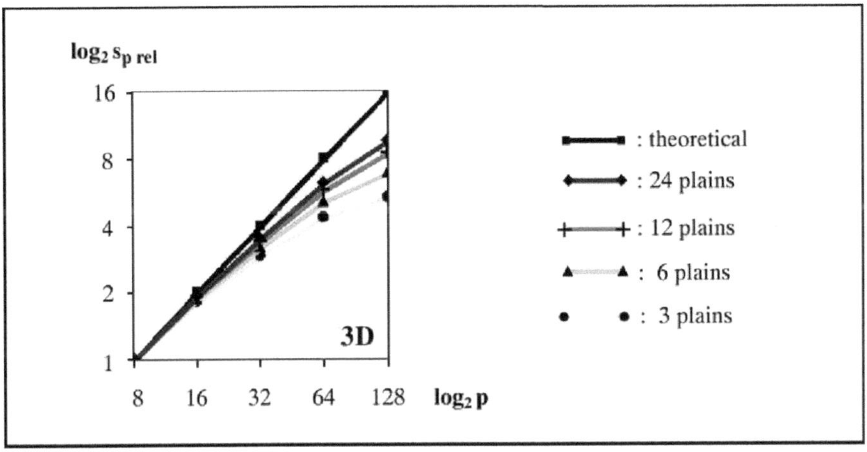

Fig. 5.62. Relative speedup for variable problem size on the nCUBE2S, after HINKELMANN (1997 [108])

Different High-Performance Computional Architectures in comparison

The aforementioned calculations with 12 planes are also performed on the workstation cluster *ParaStation* and the vector computer *S400/40* from *Siemens-Nixdorf / Fujitsu*. The ParaStation was one of the first clusters on which fine-granular problems could be treated reasonably using a special communication hardware and software (see WARSCHKO et al. (1996 [259])). The Operator-Splitting Method already described was vectorized, based on an Element-By-Element storage technique (see sec. 3.4.3) by LEPEINTRE

(1992 [166]). As a prerequisite for using the vectorized algorithm, a special element numbering must be generated in order to avoid recursions and data dependencies (see HINKELMANN (1997 [108])). A 70% usage of the vectorization unit is achieved on the S400/40. This is a reasonable number for an algorithm based on a (horizontally) unstructured grid.

Figure 5.63 compares the run times for different computer architectures. Note the different scaling of the vertical axis. The fastest run times are achieved on the parallel computer T3D. For this example, the vector computer S400/40 is approximately as fast as 16 T3D processors. The workstation cluster ParaStation has not yet been able to compete with the High-Performance Computers. Finally, it is mentioned that nowadays clusters perform much better, as discussed in secs. 3.2 and 4.2.

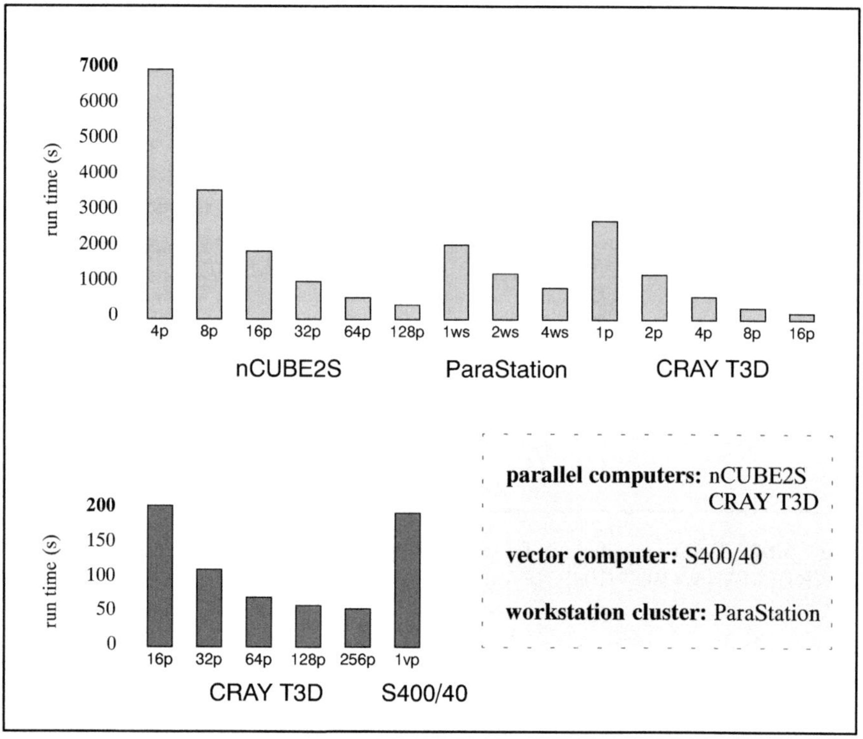

Fig. 5.63. Comparison of run times on different computer architectures, after HINKELMANN (1997 [108])

6

Summary and conclusions

In recent years, *numerical simulation models* have become indispensable in hydro- and environmental sciences and engineering, mainly for making predictions and improving the process understanding. For many problems, a physically correct and mathematically accurate simulation of the coupled complex processes requires powerfull numerical methods as well as high resolutions in space and time. In information processing, many techniques have been developed to set up systems with complex geometries and parameter distributions, and high-performance computer systems are available for fast computations. Overall, there is an urgent need for the further development and application of *efficient numerical simulation models* in hydro- and environmental sciences engineering, consisting of *efficient numerical methods* which are associated with *efficient information-processing techniques.* These topics are addressed in this work.

Chapter 1

It starts with a classification of various hydro- and environmental systems, especially in *surface waters* and the *subsurface*, and emphasizes the importance of *interaction processes.* After an illustration of the wide range of relevant *space* and *time scales*, *numerical process simulation* is introduced. Different *model concepts* and fundamental components of *modeling systems* are explained, and the need for *laboratory* and *field measurements* is demonstrated. *Deficits* in numerical simulation are mentioned, dealing with process understanding, scales, couplings, data, uncertainties and information processing. These deficits are the motivation behind this work:

- presentation of the state of the art of efficient numerical methods
- further development of numerical methods
- overview / state of the art of efficient information-processing techniques
- several extensions of the modeling system MUFTE-UG

- applications to a wide range of problems in hydro- and environmental systems
- new application fields for existing simulation models including improvements of the process understanding

Chapter 2

Physical and *mathematical models* are topics of this chapter. Depending on the geological structures, different physical model concepts for subsurface systems are chosen and explained focusing on fracture-matrix systems. Depending on the dominant processes, physical model concepts for surface-water systems are distinguished. On the basis of the *general form of the balance equation*, mathematical modeling of the following processes is briefly derived:

- groundwater flow and transport processes
- two-phase flow processes in the subsurface
- two-phase / multicomponent flow and transport processes in the subsurface
- flow and transport processes in surface water

In most cases, the resulting (systems of) partial differential equations are *(strongly) coupled*, *(highly) non-linear* and of *mixed parabolic / hyperbolic type*; this makes their further numerical treatment highly demanding.

Chapter 3

It is the most important chapter of this work and provides the state of the art of numerous *efficient numerical methods*, especially *discretization and stabilization methods*, *parallel* and *adaptive methods* as well as *fast solvers*.

- For the basic equations presented in chapter 2, different space and time *discretization* and *stabilization methods* based on the *Finite-Difference (FDM)*, *Finite-Element (FEM)* and *Finite-Volume Methods (FVM)* are discussed in detail. A method should reproduce the correct physical behavior, for example a sharp front. It should be stable and monotonous. A high order of consistency in space and time is highly desirable, at least a second order. The conservativity should be ensured; this is *globally* the case for the FEM and *locally* for the FVM. Moreover, a method should be applicable on unstructured grids to better deal with complex structures and parameter distributions. As no algorithm can fulfill all criteria satisfactorily, a good overall choice of a numerical algorithm can consist of exploiting the advantages of simple methods, such as the *Fully Upwind Box Method*, and compensating for their disadvantages with high space and time resolutions using *parallel* and *adaptive methods* as well as *fast*

solvers. Such sophisticated methods are required especially for large-scale problems because they can speed up the solution effort in orders of magnitude.
- In the context of parallel methods, the development of *High-Performance Computing*, parallelization strategies, the parallelization of basic tasks as well as *load-balancing methods* are presented. Parallel methods should use the *message-passing programming model* and provide the *Message-Passing Interface (MPI)* for the communication. Thus, maximum *portability* and the best prerequisites for *scalability* are given.
- Different adaptive methods, *error estimators* and *indicators* as well as *refinement* and *coarsening* strategies are explained. The use of *h-adaptive* and *process adaptive methods* controlled by *heuristic* error indicators is recommended for most problems considered in the context of modeling hydro- and environmental systems. If coarsement is used, the *conservation*, for example of mass, must be ensured, possibly by a mass-correction algorithm.
- Furthermore, linear single-grid solvers, such as the *Conjugate Gradient Methods*, and linear *Multigrid Methods* as well as *preconditioners* and *non-linear solvers* are introduced. For medium-scale problems, the *PCG Method* should be used for symmetric systems and the *BiCGSTAB* or the *GMRES Method* for non-symmetric ones. For large-scale problems, the Multigrid Method should be preferred, as a preconditioner of another method or as the only solution method. In most cases, the *Newton-Raphson Method* should be applied to non-linear systems.

Parallel, adaptive and Conjugate Gradient or Multigrid Methods can collaborate well and are available nowadays in several tool-boxes.

Chapter 4

An overview of various *efficient information-processing techniques* applied to the modeling of hydro- and environmental systems is given in this chapter. In a *modeling system*, preprocessors, numerical simulators and postprocessors are distinguished.

- In the context of *preprocessing tools*, *CAD systems, database-management systems (DBMS), tomography* and *scanning* and *Geographical Information Systems (GIS)* as well as *geostatistical methods* and *mesh generators* are briefly introduced. While there are only very few quasi-standard CAD systems in building and civil engineering, most of the large modeling systems in hydro- and environmental engineering have their own CAD system and standardized interfaces hardly exist. For simply structured data, relational DBMS (RDBMS) are suitable, while for complex data structures, object-orientated DBMS (OODBMS) should be preferred. The pore structure of a porous medium can be detected at a high resolution with tomography and scanning. A GIS connects geometric and other data, generally from an area on the surface of the earth. It is fast gaining importance and is applied to an increasing number of (large-scale) numerical simulations. Geostatistical methods are a means of integrating uncertainties, for example small-scale heterogeneities, into the numerical modeling. Mesh generators should use triangles in two-dimensional and tetraheda in three-dimensional simulations to enable the highest flexibility for complex structures and parameter distributions. Mesh generators which can optimize the shape of the elements are recommended.

- The processing section concentrates on *High-Performance Computers* and the modeling system *MUFTE-UG*. Among different architectures, *massively parallel systems* are the fastest computer systems installed worldwide. *Hybrid-memory systems* are becoming increasingly widespread for small- or medium-size systems. MUFTE-UG is a modeling system for the simulation of Multiphase Flow, Transport and Energy processes using parallel, adaptive Multigrid Methods as fast solvers.
- For the *visualization* of (complex) three-dimensional results, advanced visualization systems should be used.
- *WWW-based Collaborative Engineering* offers a new form of teamwork for the solution of complex global engineering problems. The collaborative work can be carried out at different locations and times and is collected with WWW-based information and communication technologies.

Chapter 5

It demonstrates the interaction of the efficient numerical methods of chapter 3 with the efficient information-processing techniques of chapter 4. A wide range of applications to problems in hydro- and environmental systems is considered.

- A new, so-called *equidimensional modeling approach* for groundwater flow and transport processes in discrete fracture-matrix systems is presented and integrated in MUFTE-UG. Fractures and the matrix are discretized with elements of the same dimension. A continuous velocity field at the fracture-matrix interface can be obtained with a few extensions; this is expected to lead to a considerable improvement of transport simulations. In this context, the high performance of h-adaptive methods is clearly demonstrated. Moreover, a basis for *coupling methods* of domains and models and for *multiphysics* is prepared.
- The two-phase flow module of the modeling system MUFTE-UG is used for new application fields dealing with the simulation of *methane-migration processes in coal-mining areas*, *gas-water flow processes in dike systems*, and a *multi-step outflow experiment.* Within the methane migration simulations, many two-dimensional test cases are investigated to determine

the dominant parameters and processes. A three-dimensional model of an abandoned coal mine is set up and, for that purpose, MUFTE-UG is extended by CAD and DBMS and the corresponding interfaces. A model calibration is envisaged in the range of available data as well as its application for several prognoses, for example dealing with optimum gas-suction measures. The dike simulations reproduce laboratory and field observations that the failure of overtopped dikes starts from the land side in areas where air is pressed out. Again, the superiority of h-adaptive methods is shown, while the multi-step outflow experiment demonstrates the efficient parallel run-time behavior.

- MUFTE-UG has been further developed to include the simulation of two-phase / three-component flow and transport processes in the interaction area groundwater - surface water. The occurrence of *gas bubbles* in the surroundings of *submarine groundwater springs* could be explained qualitatively and quantitatively with a numerical simulation taking geostatistical methods into account.
- Finally, the performance of a *parallelized Operator-Splitting Method* for free-surface flow and transport processes, which is implemented in the TELEMAC system, is illustrated. Two-dimensional simulations deal with the calibration of a model of the Baltic-Sea estuary *Darß-Zingster Boddenkette* and three-dimensional simulations with the North-Sea estuary *Weser*. An overall efficient parallelization is obtained and tested with up to 256 processors.

Further conclusions

Efficient simulation methods which are a combination of efficient numerical methods and efficient information-processing techniques should form the basis for a number of further developments. New problems of increasing size will be considered in the future. On the one hand, these require the integration of many *space* and *time scales* and, as a consequence, the (further) development of *upscaling* and *downscaling methods* as well as the treatment of more and more complex coupled processes. On the other hand, *integrated modeling* is needed to take the interaction of systems into account. For that purpose, *coupling methods* of domains and models, for example groundwater and free-surface flow, must be developed. Such couplings must go beyond merely (iteratively) exchanging simulation results via the common interfaces, for example with special coupling methods such as *Mortar Methods*.

Nowadays, as well as in the future, there will probably often be a lack of data for the numerical simulations. As the predictions of a model can only be as good as its data set, *stochastic methods* will gain importance because they enable the estimation of the *probabilities* as well as the *bandwidths* of the numerical simulation results. For that purpose, *geostatistical parameter fields* or *geostatistical fracture networks* can, for example, be used.

Finally, efficient simulation methods are essential for the necessary extension of modeling systems to *management* or *decision-support systems*, for example by means of *optimization methods*. In order to be able to implement the *European Union's Water Framework Directive*, there is an urgent need for setting up *information systems* which form the umbrella for storing, managing and analyzing large data sets, applying modeling, management and decision-support systems and providing all further problem-related information to the actors as well as the public.

References

1. Abts,D. & Mülder,W. (2000): *Aufbaukurs Wirtschaftsinformatik*, Vieweg Verlag, Braunschweig, Wiesbaden
2. AutoCAD (2001): *AutoCAD 2000 - Grundlagen*, RRZN Hannover und Herdt-Verlag, Nackenheim; http://www.autodesk.de/com
3. Ainsworth,M. & Senior,B. (1997): *Aspects of an Adaptive hp-Finite Element Method: Adaptive Strategy, Conforming Approximation and Efficient Solvers*, Comput. Methods Appl. Mech. Engrg., 150, pp. 65-87
4. Ainsworth,M., Zhu,J.Z., Craig,A.W. & Zienkiewcz,O.C. (1989): *Analysis of the Zienkiewicz-Zhu A-Posteriori Error Estimator in the Finite Element Method*, Int. J. Numer. Meth. in Engrg., Vol. 28, pp. 2161-2174
5. ArcVIEW, ArcGIS: http://www.esri.com/
6. Arya,L.M. & Paris,J.F. (1981): *A Physico Empirical Model to Predict the Soil Moisture Retention Characteristic from Particle Size Distribution and Bulk Density Data*, Soil Sci. Soc. Am. J., 45, pp. 1023-1030
7. Advanced Visualizing System AVS: http://www.avs.com/
8. Axelsson,O. & Barker,V. (1984): *Finite Element Solution of Boundary Value Problems*, Academic Press
9. Aziz,K. & Settari,A. (1979): *Petroleum Reservoir Simulation*, Applied Science Publishers, London
10. Babuska,I. & Rheinboldt,W. (1978): *Error Estimates for Adaptive Finite Element Computations*, SIAM J. Num. Anal. 15, pp. 736-754
11. Bank,R., Dupont,D.F. & Yserentant,H. (1988): *The Hierarchical Basis Multigrid Method*, Numer. Math., 52, pp. 427-458
12. Bank,R. & Xu,J. (1996): *An Algorithm for Coarsening Unstructured Meshes*, Numer. Math., 73
13. Barlag,C. (1997): *Adaptive Methoden zur Modellierung von Stofftransport im Kluftgestein*, Dissertation, Report No. 52/1997, Institut für Strömungsmechanik und Elektronisches Rechnen im Bauwesen, Universität Hannover
14. Barlag,C., Hinkelmann,R., Helmig,R. & Zielke,W. (1998): *Adaptive Methods for Modelling Transport Processes in Fractured Subsurface Systems*, 3rd International Conference on Hydroscience and Engineering, Cottbus, Center of Computational Hydroscience and Engineering, The University of Mississippi

15. Bardossy,A. (1992): *Geostatistical Methods: Recent Developments and Applications in Surface and Subsurface Hydrology*, UNESCO, Paris
16. Barrett,R., Berry,M., Chan,T., Demel,J., Donato,J., Dongarra,J., Eijkhout,V., Pozo,R., Romine,C. & van der Vorst,H. (1994): *Templates for the Solution of Linear Systems: Building Blocks for Iterative Methods*, SIAM, Philadelphia, USA; ftp netlib2.cs.utk.edu
17. Bastian,P. (1996): *Parallele Adaptive Mehrgitterverfahren*, Teubner Skripte zur Numerik, B. G. Teubner, Stuttgart
18. Bastian,P. (1999): *Numerical Computation of Multiphase Flow in Porous Media*, Habilitation, Technische Fakultät der Christian-Albrechts-Universität Kiel
19. Bastian,P., Birken,K., Johannsen,K., Lang,S., Eckstein,K., Neuss,N., Rentz-Reichert,H. & Wieners, C. (1997): *UG - A Flexible Toolbox for Solving Partial Differential Equations*, Computing and Visualization in Science (1); http://cox.iwr.uni-heidelberg.de/ ug
20. Bastian,P., Hackbusch,W. & Wittum,G. (1998): *Additive and Multiplicative Multigrid - A Comparison*, Computing 60, pp. 345-368
21. Bastian,P. & Helmig,R. (1999): *Efficient Fully-Coupled Solution Techniques for Two-Phase Flow in Porous Media. Parallel Multigrid Solution and Large Scale Computations*, Adv. Water Resour., 23, pp. 199-216
22. Bastian,P., Chen,Z., Ewing,R., Helmig,R., Jakobs,H. & Reichenberger, V. (2000): *Numerical Simulation of Multiphase Flow in Fractured Porous Media*, in Chen,Z., Ewing,R. & Shi,Z.C. (eds.): Lecture Notes in Physics, Springer, pp. 52-71
23. Bathe,K.J. (1986): *Finite-Elemente-Methoden*, Springer, Berlin
24. Baumgartner,F., Schultz,G.A. & Johnson,A.I. (1997): *Remote Sensing and Geographic Information Systems for Design and Operation of Water Resources Systems*, IAHS Publication No. 242
25. Bear,J. (1972): *Dynamics of Fluids in Porous Media*, American Elsevier, New York
26. Bear,J. & Bachmat,Y. (1990): *Introduction to Modelling of Transport Phenomena in Porous Media*, Kluwer Academic Publishers, The Netherlands
27. Bell,G. & Gray,J. (2001): *High Performance Computing: Crays, Clusters, and Centers. What Next?*, Technical Report MSR-TR2001-76, Microsoft Research Corporation, San Francisco, California
28. Bergen,O. & Forkel,C. (1999): *Seen und Reservoirs*, in Zielke,W. et al. (eds): Numerische Modelle von Flüssen, Seen und Küstengewässern, DVWK Schriften 127; see [267] in this bibliography
29. Berger,M.J. & Oliger,J. (1984): *Adaptive Mesh Refinement for Hyperbolic Partial Differential Equations*, J. Computational Physics, Vol. 53, pp. 484-512
30. Birken,K. (1998): *Ein Modell zur effizienten Parallelisierung von Algorithmen auf komplexen, dynamischen Datenstrukturen*, Dissertation, Universität Stuttgart
31. Birkhölzer,J. (1994): *Numerische Untersuchungen zur Mehrkontinuumsmodellierung von Stofftransportvorgängen in Kluftgrundwasserleitern*, Dissertation, Mitteilungen des Instituts für Wasserbau und Wasserwirtschaft, Band 93, Rheinisch-Westfälisch Technische Universität Aachen
32. Blackford,L.S., Choi,J., Cleary,A., D'Azevedo,E., Demmel,J., Dhillon,I., Dongarra,J., Hammarling,S., Henry,G., Petitet,A., Stanley,K., Walker,D. & Whaley,R.C. (1997): *ScaLAPACK User's Guide*, Report SE04, SIAM; http://www.netlib.org/scalapack

33. La Bolle,E.M., Fogg,G.E. & Tompson,A.F.B. (1996): *Random-Walk Simulation of Transport in Heterogeneous Porous Media: Local Mass-Conservation Problem and Implementation Method*, Water Resources Research, Vol. 32, No. 3, pp. 583-593
34. Bornemann,F.A., Erdmann,B. & Kornhuber,R. (1993): *Adaptive Multilevel Methods in Three Space Dimensions*, Int. J. Numer. Meth. Engrg., Vol. 36, pp. 3187-3203
35. Braess,D. (1986): *On the Combination of the Multigrid Method and Conjugate Gradients*, in Hackbusch,W. & Trottenberg,U. (eds.): Multigrid Methods II, Lecture Notes in Mathematics 960, Springer, Berlin
36. Braess,D. (1992): *Finite Elemente: Theorie, schnelle Löser und Anwendungen in der Elastizitätstheorie*, Springer, Berlin
37. Braess,D. (1996): *Towards Algebraic Multigrid for Elliptic Problems of Second Order*, Computing 55
38. Bramble,J.H., Pasciak,J.E. & Schatz,A.H. (1989): *The Construction of Preconditioners for Elliptic Problems by Substructuring*, Math. Comput., 53
39. Bramble,J.H., Pasciak,J.E., Wang,J. & Xu,J. (1991): *Convergence Estimates for Multigrid Algorithms without Regularity Assumptions*, Math. Comput., 57, pp. 23-45
40. Bramble,J.H., Pasciak,J.E. & Xu,J. (1990): *Parallel Multilevel Preconditioners*, Math. Comput., 55, pp. 1-22
41. Breiting,T., Hinkelmann,R. & Helmig,R. (2000): *Modellierung und Analyse von Mehrphasenprozessen zur Simulation von Methanausgasung im Untergrund*, Scientific Report, Institut für ComputerAnwendungen im Bauingenieurwesen, Technische Universität Braunschweig
42. Breiting,T., Hinkelmann,R. & Helmig,R. (2000): *Modeling of Hydrosystems with MUFTE-UG: Multiphase Flow and Transport Processes in the Subsurface*, Fourth International Conference on Hydroinformatics, Iowa, USA
43. Breiting,T., Hinkelmann,R., Manthey,S. & Helmig,R. (2000): *Methanausbreitungsvorgänge im Untergrund*, Darmstädter Wasserbauliches Kolloquium 1999, Mitteilungen Heft 111, Institut für Wasserbau und Wasserwirtschaft, Technische Universität Darmstadt, pp. 104-112
44. Briggs,W.L., Hensen,V.E. & McCormick,S. (1999): *A Multigrid Tutorial*, 2nd edition, SIAM, Philadelphia
45. Bronstein,I.N. & Semendjajew,K.A. (1979): *Taschenbuch der Mathematik*, Verlag Harri Deutsch, Thun und Frankfurt/Main
46. Brooks,R.H. & Corey,A.T. (1964): *Hydraulic Properties of Porous Media*, in Hydrol. Pap., Vol. 3, Fort Collins, Colorado State University
47. Brooks,A.N. & Hughes,T.J.R. (1982): *Streamline Upwind / Petrov-Galerkin Formulations for Convection Dominated Flow with Particular Emphasis on the Incompressible Navier-Stokes Equations*, Comp. Meth. Appl. Mech. Engrg., 32, pp. 199-259
48. Brussino,G. & Sonnad,V. (1989): *A Comparison of Direct and Preconditioned Iterative Techniques for Sparse, Unsymmetric Systems of Linear Equations*, Int. J. Numer. Meth. Engrg., Vol. 2
49. BSCW Basic Support for Cooperative Work: http://www.bscw.de/index.html
50. Buckley,S.E. & Leverett,M.C. (1942): *Mechanism of Fluid Displacements in Sands*, Transactions of the AIME, 146. pp. 107-116

51. Bürkle,D. & Ohlberger,M. (2002): *Adaptive Finite Volume Methods for Displacement Problems in Porous Media*, Comput. Visual. Sci., Vol. 5 (2), pp. 95-106
52. Burdine,N.T. (1953): *Relative Permeability Calculations from Pore-Size Distribution Data*, Technical Report, Petroleum Transactions, AIME
53. Burmeister,J. & Hackbusch,W. (1996): *On a Time and Space Parallel Multi-Grid Method Including Remarks on Filtering Techniques*, in Hirschel,E.H. (ed.): Flow Simulations with High-Performance Computers II, Notes on Numerical Fluid Dynamics, Vol. 52, Vieweg Verlag, Braunschweig, Wiesbaden
54. Busse,S. (2000): *Entwicklung von Finite-Differenzen-Methoden für die Flachwassergleichungen*, Studienarbeit, Institut für ComputerAnwendungen im Bauingenieurwesen, Technische Universität Braunschweig
55. Celia,M.A.,Russel,T.F., Herrera,I. & Ewing,R.E. (1990): *An Eulerian-Lagrangian Localized Adjoined Method for the Advection-Diffusion Equation*, Adv. Water Resour., 13 (4), pp. 187-206
56. Chaco: http://www.cs.sandia.gov/CRF/chaco.html
Hendrickson,B., Leland,R. et al., Sandia National Laboratories, Albuquerque, USA
57. Chandra,R., Dagum,L., Kohr,D., Maydan,D., Mc Donald,J. & Menon,R. (2000): *Parallel Programming in OpenMP*, Academic Press, San Diego; http://www.openmp.org
58. Chavent,G. & Roberts,J.E. (1991): *A Unified Physical Presentation of Mixed, Mixed Hybrid Finite Elements and Standard Finite Difference Approximations for the Determination of Velocities in Waterflow Problems*, Adv. Water Resour., 14 (6), pp. 329-348
59. Christie,I., Griffiths,D.F., Mitchell,A.R. & Zienkiewicz,O.C. (1976): *Finite Element Methods for Second Order Differential Equations with Significant First Derivatives*, International Journal for Numerical Methods in Engineering, Vol. 10, pp. 1389-1396
60. Cirpka,O.A., Frind,E.O. & Helmig,R. (1999): *Streamline-Oriented Grid-Generation for Transport Modelling in Two-Dimensional Domains Including Wells*, Adv. Water Resour., 22 (7), pp. 697-710
61. Cirpka,O. & Helmig,R. (1997): *Comparison of Approaches for the Coupling of Chemistry to Transport in Groundwater Systems*, Notes on Numerical Fluid Mechanics, Braunschweig, Vieweg Verlag, pp. 102-120
62. Class,H. (2001): *Theorie und numerische Modellierung nichtisothermer Mehrphasenprozesse in NAPL-kontaminierten porösen Medien*, Dissertation, Mitteilungen Heft 105, Institut für Wasserbau, Universität Stuttgart
63. Class,H., Helmig,R. & Bastian P. (2002): *Numerical Simulation of Nonisothermal Multiphase Multicomponent Processes in Porous Media - 1. An Efficient Solution Technique*, Adv. Water Resour., 25, pp. 533-550
64. Cordes,C.& Kinzelbach,W. (1992): *Continuous Groundwater Velocity Fields and Path Lines in Linear, Bilinear, and Trilinear Finite Elements*, Water Resources Research, Vol. 28, No. 11, pp. 2903-2911
65. COVISE: http://www.hlrs.de/organization/vis/covise
Wirse,A., Lang,U. et al., Hochleistungsrechenzentrum, Universität Stuttgart
66. Cuthill,E. & McKee,J. (1969): *Reducing the Bandwidth of Sparse Symmetric Matrices*, ACM National Conference, New York, USA
67. CVS: http://www.cvshome.org/
Cederqvist,P., Blandy,J., Vogel,K. et al.

68. Dagan,G. (1989): *Flow and Transport in Porous Formations*, Springer, Berlin
69. Davis,D.E. (2001): *GIS for Everyone*, 2nd edition, ESRI press
70. Deutsch,C. & Journel,A. (1992): *Geostatistical Software Library and User's Guide*, Oxford University Press, Oxford; http://www.gslib.com
71. Diekmann,R., Dralle,U., Neugebauer,F. & Roemke,T. (1996): *PadFEM: A Portable Parallel FEM-Tool*, International Conference on High-Performance Computing and Networking (HPCN-Europe '96), LNCS 1067, Springer, pp. 580-585; http://math-www.uni-paderborn.de/cgi-bin/frame/mathepb
72. Donea,J., Quartapelle,L. & Selmin,V. (1987): *An Analysis of Time Discretization in the Finite Element Solution of Hyperbolic Problems*, J. Comput. Phys., 70, pp. 463-499
73. Durlowsky,L.J. (1993): *A Triangle Based Mixed Finite Element - Finite Volume Technique for Modelling Two-Phase Flow through Porous Media*, J. Comput. Phys., 105, pp. 252-266
74. Ehlers,W. (1996): *Grundlegende Konzepte in der Theorie Poröser Medien*, Technische Mechanik, Band 16, Heft 1, pp. 63-76
75. Ellsiepen,P. (1999): *Zeit- und ortsadaptive Verfahren angewandt auf Mehrphasenprobleme in porösen Medien*, Dissertation, Report No. II-3, Institut für Mechanik (Bauwesen), Lehrstuhl II, Universität Stuttgart
76. Ewing,R.E. & Weekes,S. (1998): *Numerical Methods for Contaminant Transport in Porous Media*, Computational Mathematics 202, Marcel Decker, Inc., New York, USA, pp. 75-95
77. Exdasy: http://www.gup.uni-linz.ac.at/research/exdasy
Koppler,R., Kurka,G. et al., Johannes Kepler Universität Linz, Austria
78. Falta,R.W., Pruess,K., Finsterle,S. & Battistelli,A. (1995): *T2VOC User's Guide*, Lawrence Berkeley National Laboratory, University of California
79. Farhat,C. & Lesoinne,M. (1993): *Automatic Partitioning of Unstructured Meshes for the Parallel Solution of Problems in Computational Mechanics*, International Journal for Numerical Methods in Engineering, Vol. 36
80. Farthing,M.W. & Miller,C.T. (2001): *A Comparison of High-Resolution, Adaptive-Stencil Schemes for Simulating Advective-Dispersive Transport*, Adv. Water Resour., 24, pp. 29-48
81. Finlayson,B.A. (1992): *Numerical Methods for Problems with Moving Fronts*, Ravenna Park Publishing, Seattle, Washington, USA
82. Forkel,C. (2001): *Numerische Modelle für die Wasserbaupraxis*, Habilitation, Fachbereich Bauingenieurwesen, Rheinisch-Westfälisch Technische Hochschule Aachen
83. Forsyth,P.A. (1991): *A Control Volume Finite Element Approach to NAPL Groundwater Contamination*, SIAM J. Sci. Stat. Comp., Vol. 12, pp. 1029-1057
84. Fuchs,A. (1999): *Optimierte Delaunay-Triangulierungen zur Vernetzung getrimmter NURBSs-Körper*, Dissertation, Shaker Verlag
85. Fuhrmann,J. & Langmach,H. (1998): *Voronoi Box Based Finite Volume Schemes for the Numerical Solution of Subsurface Transport Problems*, International Symposium on Computer Methods for Engineering in Porous Media, Giens, France
86. Galland,J.C., Goutal,N. & Hervouet,J.M. (1991): *Telemac: A New Numerical Model for Solving Shallow Water Equations*, Adv. Water Resour., 14 (3), pp. 138-148

87. Gärtner,S. (1987): *Zur diskreten Approximation kontinuumsmechanischer Bilanzgleichungen*, Habilitation, Report No. 51/1997, Institut für Strömungsmechanik und Elektronisches Rechnen im Bauwesen, Universität Hannover
88. Garey,M.R., Johnson,D.S. & Stockmeyer,L. (1976): *Some Simplified NP-Complete Graph Problems*, Theoretical Computer Science, 1
89. Gebauer,S., Neunhäuserer,L., Kornhuber,R., Ochs,S., Hinkelmann,R., & Helmig,R. (2002): *Equidimensional Modelling of Flow and Transport Processes in Fractured Porous Systems I*, XIV International Conference on Computational Methods in Water Resources, Elsevier, Delft, The Netherlands, pp. 335-342
90. Geist,A., Beguelin,A., Dongarra,J., Jiang,W., Manchek,R. & Sunderam,V. (1994): *PVM3 User's Guide and Reference Manual*, Report ORNL/TM-12187, Oak Ridge National Laboratory, Tennessee; http://www.csm.ornl.gov/pvm
91. Gent,M.R.A. & van den Boogaard,H.F.P. (1998): : *Neural Networks and Numerical Modelling of Forces on Vertical Structures*, MAST/PROVERBS-Report, WL — Delft Hydraulics, Delft, The Netherlands
92. Van Genuchten,M.T. (1980): *A Closed-Form Equation for Predicting the Hydraulic Conductivity of Unsaturated Soils*, Soil Sci. Soc. Am. J., 44, pp. 892-898
93. Gibbs,N.E., Poole,W.G. & Stockmeyer,P.K. (1976): *An Algorithm for Reducing the Bandwidth and Profile of a Sparse Matrix*, SIAM Journal Numer. Anal. 13
94. Hackbusch,W. (1991): *Iterative Lösung großer schwachbesetzter Gleichungssysteme*,
B. G. Teubner, Stuttgart
95. Häfner,F., Sames,D. & Voigt, H.D. (1992): *Wärme- und Stofftransport: Mathematische Methoden*, Springer, Berlin, Heidelberg
96. Hansen,E. (1992): *Global Optimization Using Interval Analysis*, Monographs and Textbooks in Pure and Applied Mathematics, Marcel Dekker, Inc.
97. Helmig,R. (1993): *Theorie und Numerik der Mehrphasenströmungen in geklüftet-porösen Medien*, Dissertation, Report No. 34/1993, Institut für Strömungsmechanik und Elektronisches Rechnen im Bauwesen, Universität Hannover
98. Helmig,R. (1997): *Multiphase Flow and Transport Processes in the Subsurface - A Contribution to the Modeling of Hydrosystems*, Springer, Berlin, Heidelberg, New York
99. Helmig,R. (2000): *Einführung in die numerischen Methoden der Umweltströmungsmechanik*, Text Book, Institut für ComputerAnwendungen im Bauingenieurwesen, Technische Universität Braunschweig
100. Helmig,R., Class,H., Huber,R., Sheta,H., Ewing,J., Hinkelmann,R., Jakobs,H. & Bastian,P. (1998): *Architecture of the Modular Program System MUFTE-UG for Simulating Multiphase Flow and Transport Processes in Heterogeneous Porous Media*, Mathematische Geologie, Vol. 2, pp. 123-131
101. Helmig,R. & Cunningham,A. (2002): *Multiphase Flow, Transport and Bioremediation in the Subsurface*, Course material of IAHR-EGW short course, Lehrstuhl für Hydromechanik und Hydrosystemmodellierung, Institut für Wasserbau, Universität Stuttgart in cooperation with the Center for Biofilm Engineering, Montana State University, Bozeman

102. Helmig,R. Hinkelmann,R. & Menzel,K.: (1999) *Hydroinformatik und Hydrosystemmodellierung*, CAROLO-WILHELMINA Mitteilungen, Schwerpunktheft Informatik, Forschungsmagazin der Technischen Universität Braunschweig, pp. 106-111
103. Helmig,R. & Huber,R. (1998): *Comparison of Galerkin-Type Discretization Techniques for Two-Phase Flow in Heterogeneous Porous Media*, Adv. Water Resour., 21 (8), pp. 697-711
104. Hendrickson,B. & Leland,R. (1993): *The Chaco User's Guide, Version 1.0*, Report SAND93-0074, Sandia National Laboratories, Albuquerque, USA
105. Herrera,I., Ewing,R.E., Celia,M.A. & Russell,T.F. (1993): *Eulerian-Lagrangian Localized Adjoint Method: The Theoretical Framework*, Numer. Meth. for P.D.E.'s. 9, pp. 431-457
106. Hestenes,M. & Stiefel,E. (1952): *Methods of Conjugate Gradients for Solving Linear Systems*, J. Res. Nat. Bur. Standards 49
107. Himmelsbach,T. (1993): *Untersuchungen zum Wasser- und Stofftransportverhalten von Störungszonen im Grundgebirge (Albgranit, Südschwarzwald)*, Schriftenreihe Angewandte Geologie Karslruhe, 23
108. Hinkelmann,R. (1997): *Parallelisierung eines Lagrange-Euler-Verfahrens für Strömungs- und Stofftransportprozesse in Oberflächengewässern*, Dissertation, Report No. 51/1997, Institut für Strömungsmechanik und Elektronisches Rechnen im Bauwesen, Universität Hannover
109. Hinkelmann,R. (2000): *Hochleistungsrechnen in der Umweltströmungsmechanik*, Text Book, Institut für ComputerAnwendungen im Bauingenieurwesen, Technische Universität Braunschweig
110. Hinkelmann,R. et al. (2001): *Numerische Modellierung von Strömungs- und Transportprozessen in Oberflächengewässern*, Lecture Notes, Institut für Wasserbau, Lehrstuhl für Hydromechanik und Hydrosystemmodellierung, Universität Stuttgart
111. Hinkelmann,R. & Helmig,R. (2002): *Numerical Modelling of Transport Processes in the Subsurface*, in Wefer,G., Billett,D., Hebbeln,D., Joergensen, B.B., Schlüter,M. & van Weering,T. (eds.): Ocean Margin Systems, Springer, Berlin, pp. 269-294
112. Hinkelmann,R., Kobayashi,K., Breiting,T. & Sheta,H. (2002): *Task: Risk Assessment of Methane Emission to the Surface in Post-Mining Areas*, in Kershaw,S. (ed.): Assessment of Hazardous Gas Emission to the Surface over Former Mined Areas, Research Annual Report of the ECSC project
113. Hinkelmann,R., Malcherek,A., Jakobs,H. & Zielke,W. (1998): *A 3D Free Surface Flow and Transport Model on Different High Performance Computational Architectures*, in Emerson,D.R., Ecer,A., Periaux,J., Satofuka,N. & Fox,P. (eds.): Parallel Computational Fluid Dynamics - Recent Developments and Advances Using Parallel Computers, Elsevier Science B.V., Amsterdam, The Netherlands, pp. 407-414
114. Hinkelmann,R., Sheta,H., Class,H. & Helmig,R. (2000): *A Comparison of Different Model Concepts for Salt Water Intrusion Processes*, in Stauffer,F., Kinzelbach,W., Kovar,K. & Hoehn,E. (eds.): ModelCARE99: Calibration and Reliability in Groundwater Modelling - Coping with Uncertainties, IAHS Publications No. 265, pp. 385-391

115. Hinkelmann,R., Sheta,H., Class,H., Sauter,E.J., Helmig,R. & Schlüter,M. (2002): *Numerical Simulation of Freshwater, Salt Water and Methane Interaction Processes in a Coastal Aquifer*, in Chadam,J. Cunningham,A., Ewing,R.E., Ortoleva,P. & Wheeler,M. (eds.): IMA Volumes in Mathematics and its Applications, Vol. 131: Confinement and Remediation of Environmental Hazards and Resource Recovery, Springer, New York, pp. 263-282
116. Hinkelmann,R., Sheta,H., Helmig,R., Sauter,E.H. & Schlüter,M. (2000): *Numerical Simulation of Water-Gas Flow and Transport Processes in Coastal Aquifers*, in Sato,K. & Iwasa,Y. (eds.): Groundwater Updates, Springer, Tokyo, Berlin, New York, pp. 295-300
117. Hinkelmann,R., Stückrad,H. & Zielke,W (1994): *Hydrodynamisch-numerische Modellierungen zur Darß-Zingster Boddenkette*, Technical Report, Institut für Strömungsmechanik und Elektronisches Rechnen im Bauwesen, Universität Hannover
118. Hinkelmann,R. & Zielke,W. (1997): *A Parallel 2D Operator Splitting Method for the Navier-Stokes and Transport Equations*, Seminar on Modelling and Computation in Environmental Sciences, Notes on Numerical Fluid Mechanics, Verlag Vieweg, Braunschweig, Wiesbaden, pp. 203-214
119. Hinkelmann,R. & Zielke,W. (2000): *Parallelization of a Lagrange-Euler-Model for 3D Free Surface Flow and Transport Processes*, Computers & Fluids, Vol. 29/3, pp. 301-325
120. Holz,P., Feist,M., Nöthel,H., Lehfeldt,R., Plüß,A. & Zanke,U. (1990): *The TICAD-Toolbox Applied to Coastal Engineering Problems*, in Blain,W.R. & Ouazar,D. (eds.): Hydraulic Engineering Software Applications, Computational Mechanics Publications, Southampton, Boston, pp. 363-374
121. Hoppe,R. (1993): *A Globally Convergent Multi-Grid Algorithm for Moving Boundary Problems of Two-Phase Stefan Type*, IMA, Journal of Numerical Analysis, 13
122. Huber,R. & Helmig,R. (1998): *Multiphase Flow in Heterogenous Porous Media: A Classical Finite Element Method Versus an IMPES-based FE/FV Approach*, International Journal for Numerical Methods in Fluids, Vol. 29, pp. 899-920
123. Hughes,T.J.R. (1987): *The Finite Element Method*, Prentice-Hall, Inc., Englewood Cliffs, New Jersey
124. Hughes,J.R., Ferencz,M. & Hallquist,O. (1987): *Large Scale Vectorized Implicit Calculations in Solid Mechanics on a CRAY X-MP/48 Utilizing EBE Preconditioned Conjugate Gradients*, Computer Methods in Applied Mechanics and Engineering 61, North-Holland
125. Hunze,M. (1996): *Numerische Modellierung reaktiver Strömungen in oberflächenbelüfteten Belebungsbecken*, Dissertation, Report No. 48/1997, Institut für Strömungsmechanik und Elektronisches Rechnen im Bauwesen, Universität Hannover
126. IAHR-EGW: European Engineering Graduate School of Environment Water of the International Association of Hydraulic Engineering and Research; http://www.iws.uni-stuttgart.de/IAHR/, http://www.iahr.org
127. IBM Data Explorer: http://www.research.ibm.com/dx/
128. International Formulation Committee (1967): *A Formulation of the Thermodynamic Properties of Ordinary Water Substance*, IFC Sectretariat, Düsseldorf, Germany

129. Jacoub,G., Schultz,W. & Westrich,B. (2002): *Transport, Mixing and Fractional Deposition of Suspended Sediments in Flood Retention Reservoirs*, 27th General Assembly of the European Geophysical Society, Nice, France, 2002
130. JANET: http://www.smileconsult.de/
131. Janin,J.M. (1992): *Telemac-2D (Version 2.0): Note de Principe*, Laboratoire National d'Hydraulique, Report HE-43/92-13, Chatou, France
132. Janin,J.M., Lepeintre,F. & Pechon,P. (1992): *Telemac-3D: A Finite Element Code to Solve Free Surface Flow Problems*, Laboratoire National d'Hydraulique, Report HE-42/92-07, Chatou, France
133. Jirka,G.H. (1994): *Shallow Jets*, in Davies,P.A. & Valente Neves,M.J. (eds.): Recent Advances in the Fluid Mechanics of Turbulent Jets and Plumes, Kluwer Academic Publishers, Dordrecht
134. Jirka,G.H., von Carmer,C.F. & Wenka,T. (1999): *Nahfeld von Bauwerken und Einleitungen*, in Zielke et al.: Numerische Modelle von Flüssen, Seen und Küstengewässern, DVWK Schriften 127, pp. 343-399; see [267] in this bibliography
135. John,V. (1994): *A Comparison of some Error Estimators for Convection-Diffusion Problems on a Parallel Computer*, Preprint No. 12, Institut für Analysis und Numerik, Otto-von-Guericke-Universität Magdeburg
136. Johnson,C. (1990): *Adaptive Finite Element Methods for Diffusion and Convection Problems*, Computer Methods in Applied Mechanics and Engineering 82, North-Holland, pp. 301-322
137. Johnson,C. (1992): *Numerical Solution of Partial Differential Equations by the Finite Element Method*, Cambridge University Press, Cambridge
138. Jonasson,S.A. (1989): *Estimation of the Van Genuchten Parameters from Grain-Size Distribution*, in Indirect Methods for Estimating the Hydraulic Properties of Unsaturated Soils, Riverside, California
139. Jones,M.T. & Plassmann,P.E. (1996): *BlockSolve95 Users Manual: Scalable Library Software for the Parallel Solution of Sparse Linear Systems*, Argonne National Laboratory Report, ANL-95/48, Illinois, USA ; http://www-unix.mcs.anl.gov/sumaa3d
140. Kaiser,R. (2001): *Gitteradaption für Finite-Elemente-Modellierung gekoppelter Prozesse in geklüftet-porösen Medien*, Dissertation, Report No. 63/1997, Institut für Strömungsmechanik und Elektronisches Rechnen im Bauwesen, Universität Hannover
141. Kaleris,V., Sachinis,G., Lagas,G., Stroskidu,V., Sheta,H., Hinkelmann,R. & Helmig,R. (2001): *Task 8: Groundwater Flow Modelling*, in Sauter,E. (ed.): Executive Final Summary Report of the EU-Project SUBGATE; see [229] in this bibliography
142. Kaleris,V., Lagas,G., Marczinek,S. & Piotrowski,J.A. (2002): *Modelling Submarine Groundwater Discharge: An Example for the Western Baltic Sea*, Journal of Hydrology, Vol. 265, pp. 76-99
143. Kasper,H. (1999): *Aspekte der Datenverarbeitung*, in Zielke,W. et al. (eds): Numerische Modelle von Flüssen, Seen und Küstengewässern, DVWK Schriften 127; see [267] in this bibliography
144. Klaas,O. (1996): *Adaptive Finite-Element-Methoden für MIMD-Parallelrechner zur Behandlung von Strukturproblemen mit Anwendung auf Stabilitätsprobleme*, Dissertation, Report No. F 96/7, Forschungs- und Seminarberichte aus dem Bereich der Mechanik der Universität Hannover

145. Knödel,K., Krummel,H. & Lange,G. (1997): *Handbuch zur Erkundung des Untergrundes von Deponien und Altlasten*, Band 3: Geophysik, Berlin, Heidelberg
146. Kobayashi,K., Hinkelmann,R., Breiting,T. & Helmig,R. (2002): *Scale-Dependent Influences of Geostatistical Permeability Distributions on Gas-Water Flow Processes*, ModelCARE02 - 4th International Conference on Calibration and Reliability in Groundwater Modelling, Acta Universitatis - Geologica, Vol. 46 (2/3), pp. 571-574
147. Kobayashi,K., Hinkelmann,R., Class,H., Shoemaker,C. & Helmig,R. (2002): *Development of an Optimisation Method for Methane Extraction in Subsurface*, 4th Workshop - Porous Media, Blaubeuren
148. Kobus,H., Barczewski,B. & Koschitzky,H.P. (1996): *Groundwater and Subsurface Remediation - Research Strategies for In-situ Technologies*, Springer, Berlin, Heidelberg, New York
149. Kobus,H. & de Haar,U. (1995): *Perspektiven der Wasserforschung*, Deutsche Forschungsgemeinschaft
150. Koelbel,C.H., Loveman,B. & Schreiber,R.S. (1993): *The High Performance Fortran Handbook*, Mit Press, ISBN 0-262-61094-9
http://www.epcc.ed.ac./epcc-tec/hpf, http://www.tokyo.rist.or.jp/jahpf
151. Kopmann,R. & Jankowski,J. (2000): *Telemac auf Origin2000 - Erste Erfahrungen*, Supercomputing News, Heft 1/2000, Bundesanstalt für Wasserbau - Dienststelle Hamburg
152. Kolditz,O. (1997): *Stoff- und Wärmetransport im Kluftgestein*, Gebrüder Bornträger, Berlin, Stuttgart
153. Kolditz,O. (2000): *Numerical Methods in Environmental Fluid Dynamics*, Text Book, Institut für Strömungsmechanik und Elektronisches Rechnen im Bauwesen, Universität Hannover
154. König,C. (1991): *Numerische Berechnung des dreidimensionalen Stofftransports im Grundwasser*, Technisch Wissenschaftliche Mitteilungen der Ruhr-Universität Bochum 91/13
155. Kornhuber,R. (1997): *Adaptive Monotone Multigrid Methods for Nonlinear Variational Problems*, Teubner
156. Kinzelbach,W. (1992): *Numerische Methoden zur Modellierung des Transports von Schadstoffen im Grundwasser*, 2. Auflage, R.Oldenbourg Verlag, München, Wien
157. Krafczyk,M., Lehmann,P., Philippova,O. Hänel,D. & Lantermann,U. (2000): *Lattice Boltzmann Simulations of Complex Multi-Phase Flows*, in Sändig,A.M. et al.: Multifield Problems: State of the Art, Springer, pp. 50-57
158. Kreienmeyer,M. (1996): *Kopplung von Rand- und Finite-Element-Methoden für ebene Elastoplastizität mit Implementierung auf Parallelrechnern*, Dissertation, Report No.
F 96/5, Forschungs- und Seminarberichte aus dem Bereich der Mechanik der Universität Hannover
159. Kröhn,K.P. (1991): *Simulation von Transportvorgängen im klüftigen Gestein mit der Methode der Finite Elemente*, Dissertation, Report No. 29/1991, Institut für Strömungsmechanik und Elektronisches Rechnen im Bauwesen, Universität Hannover
160. Kunert,G. (2001): *A Posteriori L_2 Error Estimation on Anisotropic Tetrahedral Finite Element Meshes*, IMA J. Numer. Anal., 21, pp. 503-523

161. Lämmer,L. (1997): *Parallelisierung von Anwendungen der Finite-Elemente-Methode im Bauingenieurwesen*, Habilitation, Report No. 1/97, Institut für Numerische Methoden und Informatik im Bauwesen, Technische Hochschule Darmstadt
162. Lang,S. (2001): *Parallele numerische Simulation instationärer Probleme mit adaptiven Methoden auf unstrukturierten Gittern*, Dissertation, Mitteilungen Heft 110, Institut für Wasserbau, Universität Stuttgart
163. Lang,U. (1995): *Simulation regionaler Strömungs- und Transportvorgänge in Karstaquiferen mit Hilfe des Doppelkontinuum-Ansatzes: Methodenentwicklung und Parameteridentifikation*, Dissertation, Mitteilungen Heft 85, Institut für Wasserbau, Universität Stuttgart
164. Lege,T., Kolditz,O. & Zielke,W. (1996): *Strömungs- und Transportmodellierung*, Handbuch zur Erkundung des Untergrundes von Deponien und Altlasten, Band 2, Springer, Berlin, Heidelberg, New York
165. Lehfeldt,R. (1991): *Ein algebraisches Turbulenzmodell für Ästuare*, Dissertation, Report No. 30/1991, Institut für Strömungsmechanik und Elektronisches Rechnen im Bauwesen, Universität Hannover
166. Lepeintre,F. (1992): *Bibliothque BIEF - Note de Principe et Descriptif Informatique*, Laboratoire National d'Hydraulique, Report HE-43/92-16, Chatou, France
167. Leverett,M.C. (1941): *Capillary Behavior in Porous Solids*, Transactions of the AIME, 142, pp. 152-169
168. Leydag,S. (2001): *Entwicklung einer adaptiven Finite-Volumen-Methode für die Flachwassergleichungen*, Studienarbeit, Institut für ComputerAnwendungen im Bauingenieurwesen, Technische Universität Braunschweig
169. Leydag,S., Busse,S. & Hinkelmann,R. (2001): *Entwicklung einer adaptiven Finite-Volumen Methode für die Flachwassergleichungen*, Forum Bauinformatik 2001, Fortschrittsberichte VDI Verlag, Reihe 4: Bauingenieurwesen, No. 169, pp. 192-200
170. Malcherek,A. (2000): *Physik und Numerik der Fließgewässer*, Technical Documentation, Bundesanstalt für Wasserbau, Außenstelle Küste, Hamburg
171. Maloszewski,P. & Zuber,A. (1993): *Tracer Experiments in Fractured Rocks: Matrix Diffusion and the Validity of Models*, Water Resources Research, 29 (8), pp. 2723-2735
172. Manthey,S., Ölmann,U., Helmig,R. & Hinkelmann,R. (2003): *MUSKAT, MUltiSKAlenTransport, Teilprojekt C: Identifikation wesentlicher Prozesse zur Bestimmung effektiver Parameter in porösen Mehrphasen-Mehrkomponentensystemen*, Scientific Report, Institut für Wasserbau, Lehrstuhl für Hydromechanik und Hydrosystemmodellierung, Universität Stuttgart
173. Markofsky,M. (1980): *Strömungsmechanische Aspekte der Wasserqualität*, Schriftenreihe gwf Wasser · Abwasser, Oldenbourg Verlag, München, Wien
174. Mase,H., Sakamoto,M. & Sakai,T. (1995): *Neural Network for Stability Analysis of Rubble-Mound Breakwaters* , Journal of Waterway, Port, Coastal and Ocean Engineering, ASCE, November/December, pp. 294-299
175. McDonald,M.G. & Harbaugh,A.W. (1984): *A Modular Three-Dimensional Finite-Difference Groundwater Flow Model*, USGS, Reston
176. Meister,A. (1999): *Numerik linearer Gleichungssysteme - Eine Einführung in moderne Verfahren*, Vieweg, Braunschweig, Wiesbaden

177. Meuer,H.W., Strohmaier,E., Dongarra,J. & Simon,H.D (2001): *TOP500 Supercomputer Sites*, 17th Edition, Rechenzentrum Universität Mannheim, Report RUM 62/2001, Lawrence Berkeley National Laboratory, California, Report LBNL 4281; http://www.top500.org
178. Miller,C.T., Christakos,G., Imhoff,P.T., McBride,J.F., Pedit,J.A. & Trangenstein,J. (1998): *Multiphase Flow and Transport Modelling in Heterogeneous Media: Challenges and Approaches*, Adv. Water Resour., 21 (2), pp. 77-120
179. MODFLOW: http://www.modflow.com/
180. Molenaar,J. (1995): *Multigrid Methods for Fully Implicit Oil Reservoir Simulation*, Copper Mountain Conference on Multigrid Methods
181. Molkenthin,F. (2000): *WWW based Hydroinformatics Systems*, Habilitation, Progress in Bauinformatik, Institut für Bauinformatik, Brandenburg University of Technology
182. Mualem,Y. (1976): *A New Model for Predicting the Hydraulic Conductivity of Unsaturated Porous Media*, Water Resources Research, 12, pp. 513-522
183. Myrinet: http://www.myri.com/myrinet
184. MySQL: http://www.mysql.com/
185. Narasimhan,T.N. & Pruess,K. (1987): *MINC: An Approach for Analyzing Transport in Strongly Heterogeneous Systems*, in Custodio,E., Gurgui,A. & Lobo Ferreira,J.P. (eds.): Groundwater Flow and Quality Modelling, D.Reidel Publishing Company, Dordrecht, pp. 375-391
186. Neunhäuserer,L. (2003): *Diskretisierungsansätze zur Modellierung von Strömungs- und Transportprozessen in geklüftet-porösen Medien*, Dissertation, Mitteilungen Heft 119, Institut für Wasserbau, Universität Stuttgart
187. Neunhäuserer,L., Fuchs,A., Hemminger,A. & Helmig,R. (1998): *Flow and Transport Processes in Fractured Porous Media*, in Burganos,V.N., Karatzas,G.P., Payatakes,A.C., Brebbia,C.A. & Gray,W.G. (eds.): XII International Conference on Computational Methods in Water Resources, Vol. 2, Kluwer Academic Publishers, Dordrecht, Boston, London, pp. 101-108
188. Neunhäuserer,L., Gebauer,S., Ochs,S., Hinkelmann,R., Kornhuber,R. & Helmig,R. (2001): *Mehrgittermethoden und adaptive Euler-Lagrange-Verfahren zur Simulation von Strömungs- und Transportvorgängen in Kluftaquifersystemen*, Scientific Report, Institut für Mathematik, Freie Universität Berlin and Institut für Wasserbau, Lehrstuhl für Hydromechanik und Hydrosystemmodellierung, Universität Stuttgart
189. Neunhäuserer,L., Gebauer,S., Ochs,S., Hinkelmann,R., Kornhuber,R., & Helmig,R. (2002): *Equidimensional Modelling of Flow and Transport Processes in Fractured Porous Systems II*, XIV International Conference on Computational Methods in Water Resources, Elsevier, Delft, The Netherlands, pp. 343-350
190. Neunhäuserer,L., Hemminger,A. & Helmig,R. (2000): *Einsatz von diskreten Modelansätzen*, Scientific Report of the DFG Project: Festgesteins - Aquiferanalog: Experimente und Modellierung, Institut für ComputerAnwendungen im Bauingenieurwesen, Technische Universität Braunschweig
191. Nitsche,G. (1985): *Explizite Finite-Element-Modelle und ihre Naturanwendungen auf Strömungsprobleme in Tidegebieten*, Dissertation, Report No. 19/1997, Institut für Strömungsmechanik und Elektronisches Rechnen im Bauwesen, Universität Hannover

192. Noorishad,J., Tsang,C.F., Perrochet,P. & Musy,A. (1992): *A Perspective on the Numerical Solution of Convection-Dominated Transport Problems: A Price to Pay for the Easy Way Out*, Water Resources Research, 28(2), pp. 551-561
193. Ochs,S., Hinkelmann,R., Neunhäuserer,L., Helmig,R., Gebauer,S. & Kornhuber,R. (2002): *Adaptive Methods for the Equidimensional Modeling of Flow and Transport Processes in Fractured Aquifers*, 5th International Conference on Hydrosciences & Engineering, Warsaw, Poland
194. Ölmann,U., Hinkelmann,R., Manthey,S., Lang,S. & Helmig,R. (2002): *Parallel Numerical Simulation with MUFTE-UG: Two Phase Flow Processes in the Subsurface*, in Cluckie,I.D., Han,D., Davies,J.P.& Heslop,S. (eds.): Hydroinformatics 2002, Volume 2: Software Tools and Management Systems, IWA Publishing, London, pp. 1441-1446
195. Ogata,A. & Banks,R.B. (1961): *A Solution of the Differential Equation of Longitudinal Dispersion in Porous Media*, Professional Paper 411, A.U.S. Geological Survey, Washington, D.C.
196. Ohlberger,M. (2001): *A Posteriori Error Estimates and Adaptive Methods for Convection Dominated Transport Processes*, Dissertation, Mathematische Fakultät, Universität Freiburg
197. Ohlberger,M. (2001): *A Posteriori Error Estimates for Vertex Centered Finite Volume Approximations of Convection-Diffusion-Reaction Equations*, ESAIM: M2AN Mathematical Modelling and Numerical Analysis, Vol. 35, No. 2, pp. 355-387
198. Ohnimus,S. (1996): *Theorie und Numerik dimensions- und modelladaptiver Finite-Elemente-Methoden von Flächentragwerken*, Dissertation, Report No. F 96/6, Forschungs- und Seminarberichte aus dem Bereich der Mechanik der Universität Hannover
199. Oldenburg,C.M. & Pruess,K. (1995): *Dispersive Transport Dynamics in a Strongly Coupled Groundwater-Brine Flow System*, Water Resources Research, 31 (2), pp. 289-302
200. ORACLE: http://www.oracle.com/
201. Papastavrou,A. (1998): *Adaptive Finite Element Methoden für Konvektions-Diffusionsprobleme*, Graduiertenkolleg Computational Structural Dynamics, Ruhr-Universität Bochum
202. ParMetis: http://www-users.cs.umn.edu/ karypis/metis
Karypis,G., Kumar,V. et al., University of Minnesota, Minneapolis, USA
203. Paul,M.(2003): *Simulation of Two-Phase Flow Processes in Porous Media with Adaptive Methods*, Dissertation, Mitteilungen Heft 120, Institut für Wasserbau, Universität Stuttgart
204. Paul,M., Hinkelmann,R., Sheta,H. & Helmig,R. (1999): *Two-Phase Flow Modelling of Flood Defense Structures*, in Ehlers,W. (ed.): IUTAM Symposium on Theoretical and Numerical Methods in Continuum Mechanics of Porous Materials, Stuttgart, 1999, Kluwer Academic Publisher, Dordrecht, Boston, London, pp. 163-168
205. Paul,M., Stauffer,F., Hinkelmann,R. & Helmig,R. (2000): *Calibration of a Water-Gas Flow Model for Dike Systems*, in Bentley,L.R., Sykes,J.F., Brebbia,C.A., Gray,W.C. & Pinder,G.F. (eds.): XIII International Conference on Computational Methods in Water Resources, A.A. Balkema, Rotterdam, Brookfield, pp. 201-208

206. Perrochet,P. & Berod,D. (1993): *Stability of Standard Crank-Nicholson-Galerkin Scheme Applied to the Diffusion-Convection Equation: Some New Insights*, Water Resources Research, Vol. 29, No. 9, pp. 3291-3297
207. Pfingsten,W. (1990): *Stofftransport in Klüften mit porösen Gesteinen*, Mitteilungen Institut für Wasserwirtschaft, Hydrologie und Landwirtschaftlichen Wasserbau, Universität Hannover
208. Press,W.H.,Teukolsky,S.A., Vetterling,W.T. & Flannery,B.P. (1992): *Numerical Recipes in Fortran - The Art of Scientific Computing*, 2. Edition, Cambridge University Press, USA
209. PRO/ENGINEER: http://www.ptc.com/products/proe/index.htm
210. Pruess,K. (1991): *TOUGH 2, A General-Purpose Numerical Simulator for Multiphase Fluid and Heat Flow*, Lawrence Berkeley National Laboratory, University of California, USA
211. Pruess,K. & Tsang,Y.W. (1990): *On Two-Phase Relative Permeability Capillary Pressure of Rough-Walled Rock Fractures*, Water Resources Research, 10
212. Rank,E. (1987): *An Adaptive hp-Version in the Finite Element Method*, Numerical Methods in Engineering Theory and Applications
213. Rank,E. & Düster,A. (2001): *H-versus P-version Finite Element Analysis for J2 Flow Theory*, First M.I.T Conference on Computational Fluid and Solid Mechanics, Cambridge, Massachusetts, USA
214. Rank,E. & Zienkiewic,O.C. (1987): *A Simple Error Estimator in the Finite Element Method*, Comm. Appl. Num. Meth. 3, pp. 449-454
215. Reid,J. (1971): *On the Method of Conjugate Gradients for the Solution of Large Sparse Systems of Linear Equations*, in Reid,J. (ed.): Large Sparse Sets of Linear Equations, Academic Press, London, pp.231-254
216. Reid,R.C., Prausnitz,J.M. & Poling,B.E. (1987): *The Properties of Gases and Liquids*, MacGraw-Hill
217. Rentz-Reichert,H. (1996): *Mehrgitterverfahren zur Lösung der inkompressiblen Navier-Stokes Gleichungen: Ein Vergleich*, Dissertation, Institut für Computeranwendungen, Universität Stuttgart
218. Riegger,J. (2001): *Hydroinformatics*, Lecture Notes, Institut für Wasserbau, Universität Stuttgart
219. Van Rijn,L.C. (1990): *Principles of Fluid Flow and Surface Waves in Rivers, Estuaries, Seas and Oceans*, Aqua Publications, Amsterdam
220. Riviere,B. & Wheeler,M.F. (2000): *Discontinuous Galerkin Methods for Flow and Transport Problems in Porous Media*, Commun. Numer. Meth. Engrg, 00, pp. 1-6
221. Riviere,B., Wheeler,M.F. & Banas,K. (2000): *Discontinuous Galerkin Method Applied to Single Phase Flow in Porous Media*, Computational Geosciences 4, pp. 337-346
222. Rodi,W. (1984): *Turbulence Models and their Applications in Hydraulics*, IAHR Publication, Delft
223. Romm,E.S. (1966): *Fluid Flow in Fractured Rocks*, Moscow (translated by W.R. Blake, Bartlesville, OK)
224. Rosen,B. (2000): *Numerische Modellierung der Wechselwirkung zwischen gesättigt-ungesättigter Grundwasserströmung und Verformung im geklüfteten Gestein*, Fortschritt-Berichte VDI, Reihe 15 Umwelttechnik, No. 222

225. Rother,T. (2001): *Geometric Modelling of Geo-Systems*, Dissertation, Report No. 64/2001, Institut für Strömungsmechanik und Elektronisches Rechnen im Bauwesen, Universität Hannover
226. Ruge,J. & Stüben,K. (1986): *Algebraic Multigrid (AMG)*, Arbeitspapiere der GMD 210, St. Augustin
227. Rust,W. (1991): *Mehrgitterverfahren und Netzadaption für lineare und nichtlineare statische Finite-Elemente-Berechnungen von Flächentragwerken*, Dissertation, Report No. F 91/2, Forschungs- und Seminarberichte aus dem Bereich der Mechanik der Universität Hannover
228. Saad,Y. & Schulz,M. (1985): *Conjugate Gradient-Like Algorithms for Solving Non-Symmetric Linear Systems*, Mathematics of Computation, Vol. 44, pp. 417-424
229. Sauter,E. (2001): *Executive Final Summary Report of the EU-Project SUBGATE*, Environment & Climate Research Programme (1994-1998), Contract ENV4-CT97-0631
230. Sheta,H. (1999): *Simulation von Mehrphasenvorgängen in porösen Medien unter Einbeziehung von Hysterese-Effekten*, Dissertation, Mitteilungen Heft 100, Institut für Wasserbau, Universität Stuttgart
231. Schönung,B.E. (1990): *Numerische Strömungsmechanik*, Springer, Berlin, Heidelberg, New York
232. Schotting,R.J. (1998): *Mathematical Aspects of Salt Transport in Porous Media*, Proefschrift, Centrum voor Wiskunde en Informatica, Technische Univeriteit Delft, The Netherlands
233. Schüle,J. (1998): *Paralleles Rechnen - Trends und Algorithmen*, Lecture Notes SS 98, Institut für Wissenschaftliches Rechnen, Technische Universität Braunschweig
234. Schultz,G.A. & Engman,E.T. (2000): *Remote Sensing in Hydrology and Water Management*, Springer, Berlin, Heidelberg, New York
235. Schwarz,H.R. (1991): *Methode der finiten Elemente*, 3rd edition, B.G. Teubner Stuttgart
236. Sellerhof,F. (2002): *Ein punktbasiertes Informationsmodell für das Küsteningenieurwesen*, Dissertation, Publication No. 1/2002, Institutsreihe des Instituts für Bauinformatik, Universitt Hannover
237. Silberhorn-Hemminger,A. (2002): *Modellierung von Kluftaquifersystemen: Geostatistische Analyse und deterministisch-stochastische Kluftgenerierung*, Dissertation, Mitteilungen Heft 114, Institut für Wasserbau, Universität Stuttgart
238. Siebert,K.G. (1996): *An A Posteriori Error Estimator for Anisotropic Refinement*, Numer. Math. 73, pp.373-398
239. Simon,H.D., Farhat,C. & Lanteri,S. (1993): *TOP/DOMDEC - a Software Tool for Mesh Partitioning and Parallel Processing*, NAS Technical Report RNR-93-011, NASA Ames Research Center, Moffett Field, California
240. Smith,G.D. (1970): *Numerische Lösung von partiellen Differentialgleichungen*, Vieweg, Braunschweig
241. Sonnenborg,T.O., Tofteng,C., Jensen, K.H. & Butts, M.B. (1997): *Flow and Solute Transport in Laboratory Fracture-Matrix Systems*, in Holly Jr.,F.M., Alsaffar,A., Findikakis,A.N. & Stauffer,F. (eds.): Groundwater: An Endangered Resource, 27th IAHR Congress, San Francisco

242. SPRING (2000): *Program System for the Computation of Two- and Three-Dimensional Groundwater Flow and Transport Models - User's Guide version 1.1 GKW GmbH*, Bochum; http://www.gkw-gmbh.de/
243. Stückrad,H. & Hinkelmann,R. (1995): *Numerische Modellrechnungen zur Darß-Zingster Boddenkette*, Deutsche Hydrographische Zeitschrift, Vol. 47, No. 2, Hamburg, Rostock, pp. 93-107
244. Tang,D.H., Frind,E.O. & Sudicky,E.A. (1981): *Contaminant Transport in Fractured Porous Media: Analytical Solution for a Single Fracture*, Water Resources Research, Vol. 17, pp. 555-564
245. Taniguchi,T., Goda,T., Kasper,H., Zielke,W. & Kosakowski,G. (1996): *Hexahedral Mesh Generation of Complex Composite Domain - Groundwater Flow and Transport Analysis in Fractured Rock*, 5th International Conference on Num. Grid Generation in Comp. Field Simulations, Starkville, Mississippi
246. TELEMAC Modelling System, Electricité de France, Département Laboratoire National d'Hydraulique, SOGREAH Consultants; http://www.telemac-system.com/
247. Thiele,K. (1999): *Adaptive Finite Volume Discretization of Density Driven Flows in Porous Media*, Dissertation, Naturwissenschaftliche Fakultäten der Friedrich-Alexander-Universität Erlangen-Nürnberg
248. Thorenz,C. (2001): *Model Adaptive Simulation of Multiphase and Density Driven Flow in Fractured and Porous Media*, Dissertation, Report No. 62/2001, Institut für Strömungsmechanik und Elektronisches Rechnen im Bauwesen, Universität Hannover
249. Toro,E.F. (2001): *Shock-Capturing Methods for Free-Surface Shallow Flows*, John Wiley & Sons, Ltd, Chichester, New York, Weinheim, Bisbane, Singapore, Toronto
250. UG, DDD: http://cox.iwr.uni-heidelberg.de/ ug/index.html
Wittum,G., Bastian,P., Birken,K., Institut für Wissenschaftliches Rechnen, Universität Heidelberg
251. UNESCO (1987): *International Oceanographic Tables*, Vol. 4, UNESCO Technical Papers in Marine Sciences, No. 40
252. Le Veque,R. (1992): *Numerical Methods for Conservation Laws*, Birkhäuser Verlag, Basel, Boston, Berlin
253. Verfürth,R. (1996): *A Review of A Posteriori Error Estimation and Adaptive Mesh Refinement Techniques*, John Wiley & Sons and B.G. Teubner Publishers, Stuttgart
254. Vogel,H.J. (1998): *RöFo, Fortschritte auf dem Gebiet der Röntgenstrahlen und der bildgebenden Verfahren*, Title Page, Vol. 169
255. Vogel,M. & Bunte,P. (2001): *Pro/ENGINEER und Pro/MECHANICA - Konstruieren, Berechnen und Optimieren*, Edition 2000i2, Carl Hanser Verlag, München
256. Vogel,H.J. & Roth,K. (1998): *A New Approach for Determining Effective Soil Hydraulic Functions*, Europ. J. Soil Sci., Vol. 49, pp. 547-556
257. Wackernagel,H. (1998): *Multivariate Geostatistics*, Springer, Berlin, Heidelberg
258. Walker,D.W. & Dongarra,J,J. (1996): *MPI: A Standard Message Passing Interface*, Supercomputer, Vol. XII, No. 1; http://www.mpi-forum.org
259. Warschko,T.M., Blum,J.M., & Tichy,W.F. (1996): *The ParaPC/ParaStation Project: Efficient Parallel Computing by Clustering Workstations*, Report 13/96, Fakultät für Informatik, Universität Karlsruhe

260. Westrich,B., Salden,D., Siebel,R. & Zweschper,B. (2002): *Naturnahe Bauweisen für überströmbare Dämme*, BWPLUS Interim Report, Institut für Wasserbau, Universität Stuttgart
261. White, M.D. & Oostrom,M. (2000): *STOMP, Subsurface Transport over Multiple Phases*, Pacific Northwest National Laboratory
262. Witte,B. (2000): *Parallelisierung eines auf der Integralen-Finite-Differenzen-Methode basierenden Grundwasserströmungsmodells*, Diplomarbeit, Institut für ComputerAnwendungen im Bauingenieurwesen, Technische Universität Braunschweig
263. Witte,B., Hinkelmann,R. & Helmig,R. (2001): *Development of a Parallel FVM Based Groundwater Flow Model*, in Breuer,M., Durst,F. & Zenger,C. (eds.): High Performance Scientific and Engineering Computing - Methods, Developmemts, and Applications, Lecture Notes in Computational Science and Engineering, Springer, Berlin, Heidelberg, New York, Vol. 21, pp. 29-36
264. Wollrath,J. (1990): *Ein Strömungs- und Transportmodell für klüftiges Gestein und Untersuchung zu homogenen Ersatzsystemen*, Dissertation, Report No. 28/1990, Institut für Strömungsmechanik und Elektronisches Rechnen im Bauwesen, Universität Hannover
265. Wollschläger,A. (1996): *Ein Random-Walk-Modell für Schwermetallpartikel in natürlichen Gewässern*, Dissertation, Report No. 49/1996, Institut für Strömungsmechanik und Elektronisches Rechnen im Bauwesen, Universität Hannover
266. Yserentant,H. (1993): *Two Preconditioners Based on Multilevel Splitting of Finite Element Spaces*
267. Zielke,W., Bergen,O., Bloß,S., von Carmer,C.F., Kasper,H., Malcherek,A., Mayerle,R., Schröder,P.M. & Wenka,T. (1999): *Numerische Modelle von Flüssen, Seen und Küstengewässern*, DVWK Schriften 127
268. Zielke,W. & Mayerle,R. (1999): *Küstengewässer*, in Zielke,W. et al. (eds): Numerische Modelle von Flüssen, Seen und Küstengewässern, DVWK Schriften 127; see [267] in this bibliography
269. Zienkiewicz,O.C. & Taylor,R.L. (1991): *The Finite Element Method*, Vol. 2, McGraw-Hill Book Company, London
270. Zienkiewicz,O.C. & Zhu,J.Z. (1987): *A Simple Error Estimator and Adaptive Procedure for Practical Engineering Analysis*, International Journal for Numerical Methods in Engineering, Vol. 24, pp. 337-357

Index

Printing: Krips bv, Meppel
Binding: Litges & Dopf, Heppenheim

GPSR Compliance

The European Union's (EU) General Product Safety Regulation (GPSR) is a set of rules that requires consumer products to be safe and our obligations to ensure this.

If you have any concerns about our products, you can contact us on ProductSafety@springernature.com

In case Publisher is established outside the EU, the EU authorized representative is:

Springer Nature Customer Service Center GmbH
Europaplatz 3
69115 Heidelberg, Germany

Batch number: 10755329

Printed by Printforce, the Netherlands